MINING IN ECOLOGICALLY SENSITIVE LANDSCAPES

To my parents, for their love and support

MINING IN ECOLOGICALLY SENSITIVE LANDSCAPES

EDITOR: MARK TIBBETT

© Mark Tibbett 2015

All rights reserved. Except under the conditions described in the *Australian Copyright Act 1968* and subsequent amendments, no part of this publication may be reproduced, stored in a retrieval system or transmitted in any form or by any means, electronic, mechanical, photocopying, recording, duplicating or otherwise, without the prior permission of the copyright owner. Contact CSIRO Publishing for all permission requests.

National Library of Australia Cataloguing-in-Publication entry

Mining in ecologically sensitive landscapes / Mark Tibbett, editor.

9780643106352 (hardback)
9780643106369 (epdf)
9780643106376 (epub)

Includes bibliographical references and index.

Mines and mineral resources – Environmental aspects.
Biotic communities – Effect of mining on.
Conservation of natural resources.
Ecosystem management.
Environmental responsibility.
Social responsibility of business.

Tibbett, Mark, editor.

333.8517

Published exclusively in Australia and New Zealand by

CSIRO Publishing
Locked Bag 10
Clayton South VIC 3169
Australia

Telephone: +61 3 9545 8400
Email: publishing.sales@csiro.au
Website: www.publish.csiro.au

Published exclusively throughout the world (excluding Australia and New Zealand) by CRC Press/Balkema, with ISBN 978-0-415-62091-8

CRC Press/Balkema
P.O. Box 11320
2301 EH Leiden
The Netherlands
Tel: +31 71 524 3080
Website: www.crcpress.com

Front cover: Mining and restoration in banded ironstone formation geology in Western Australia (photo: Rob Howard).

Set in 10.5/14 Palatino & Optima
Edited by Jane Kunstler
Cover design by James Kelly
Typeset by Thomson Digital
Index by Bruce Gillespie
Printed in China by 1010 Printing International Ltd

CSIRO Publishing publishes and distributes scientific, technical and health science books, magazines and journals from Australia to a worldwide audience and conducts these activities autonomously from the research activities of the Commonwealth Scientific and Industrial Research Organisation (CSIRO). The views expressed in this publication are those of the author(s) and do not necessarily represent those of, and should not be attributed to, the publisher or CSIRO. The copyright owner shall not be liable for technical or other errors or omissions contained herein. The reader/user accepts all risks and responsibility for losses, damages, costs and other consequences resulting directly or indirectly from using this information.

Original print edition:
The paper this book is printed on is in accordance with the rules of the Forest Stewardship Council®. The FSC® promotes environmentally responsible, socially beneficial and economically viable management of the world's forests.

CONTENTS

List of contributors — vii
Technical reviewers — ix

Introduction, concepts and challenges — 1

1 Mining in ecologically sensitive landscapes: concepts and challenges — 3
   Mark Tibbett

2 Mined land reclamation in the Appalachian coalfields: a case for an ecosystem reclamation approach — 7
   James A Burger

3 Hidden costs of mining in ecologically sensitive areas — 29
   Howard D Smith

Endemism in ironstone geosystems — 53

4 Iron geosystems: priority areas for conservation in Brazil — 55
   Claudia M Jacobi, Flávio F do Carmo, Felipe F do Carmo and Iara C de Campos

5 Flora and vegetation surveys of the ironstone ranges of the Yilgarn Craton — 79
   Neil Gibson, Rachel Meissner, Adrienne S Markey and Wendy A Thompson

6 Lessons drawn from iron ore mining in the Yilgarn region of Western Australia — 95
   Robert K Howard

Progress in bauxite mining — 109

7 Mining and ecological restoration in the jarrah forest of Western Australia — 111
   John M Koch

8	Recovery of terrestrial orchids in natural ecosystems after severe disturbance	141
	Margaret Collins and Mark Brundrett	
9	The mining-restoration system and ecosystem development following bauxite mining in a biodiverse environment of the seasonally dry tropics of Australia	159
	Alister V Spain, Mark Tibbett, Dieter A Hinz, John A Ludwig and David J Tongway	

Ways forward — 229

10	Key values of metallophytes for the minerals industry in Australasia	231
	Antony van der Ent	
11	A collaborative approach for mining, environment organisations and Traditional Owners to manage and conserve biodiversity in Australia's Great Western Woodlands	251
	Alexander WT Watson, Simon Judd, Wayne O'Sullivan and James EM Watson	
	Index	265

LIST OF CONTRIBUTORS

Mark Brundrett
Department of Parks and Wildlife, Swan Region, Australia II Drive, Crawley, WA 6000; Post: Locked Bag 104, Bentley Delivery Centre, WA, 6983 Australia

James A Burger
Professor Emeritus, Virginia Polytechnic Institute and State University, Blacksburg, VA 24061 USA

Margaret Collins
Faculty of Science, The University of Western Australia, Crawley, WA 6009 Australia

Iara C de Campos
Departamento de Biologia Geral, Instituto de Ciências Biológicas, Universidade Federal de Minas Gerais, Brazil

Flávio F do Carmo
Departamento de Biologia Geral, Instituto de Ciências Biológicas, Universidade Federal de Minas Gerais, Brazil

Felipe F do Carmo
Departamento de Biologia Geral, Instituto de Ciências Biológicas, Universidade Federal de Minas Gerais, Brazil

Neil Gibson
Keiran McNamara Conservation Science Centre, Science and Conservation Division, Department of Parks and Wildlife, Locked Bag 104, Bentley Delivery Centre, WA 6983 Australia

Dieter A Hinz
Ecological Restoration, PO Box 92, Crows Nest, QLD 4355 Australia

Robert K Howard
Cliffs Asia Pacific Iron Ore Pty Ltd

Claudia M Jacobi
Departamento de Biologia Geral, Instituto de Ciências Biológicas, Universidade Federal de Minas Gerais, Brazil

Simon Judd
School of Natural Sciences, Edith Cowan University, Joondalup, WA 6027 Australia

John M Koch
Alcoa Australia Ltd

John A Ludwig
CSIRO Ecosystem Sciences, Atherton, QLD

Adrienne S Markey
Science and Conservation Division, Department of Parks and Wildlife, Locked Bag 104, Bentley Delivery Centre, WA 6983 Australia

Rachel Meissner
Science and Conservation Division, Department of Parks and Wildlife, Locked Bag 104, Bentley Delivery Centre, WA 6983 Australia

Wayne O'Sullivan
The Great Western Woodland Collaboration, 2 Delhi St, West Perth, WA 6005 Australia

Howard D Smith
Northern Land Council, Darwin, NT Australia

Alister V Spain
School of Earth and Environment, The University of Western Australia, Perth, WA Australia

Wendy A Thompson
Science and Conservation Division, Department of Parks and Wildlife, Locked Bag 104, Bentley Delivery Centre, WA 6983 Australia

Mark Tibbett
Centre for Agri-Environmental Research, School of Agriculture Policy and Development, University of Reading, Berkshire, RG6 6AR, UK

David J Tongway
School of Environmental and Rural Science, University of New England, Armidale, NSW, Australia

Antony van der Ent
Centre for Mined Land Rehabilitation, Sustainable Minerals Institute, The University of Queensland, Brisbane, QLD 4072, Australia

Alexander WT Watson
School of Natural Sciences, Edith Cowan University, Joondalup, WA 6027 Australia

James EM Watson
Global Conservation Programs, Wildlife Conservation Society, 2300 Southern Blvd, Bronx NY 10460 USA; and The University of Queensland, The Ecology Centre, QLD 4072 Australia

TECHNICAL REVIEWERS

The editor thanks the following people who contributed their time and expertise as reviewers of the manuscripts for *Mining in Ecologically Sensitive Landscapes*. A technical and critical review of each chapter was undertaken by a minimum of two independent reviewers for the production of this book.

Jim Burger
Virginia Polytechnic Institute and State University

Mark Brundrett
The University of Western Australia

Mark A Clements
CSIRO

Augustine Dornilla
University of Melbourne

Sonia Finucane
Bioscope Environmental Consulting Pty Ltd

Neil Gibson
Department of Environment and Conservation

David Jasper
Outback Ecology

Tim Lardner
The University of Western Australia

Pieter Poot
The University of Western Australia

David Roberts
University of Kent

Juliet Rose
Post-mining Alliance, Eden Project

Robert Simmons
Cranfield University

Howard Smith
Cliffs Natural Resources

Alister Spain
The University of Western Australia

David Tongway
CSIRO

Peter Whitbread-Abrutat
Wardell Armstrong International

INTRODUCTION, CONCEPTS AND CHALLENGES

1

Mining in ecologically sensitive landscapes: concepts and challenges

Mark Tibbett

Summary

Many areas of unusual geology that contain ore bearing bodies also support unique ecological communities of plants and animals. In such landscapes, increasing demand to exploit rich mineral deposits can lead to a conflict between mining and conservation interests. This book brings together expert commentary in the fields of mining and conservation to grapple with this pressing issue and to work towards a positive outcome for all. The book is divided into four sections, where issues are dealt with in a series of case studies spanning four continents. The first section provides an introduction to concepts and challenges of mining in ecologically sensitive landscapes. The second section deals with rarity and endemism in plant species found on highly sought after (for mining) ironstone geosystems. The third section focuses on progress that has been made in restoring biodiverse forest after bauxite mining. The final section considers ways forward, away from conflict. *Mining in Ecologically Sensitive Landscapes* aims to generate an awareness of some of the key issues through a series of case study exemplars of this complex yet globally critical issue, one that wider society rarely considers in informed terms.

Introduction

Since human settlement began we have always been dependent on two sources of materials for our activities and endeavours: things that we grow and things that we mine. With a mining heritage over three millennia, what could be more human? For humans, what could be more natural? Mining has provided us with the raw materials and fuel to build towns and cities, to industrialise, and to hunt and farm throughout our history.

Today, with higher than ever demand for commonly mined elements (iron, aluminium, coal, gold and copper) and greater supplies of rare earth and other critical elements demanded (European Commission 2010; United States Department of Energy 2011) – and competing national agendas for resource security – mining has never been so important to humanity. It is at the very foundation on which our societal needs are met and is the base of almost all non-grown material supply chains. Yet mining has come into conflict with other human needs in the form of competition for land resources and particularly environmental degradation that can have serious and long-term effects on ecosystem and human health and wellbeing. Mines of all sizes and types should now be compelled to show how their activity meets the triple bottom line of being economically viable, environmentally benign and socially responsible, a widely accepted definition of sustainability (Brundtland 1987).

Many areas of unusual geology that contain ore bearing bodies also support unique ecological communities of plants and animals. Increasing demand to exploit rich mineral deposits can lead to a conflict between mining and conservation interests in such landscapes. This book brings together expert commentary in the fields of mining and conservation to grapple with this pressing issue and to work towards a positive outcome for all.

Concepts and conflicts

Mining may come into conflict with other sectors of society for any number of reasons. The benefits of greater employment and infrastructure development are often seen as being offset by transient workforces, environmental destruction and negative post-closure legacies. There are plenty of examples where these concerns are not realised and innovative solutions have been found (e.g. Pearman 2009). However, too often in some or all these matters, miners are found wanting.

Environmental destruction is not a pejorative term in this context. A mine completely obliterates its indigenous ecosystem and over much of the mine's footprint it is unlikely that the precise ecosystem, with its symbionts, hydrology, soil profile, pollinators, nutrient cycling, and precise floristic and faunal community structure is able to be replaced as though undisturbed. Even in the best examples, after the most benign strip-mining, not all is as it once was (Koch and Hobbs 2007; Tibbett 2010). It is for such reasons that the mining industry rarely uses the term 'restoration' or 'restoration ecology' as a discipline base, the term 'reclamation' or 'rehabilitation' are more likely to be used. Nonetheless, for some of the chapters in this book, particular related to the 'Progress in Bauxite Mining' section, restoration is probably a more appropriate term and is used accordingly.

The environmental consequences during mining and post-closure can be expected, regardless of well intended plans, and can leave considerable, long-term off-site legacies. Examples are many, with the more easily reviewed cases including Rum Jungle in Australia (Taylor *et al.* 2003) and Ok Tedi in Papua New Guinea (Ghazi 2003) to name but two. As in many of these cases, the worst conflicts occur when mining impacts on sensitive ecosystems and landscapes. As almost any ecosystem might be described as sensitive for any number of reasons, boundaries must be set on what counts as sensitive. For the purposes of this book, sensitive will mean a landscape 'quick or easy to respond to even slight changes or influences'; so it will not take much to upset or disturb the ecological balance in some way. This may range from a loss of hydrological functioning to the loss of key pollinators and everything in-between. In addition, and implicit in this, is also a landscape of high value for the rarity of its flora or fauna or its uniqueness as an endemic ecosystem or ecosystem service provision. For example, the iron geosystems such as the banded ironstone formations in Australia and Brazil possess both unique plants in unique ecosystems embedded in some of the richest iron ore deposits on earth.

Sections

The book is divided into four sections, where issues are dealt with in a series of case studies spanning four continents. The first section provides an introduction to the concepts and challenges of mining in ecologically sensitive landscapes. After the current introduction (Chapter 1), Chapters 2 and 3 focus on the removal of earthbound resources in the most remarkable manner (sometimes called mountaintop mining) and the human and ecological communities that dwell in effected landscapes, respectively. The concepts here are exemplars of the most extreme ecological and ecosystem damage, and of the most sensitive ecological and cultural landscapes one can find. The chapters lay out the conflicts and address how these have been and might be dealt with in the future.

Specifically, Chapter 2 (Burger) deals with the challenges posed by mountaintop mining in the Appalachians in the USA. Here, coal mining techniques have evolved to exploit ever larger volumes of land and the chapter deals with the response to this as new methods of reclamation and restoration are needed to minimise impacts on human and natural ecologies. We learn how a reforestation approach is superseding previous grassland-based approaches, where forests are now seen as the rational and sustainable post-mining land use. Public engagement, requiring improved environmental outcomes, may yet require a more complete ecosystem-based approach that is grounded in science. This could be applied through a process of adaptive management to help the coal industry in the USA maintain its social licence to operate. In Chapter 3 (Smith), using the Ranger uranium mine as an example, some of the hidden socio-cultural and ecological costs relevant to Australia's ecologically sensitive Kakadu National Park are described. These concepts are then applied to other mining situations in Tanzania and Namibia, and discussed in terms of those countries' regulatory systems. While there is a special focus on uranium in this chapter, many concepts and challenges that are identified are clearly applicable to other types of mining, landscapes and communities.

Australia and Brazil are blessed with some of the most geodiverse and biodiverse regions in the world and Section Two of the book deals with endemism (of plant communities) in the distinctive ironstone geosystems found as banded ironstone formations (BIF). Here these distinct geologies have evolved the most unique and rare plant communities found on earth, where the conflicts between mining and ecology could not be more clear or stark. Yet despite this, solutions are offered to even these seemingly intractable problems and it often becomes a matter of moral judgement and philosophical evaluation, as much as technical feasibility, to determine if these are acceptable. The first two chapters in this section (Chapters 4 and 5 – Jacobi *et al.*; Gibson *et al.*) illustrate the high biodiversity and ecosystem values of ironstone systems, and point to the importance of understanding the unique natural histories of these regions in assessment of current future resource development proposals. The vulnerability of BIF is inherently high and the final chapter in this section (Howard) presents information on the management of current mining impacts on rare flora and considers the challenges for prospective mining projects on BIF landscapes. The chapter also considers the strategic planning in balancing resource extraction and biodiversity conservation at a regional level.

The third section of the book discusses progress that has been made in biodiverse forest restoration after bauxite mining, focusing on examples from Australia. Here large tracts of forest are completely destroyed each year by strip-mining in areas that are among the most floristically diverse on the planet (Hopper and Gioia 2004). However, techniques and methods have been developed over many years to restore the former ecosystem with as many functional attributes rehabilitated as best as possible (Koch 2007a, b; Koch and Samsa 2007; Tibbett 2010). Notably, the relative success achieved here has been largely dependent on, and underpinned by, a significant research effort. The section deals with state-of-the-art restoration practices and the ecological outcomes over four decades of work, and provides overviews from two large bauxite mine restoration projects: one in the Jarrah (*Eucalyptus marginata* Donn ex Sm) forest of Western Australia and the other in the Darwin Stringybark (*Eucalyptus tetrodonta* F. Muell.) forest, northern tropical Australia (Chapters 7 and 9 – Koch; Spain *et al.*). Chapter 8 (Collins and Brundrett) provides an insight into the recovery of one diverse group of organism, the terrestrial orchids and their critical relationship with soil (mycorrhizal) fungi.

While there are clearly conflicts between mining, ecology and local communities there are ways forward from seemingly intractable issues. The future must build on what we have learnt from past mistakes and also employ novel and innovative ideas in terms of finding positives from the particular points of conflict. Too often we see win–lose or lose–lose scenarios where conflicts arise. We need to find win–win outcomes to deal with the inevitability of mining in ecologically sensitive areas. Easier said than done, but the final section (Ways forward) contains two chapters that take steps in this direction. Chapter 10 (Van der Ent) deals with offsetting and adding value to rare species in ways that could lead to conservation, not extinction. The final chapter (Chapter 11 – Watson *et al.*) considers moving away from conflict towards new paradigm for industry, regional communities, environmental organisations and traditional land owners. These are examples of ways forward and hope for the future.

Concluding remarks

Mining in Ecologically Sensitive landscapes should give the reader an awareness of some of the key issues in this complex yet globally critical issue that wider society rarely considers in informed terms. Some of the outrage against mining may be well justified but too often the essential nature of mining to our lives is misjudged in a clamour to retain an unspoilt Eden. In some cases this should be a clearly desired outcome; in others we may be best to accept new land uses or novel ecosystems (Hobbs *et al.* 2009; Pearman 2009). Compromises will have to be made and where this is so, it is how we make these compromises, and deliver on their execution, that will define our success in mining in ecologically sensitive landscapes in the future.

References

Brundtland GH (Ed.) (1987) *Our Common Future: The World Commission on Environment and Development*. Oxford University Press, Oxford, UK.

European Commission (2010) *Critical Raw materials for the European Union*. EU publication, Brussels, Belgium.

Ghazi P (2003) *Unearthing Controversy at the Ok Tedi Mine*. WRI Features, World Resources Institute.

Hobbs RJ, Higgs E, Harris JA (2009) Novel ecosystems: implications for conservation and restoration. *Trends in Ecology & Evolution* **24**, 599–605. doi:10.1016/j.tree.2009.05.012

Hopper SD, Gioia P (2004) The southwest Australian floristic region: evolution and conservation of global hotspot of biodiversity. *Annual Review of Ecology Evolution and Systematics* **35**, 623–650. doi:10.1146/annurev.ecolsys.35.112202.130201

Koch JM (2007a) Bauxite mining restoration by Alcoa World Alumina Australia in Western Australia: social, political, historical, and environmental contexts. *Restoration Ecology* **15**(4), S3–S10.

Koch JM (2007b) Alcoa's mining and restoration process in south Western Australia. *Restoration Ecology* **15**, S11–S16. doi:10.1111/j.1526-100X.2007.00288.x

Koch JM, Hobbs RJ (2007) Synthesis: Is Alcoa Successfully Restoring a Jarrah Forest Ecosystem after Bauxite Mining in Western Australia? *Restoration Ecology* **15**, S137–S144. doi:10.1111/j.1526-100X.2007.00301.x

Koch JM, Samsa G (2007) Restoring jarrah forest trees after bauxite mining in Western Australia. *Restoration Ecology* **15**(4), S17–S25. doi:10.1111/j.1526-100X.2007.00289.x

Pearman G (2009) *101 Things to do with a hole in the ground*. Post mining Alliance, Cornwall, England.

Taylor G, Spain A, Nefiodovas A, Timms G, Kuznetsov V, Bennett J (2003) *Determination of the reasons for deterioration of the Rum Jungle waste rock cover*. Australian Centre for Mining Environmental Research, Brisbane.

Tibbett M (2010) Large-scale mine site restoration of Australian eucalypt forests after bauxite mining: soil management and ecosystem development. In *Ecology of industrial pollution*. (Eds LC Batty & K Hallberg). Cambridge University Press, UK.

United States Department of Energy (2011) *Critical Metals Strategy*. USDE Publication.

2

Mined land reclamation in the Appalachian coalfields: a case for an ecosystem reclamation approach

James A Burger

Summary

The demand for energy throughout the world grows each day, and coal will be needed to meet a large portion of that demand. Coal mining techniques in the Appalachian and Midwestern coalfields have evolved to mine larger land areas and multiple seams at greater depths. New reclamation methods and approaches must also evolve to minimise cumulative impacts on aquatic, terrestrial, and human resources. Mined land reforestation practices over the past 80 years illustrate the evolution of reclamation in the eastern coalfields of the USA Prior to the implementation of the Surface Mining Control and Reclamation Act (SMCRA), tree planting was synonymous with reclamation. Grassland reclamation became the dominant approach post-SMCRA. As stakeholders of the mining and reclamation process have begun to appreciate the value of forest ecosystems, there is greater emphasis on ensuring land and forest restoration and proper ecosystem functioning on reclaimed mined land. A forestry reclamation approach is supplanting grassland reclamation where forests are the logical post-mining land use. Restoring forestland capability, native species, and watershed protection are positive outcomes. However, greater public demand for stream protection, water quality, biodiversity, carbon sequestration, native wildlife habitat, and human protection may require a more comprehensive *ecosystem reclamation approach*. In my view, the components of such an approach already have a good basis in science and could be applied through a process of adaptive management to help the coal industry in the USA maintain its social licence to operate.

Introduction

The Appalachian coalfields cover a broad area from western Pennsylvania to northern Alabama. Portions of the region also lie in Ohio, West Virginia, Maryland, Virginia, Kentucky, and Tennessee (Fig. 2.1). The Midwestern coalfields cover parts of Illinois, Indiana, and Kentucky. Over 600 000 ha have been disturbed by surface mining in the region, and more than 15 000 ha are surface mined each year (US OSM 2008). Coal from the region supplies ~15% of the total used in the US for power production, and high-quality metallurgical coal is shipped regionally and abroad. Mining in the region over the past century has evolved from mostly deep, post-and-pillar exploitation of coal seams 1 m thick or more, to a process of open pit and long-wall mining in the Midwest, and to mountaintop removal surface mining in the Appalachians that removes as much as 100 m of overburden for multiple seams of coal. Several hundred contiguous hectares can be disturbed in a single mine operation resulting in environmental, economic and societal impacts that are controversial. Given the current and projected demand for coal nationally and internationally, it is highly likely that land disturbance associated with surface mining will become even more extensive.

Largely unregulated mining at a smaller scale than practiced today prompted the enactment of the Surface Mining Control and Reclamation Act of 1977 (SMCRA 2006) with mining and reclamation provisions appropriate for the time and scale of mining. Three decades later, the spatial scale of mining and the techniques used are quite different and require a commensurate change in reclamation techniques to meet the intent of the SMCRA provisions. Another significant change in the past three decades is society's awareness of, and demand for, processes that assure sustainable environments, economies and human communities. Businesses of all sizes and types are now compelled to show how their activity meets the triple bottom line of being economically viable, environmentally benign and socially responsible, a widely accepted definition of sustainability (Brundtland 1987). The purpose of this paper is to present the evolution of mined land

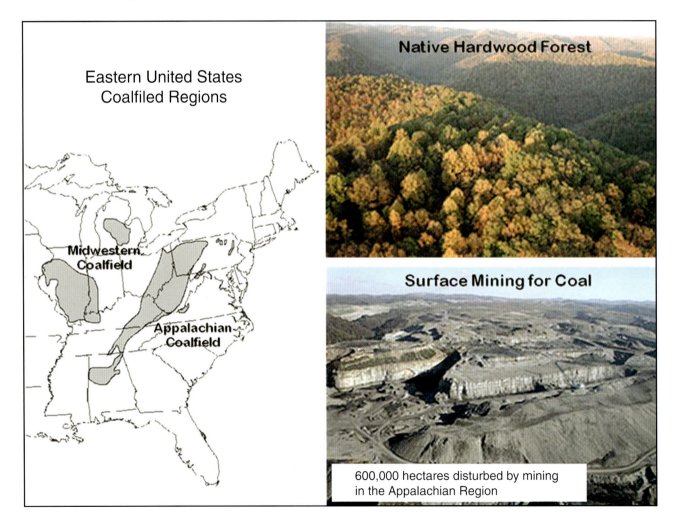

Figure 2.1: The Appalachian and Midwestern Coalfields region in the eastern United States.

reforestation in the eastern USA and to suggest an ecosystem reclamation approach that better addresses the impact of surface mining within a sustainable mining and reclamation paradigm.

Value of native forests of the Eastern Coalfields

The Appalachian and Midwestern coalfields are within the mountainous, largely intact part of the eastern deciduous forest biome of the United States; most land was forested before mining. The forests are Mixed-Mesophytic, Appalachian Oak and Oak-Hickory forest associations (Braun 1964), some of the most diverse, contiguous, temperate, broad-leaved forests in the world (Ricketts *et al.* 1999) (Fig. 2.1). These forests have provided products and services to their human inhabitants for centuries. Over the past 150 years, virgin timber and second-growth forests supported an important forest products industry. According to the American Forest and Paper Association (www.afandpa.org), the eastern coal producing states (West Virginia, Kentucky, Maryland, Indiana, Illinois, Ohio, Pennsylvania, Tennessee and Virginia) have more than 6300 wood products manufacturing facilities that employ over 300 000 workers with an annual payroll over US$15 billion. The overall value of annual industry shipments exceeds US$62 billion. The services provided by forests, which include watershed control, water quality, carbon sequestration, biodiversity and habitat, are reportedly more valuable than the wood products they provide. Forest services are poorly monetised, but their collective value has been estimated at three to ten times the value of wood products extracted (Costanza *et al.* 1997; Heal 2000). There is a rich history of human communities using these diverse forests for sustenance and shelter, and the use of non-timber forest products is an important part of local culture (Bolgiano 1998). On the steep terrain of the Appalachian Mountains where mean annual rainfall exceeds 1300 mm, much of it as intense, high-energy storms, forest cover is especially important for sediment and flood control and water quality.

Therefore, there are many compelling social, economic and environmental reasons to restore the forest on land disturbed by mining.

Early reports and records summarised by Utley (2008) show the evolution of strip mining for coal in the United States. In 1804, coal was reportedly dug by hand and with mule scrappers from the bottom of a creek near Lancaster, Pennsylvania. Mechanised strip mining was accomplished with Otis steam shovels and Vulcan railroad shovels by 1907 in Kentucky and other regions of the Appalachian and Midwestern coalfields. Electric stripping shovels were used as early as the 1920s and modern versions continue to be used nearly a century later. Early on, stripping followed coal outcrops, but electric and diesel-powered shovels allowed removal of large amounts of overburden in area and contour mines. By 1980, the scale of mechanisation was such that entire ridgelines and mountaintops consisting of several hundred contiguous hectares were being removed in the process of harvesting multiple coal seams lying approximately horizontally with the landscape. This complete removal of landscape features is called mountaintop/valley-fill mining and is now quite common (and controversial) in the Appalachian region. Explosives are used to break rock layers and large shovels and draglines remove the overburden and interburden between coal seams. Excess spoil is typically placed in valley fills, some of which may be 1 km wide and 3 km long. Despite the tremendous change in scale and method since the SMCRA was implemented three decades ago, reclamation methods used on most surface mines have changed little.

Mined land reforestation history

There are four periods of mined land reforestation over the past 80 years that influenced the composition and productivity of woody vegetation on surface mined sites in the eastern coalfields: 1) pre-federal law tree planting period; 2) grassland period; 3) woody shrub period; and 4) native forest restoration period (Fig. 2.2). Except for tree planting, rehabilitation of areas disturbed by mining was negligible until the passage of the SMCRA in 1977. Most Appalachian and Midwestern states had rudimentary reclamation laws with provisions for revegetation, but few required highwall and pit backfilling, landscape reshaping to approximate original contour (AOC), or replacement of topsoil. According to Medvick (1980), Indiana was the first state to organise a tree planting program for mined land in 1928. Over the next four decades, land reclamation and tree planting became synonymous terms in most eastern coalfield states. State coal operator associations enlisted the aid of state, federal and academic forestry researchers which resulted in reforestation guidelines for pre-law mined sites (Davis 1965; Vogel 1981). Prescribed tree mixes, including valuable native hardwood species, generally grew well on loose, deep, overburden materials and many of these forest stands have become quite valuable for saw timber (Ashby 1996; Rodrigue *et al*. 2002).

Rodrigue *et al*. (2002) documented the productivity, diversity and value of these new pre-law forests. They measured 14 stands on mined sites across a seven-state area in both the Appalachian and Midwestern coalfields and compared them with adjacent, mature stands on non-mined areas with similar pre-mining conditions. The forest stands were mixes of native hardwoods, native conifers, or a mix of hardwoods and conifers, and ranged in age from 18 to 71 years. Growth rates, expressed as mean annual increment, were greater

Figure 2.2: Eighty-year history of reclamation in the Appalachian Coalfields.

on 10 of the 14 mined sites than on adjacent non-mined sites (Rodrigue *et al.* 2002). In most cases, depending on the value of the species mix, the wood product value on mined sites exceeded the value of that on non-mined sites. Carbon sequestration on the reforested sites ranged from 3 to 4 Mg ha^{-1} yr^{-1}, and based on a sequestration projection, carbon accumulation will reach the average level in adjacent, non-mined stands by age 80 (Amichev *et al.* 2008). Despite widespread tree planting during the decades of the 1950s, '60s and '70s, and the overall success of tree planting programs, environmental and human safety issues remained. Mass wasting of debris slopes, dangerous highwalls, acid mine drainage and sediments in streams ultimately led to passage of the SMCRA.

When the SMCRA was implemented in 1978, it greatly changed the way coal surface mines were reclaimed. Toxic materials that could lead to degraded water quality were buried and isolated to prevent oxidation and seepage of acids and minerals into downstream aquifers; pits and highwalls were backfilled and the landscape was returned to approximate original contour; strict sedimentation and air quality provisions prevented off-site pollution; and the graded mined sites were sown with agricultural grasses and legumes for erosion and sediment control for improved water quality. Regulations based on the SMCRA require native soil replacement on row-cropped agricultural land after mining and reclamation; however, there is very little crop land in the Appalachian region. For the predominantly forest and pasture land, alternative overburden materials are allowed in lieu of native soil if the coal miner demonstrates equal or better capability shown by certain performance standards such as herbaceous cover or number of woody stems per unit area. With this regulatory provision, virtually no native soil is salvaged and replaced on mined surfaces in the Appalachian region, despite concerns that current performance standards are inadequate measures of land capability, site quality or pre-mining forest productivity.

Because of their previous pre-law success with tree planting, coal operators planted trees after the implementation of the SMCRA, but with little to no success. The extensive grading with heavy equipment needed to achieve AOC severely compacted the mine surfaces; mostly alkaline, unweathered rock was used as soil-substitute growth media; and the herbaceous ground cover sown for erosion control overtopped planted seedlings, causing high mortality rates, stunted growth and low species diversity (Groninger *et al.* 2006; Holl 2002). Where and when trees survived, they grew poorly, demonstrating that only part of the original pre-mining capability was restored (Burger and Evans 2010; Burger and Fannon 2009). As a result, within two years after implementation of the new federal law, coal operators reclaimed most mined lands to alternative, approved uses such as *hayland/pasture* (italics indicates a defined, post-mining land use (PMLU) in the federal regulations) for the 15-year period from 1980 to 1995. Very few of the ~300 000 ha reclaimed in the Appalachian region during this time were managed and used for their prescribed PMLU. There is no significant livestock industry in the central Appalachians and most mined sites are remote, without water for livestock, and difficult to fence and manage. Without management these *hayland/pastures* were quickly (within 10 years) overcome with invasive species such as serecia lespedeza (*Lespedeza cuneata*), Japanese honeysuckle (*Lonicera japonica*) and autumn olive (*Elaeagnus umbellata Thunb.*).

By the mid-1990s, landowners, ecologists, and the general public expressed concerns to coal operators and regulators about the low value and poor condition of reclaimed land. This initiated a 10-year period (1996–2005) during which the predominant PMLUs in most areas of the Appalachian coalfields were permitted *wildlife habitat* and *unmanaged forest* (GAO 2009). In actual practice, coal operators simply added the additional step of planting shrubs and tolerant tree species to grassland reclamation to achieve these PMLUs; the mined land was still heavily graded, compacted and sown with agricultural grasses and legumes as required by most federal and state regulators. Within 10 years, the *wildlife habitat* prescription was largely overtaken by invasive species that were more competitive than the planted wildlife shrubs, which is preventing bond release due to unmet performance standards. *Unmanaged forest* consisted of a few early-successional species such as black locust (*Robinia psuedoacacia*) and *Pinus* spp. that could tolerate the inhospitable, compacted mine soils covered with sown, competitive herbs. By the year 2000, on a half million hectares reclaimed under the provisions of the new law in the Appalachian region, approximate post-mining coverage consisted of 35% *hayland/pasture*, 45% *wildlife habitat*, 15% *unmanaged forests* and 5% other uses (OSM 2010). Based on anecdotal reports and surveys, very little reclaimed land was in productive use, and regardless of the prescribed PMLU, most of it was in a state of arrested succession (Holl 2002; Putz and Canham 1992; Williamson and Gray 1996) and heavily infested with invasive and exotic species. By the turn of the century, it was clear to many in the mining and reclamation community that a different approach to rehabilitation of mined land was needed.

A forestry reclamation approach for reclaiming mined land

In 1980, shortly after the implementation of SMCRA, Virginia Polytechnic Institute and State University established the Powell River Project (PRP) (www.prp.cses.vt.edu) with a mission to conduct research and education programs to enhance restoration of mined land for the benefit of communities and businesses in the region. The PRP, funded by government and industry, supported research on AOC methods, mine soil suitability, forage and livestock production, acid mine drainage remediation, water quality, horticultural and tree crops, and forest restoration. For over 30 years, reforestation research through the PRP has been conducted in a seven-state region on issues including selection and placement of mine soils suitable for native forest restoration; development of erosion-control ground covers compatible with native plant colonisation; and researching methods for increasing diversity, productivity and value of restored forests. The PRP reforestation research program was built upon several decades of pre-law research by agency, university and industry scientists. Because the federal SMCRA dramatically changed the way mined land was reclaimed, additional research was needed after 1977 to assess the impact of AOC grading, heavily compacted mine soils, suitability of overburden as plant growth media, compatibility of traditionally used reclamation species with native forest trees, and the integration of the biology of reforestation with site preparation, regulatory requirements, and cost optimisation. In many respects, the SMCRA created a whole new landscape that required new reclamation approaches for which there was little science for guidance.

The culmination of this long-term PRP reforestation research program, and that of several others established in adjacent states a decade later, was a Forestry Reclamation Approach (FRA) (Burger and Zipper 2009a; Burger *et al.* 2005; Zipper *et al.* 2011) very different from common practices used previously. The FRA entails: 1) selection of soil and overburden growth media specifically for forest trees; 2) placement of mine soils 1.2 m deep in a loose, uncompacted condition; 3) use of tree-compatible ground cover that minimises competition and allows germination, emergence and establishment of native, volunteer vegetation; and 4) planting by professional tree planters of valuable, native trees and shrubs that enhance biodiversity and other forest services. Thorough reviews of the research on which the steps of this FRA are based have been published (Franklin *et al.* 2012; Zipper *et al.* 2011).

Native forest soils in the Appalachians range in depth from 30 to 150 cm, have surface soil densities ranging from 0.8 to 1.3 g cm^{-3}, and are generally moderately acidic, low in soluble salts and well drained. Forest productivity can be restored by salvaging and returning soil and regolith materials with physical and chemical properties similar to native soils. However, high soil density caused by mining equipment often impairs plant productivity on Appalachian mine soils (Daniels and Amos 1985; Davidson *et al.* 1984; Haering *et al.* 2004). Mine reforestation studies have consistently revealed negative relationships between soil density and tree survival (Burger and Evans 2010; Conrad *et al.* 2002; Torbert and Burger 1990) and between soil density and tree growth (Andrews *et al.* 1998; Ashby 1990, 1997; Burger and Evans 2010; Jones *et al.* 2005; Skousen *et al.* 2009; Torbert and Burger 1990; Torbert *et al.* 1988). Dense soils impede root growth and water infiltration, and limit soil water holding capacity. The lack of air spaces within dense soils limits oxygen availability, further inhibiting the growth and function of roots and soil microbes, leading to poor water and nutrient uptake. Practices that avoid or remedy these limitations are outlined in 'best practices' publications by Sweigard *et al.* (2007a) and Sweigard *et al.* (2007b) located at: http://arri.osmre.gov/Publications/Publications.shtm.

Mine soil chemical properties can also limit tree survival and growth. Soil pH, soluble salts and base saturation are consistently correlated with the performance of Appalachian forest species (Rodrigue and Burger 2004; Showalter *et al.* 2007). Other soil properties such as soil N, organic C and P have been reported as growth limiting on mined sites but usually within the first 10 years after disturbance (Andrews *et al.* 1998; Czapowskyj 1978; Torbert *et al.* 1988). Skousen *et al.* (2011) describe methods for selecting soils and alternative materials with desirable chemical properties for the Appalachian coal region.

Using tree-compatible herbaceous ground cover for initial erosion control or using herbicides to reduce competitive vegetation aids establishment of planted trees on Appalachian coal mine sites (Ashby 1997; Chaney *et al.* 1995; Torbert *et al.* 2000). On a Virginia mine planted with native hardwoods, Burger *et al.* (2008) demonstrated that reducing herbaceous competition accelerates both survival and growth of planted trees. Skousen *et al.* (2006) found that native Appalachian hardwood trees were able to volunteer on older mine sites that were not seeded with herbaceous vegetation, while seeded areas supported only sparse tree cover several decades after reclamation. Some herbaceous vegetation; however, is often required to control erosion, especially on steep slopes (Jeldes *et al.* 2010), and the SMCRA is explicit in requiring erosion control. Hence, the FRA's third step emphasises establishment of tree-compatible herbaceous vegetation as an important complement to planted trees (Burger *et al.* 2009).

Short-term reforestation success using the FRA has been documented in several research studies (Burger *et al.* 2008; Emerson *et al.* 2009; Fields-Johnson *et al.* 2010). When the FRA is used, hydrologic flow paths and runoff and infiltration processes are more like native forests than mined land reclaimed using traditional methods (Taylor *et al.* 2009). Field surveys show diverse (10 planted woody species and as many naturally colonised), fast-growing stands of native trees intermixed with native herbs and grasses (Angel *et al.* 2005). The extent to which native biodiversity will be restored using the FRA is still unknown, but initial observations suggest that many FRA sites planted with a variety of native species are on a normal, natural-succession trajectory.

Long-term evidence showing that forest productivity can be restored using the FRA was reported by Burger and Zipper (2009b) and Casselman *et al.* (2007) in studies of an eastern white pine (*Pinus strobus* L.) stand established in 1978 using FRA procedures (Fig. 2.3A). White pine is a native species sensitive to gradients of soil quality; therefore, it provides a good bioassay of forest land capability. Forest site index projected to age 50 was 34 m, which is excellent land capability on any site, disturbed or not. Figure 2.3B shows white pine cross sections, all 17 years old, growing on soils of different quality. The intermediate sized cross-section of 17-year-old white pines shown in Fig. 2.3B depicts the average growth rate of white pines across the Appalachian region (Doolittle 1962; Vimmerstedt 1962). The smallest cross-section depicts growth on compacted, alkaline overburden prepared for grassland PMLUs. The largest cross-section depicts growth on mine soils prepared using FRA techniques (Fig. 2.3A). Burger and Zipper (2009b) summarised data comparing wood production and

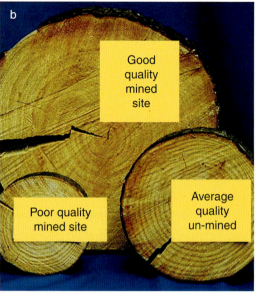

Figure 2.3: (a) Twenty-eight-year-old forest stand on mined land reclaimed using FRA procedures. (b) Comparative growth of 17-year-old white pine for average unmined site, mined site with poor soil quality and mined site with excellent soil quality.

value of stands growing at these different rates (Table 2.1). Wood volume increases disproportionately faster as site index increases, and wood value at rotation age increases exponentially with site quality because large trees have greater value for more diverse products. Pre-mining capability in terms of tree growth and forest economic value can be restored using the FRA. Preliminary studies indicate that other ecosystem services provided by forests such as biodiversity, watershed control, and wildlife habitat can be restored using the FRA, but more research is needed to determine the extent of recovery towards pre-mining conditions.

Adoption of the forestry reclamation approach

After much public criticism of the extensive amounts of unused and unproductive mined land being created throughout the Appalachians, the

Table 2.1. The effects of reclamation technique on white pine productivity and stand value at 30 years with average management

Case	White pine site type	Site index[a]	Timber volume (m³/ha)[b]	Harvestable wood products	Total value (US$/ha)[c]
I	Average quality of an undisturbed Appalachian forest site (Doolittle 1962)	17	500	small saw timber	$3460
II	Average quality of a post-SMCRA non-FRA mine soil (Torbert et al. 1988)	14	90	pulp	$250
III	Actual quality of a white pine stand on a good mine soil in Virginia (Kelting et al. 1997)	21	670	large saw timber	$6900

[a] Site index = expected eastern white pine height after 25 years, in metres.
[b] Harvestable timber volume at age 30, expressed in cubic metres per hectare, based on yield tables prepared by Vimmerstedt (1962).
[c] Harvest price, estimated as an average of what was typical over the 1999–2009 period. Actual harvest prices vary with market conditions.

USA Office of Surface Mining established the Appalachian Regional Reforestation Initiative (ARRI) in 2004. Its purpose was communicating and encouraging mine reclamation methods that plant more high-value hardwood trees, increase the survival and growth rates of planted trees, and expedite the establishment of forest habitat through natural succession (Angel *et al.* 2005; ARRI 2010). ARRI is organised with teams of federal and state regulators, and a science team consisting of academic and agency reclamation researchers. The ARRI adopted and encouraged the application of the FRA to improve mine soil and forest productivity and to restore forest services including watershed control, water quality, biodiversity, carbon sequestration and habitat. The ARRI program has achieved modest success: from 2005 to 2010 ~40 million trees were planted on 30 000 ha using FRA practices. This represents ~15% of the land mined and reclaimed. Based on reported tree planting data (Zipper *et al.* 2011), use of the FRA has levelled off short of expectations despite evidence showing that use of the FRA is not significantly more costly than reclamation practiced traditionally to produce grassland (Baker 2009), especially when including lower costs for maintaining erosion control structures due to reduced runoff and sedimentation.

The reasons for limited adoption of the FRA are unclear, but it may be a function of the institutional context within which mining and reclamation is done (Zipper 1987). Reforestation to pre-mining capability is a function of the quality of the post-mining landscape (Fig. 2.4). The biology and 'best management practices' for the FRA are well established by science and experience (Zipper *et al.* 2011), but applying the FRA is voluntary. The cost is not significantly different, but the coal operator has little economic incentive after final bond release given no long-term interest in the PMLU. From an economic perspective, the landowner is most concerned with coal royalties and is unlikely to

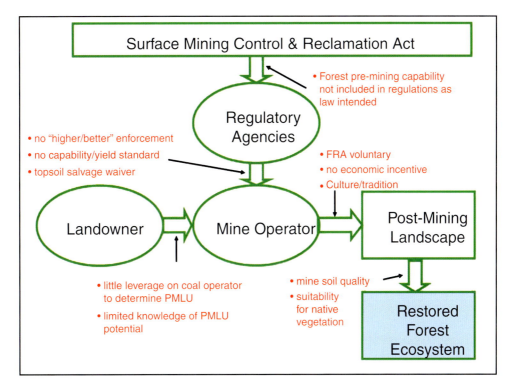

Figure 2.4: The institutional context for mining and reclamation constraining the adoption of the Forestry Reclamation Approach.

participate in the permitting process and selection of the PMLU; it is the coal operator's responsibility to meet reclamation standards required by regulations. In many cases the landowners do not appreciate the potential for post-mining income through establishment of productive forest land. State and federal regulations do not *prevent* the use of the FRA in any way, but current interpretations do not *require* its use even when the pre-mining condition consisted of native forest. Wider adoption of the FRA and restoration of pre-mining forest capability would be enhanced if restored forests were properly categorised as a 'higher and better' PMLU than abandoned, unproductive grassland of largely non-native species. Requiring topsoil replacement and a capability standard equivalent to pre-mining forest site index would also enhance biodiversity and forest productivity (capability or yield standards are required for other PMLUs, but no yield or productivity standard is required for forest land). A final institutional constraint for adopting the FRA is little expressed support by leading environmental groups whose primary agenda is restricting surface mining through litigation based on interpretations of the SMCRA in federal and state regulations. Expressed support for better reclamation using the FRA would undermine their position that surface mining causes irreversible damage to terrestrial ecosystems, meaning that restricting surface mining is the only solution.

In any case, restoration of the native forest and all its services, in lieu of other PMLUs that add little or no value to the mined landscape, would appear to be in the best interest of all current stakeholders of this institutional arrangement. This is the case given that recent rulings associated with litigation dealing with disposal of mining spoil in valley streams (Bragg v. Robertson 1999) have created even higher expectations for good reclamation, especially for mountaintop/valley-fill mining (Palmer *et al.* 2010). Given current expectations, a sustainable mining and reclamation philosophy and a holistic, ecosystem approach for reclamation is needed to deal with aquatic, terrestrial and human justice issues associated with surface mining in the eastern USA coalfields. Offered below are suggestions for 'what' an ecosystem reclamation approach (ERA) would consist of, 'how' an ERA could be applied and adopted by all involved, and 'why' it may be important for the coal industry to pursue an ERA.

Improved reclamation approaches

Mountaintop mining (MTM) has become increasingly common since the early 1990s. Even when constructed to AOC, excess spoil, steep terrain and large scale of the operations create additional reclamation challenges. Because of the cumulative impacts of MTM on terrestrial, aquatic and human conditions, citizen groups have been using litigation to challenge the practice, particularly the process of placing excess spoil in valley fills which reportedly violates the SMCRA buffer zone rule and provisions of the Clean Water Act (CWA). Litigation (Bragg v. Robertson 1999) settlements ultimately led to 1) a multi-agency environmental impact statement (EIS) on the effects of MTM with alternative recommendations for reducing its impacts; and 2) a requirement that OSM revise its stream buffer rules to be consistent with CWA provisions. The final programmatic EIS (US EPA 2005) required multi-agency actions addressing both aquatic and terrestrial impacts. On 11 June 2009, the USA Department of the Army, USA Department of the Interior and USA Environmental Protection Agency signed a memorandum of understanding implementing the interagency action plan on Appalachian surface coal mining designed to 'significantly reduce the harmful environmental consequences of Appalachian surface coal mining operations, while ensuring that future mining remains consistent with federal law' (US EPA 2009). Impacts identified in the EIS included loss of terrain features, loss of headwater streams, modified hydrologic flow paths, degraded stream water quality, loss of forest and interior forest, reduced nutrient and carbon cycling, reduced soil and forest productivity, and loss of required habitat and native biodiversity. In an article in the journal

Science, 12 renowned scientists (Palmer *et al*. 2010) charged that MTM '...impacts are pervasive and irreversible and that mitigation cannot compensate for losses.' They concluded that 'MTM permits should not be granted unless new methods—... remedy these problems.'

A new method to remedy these problems might include a holistic, ecosystem reclamation approach (ERA) that includes the FRA as a component but also incorporates practices such as geomorphic landscape design, stream restoration, soil building and techniques for restoring biodiversity. This approach is consistent with the reclamation model proposed by Bradshaw (1984) showing reclamation and restoration of ecosystems as a process of returning ecosystem function (y-axis: biomass/carbon accumulation, hydrologic function, flood control, etc.) and structure (x-axis: species diversity, habitat, water quality) from a degraded condition towards its original condition (Fig. 2.5). According to Bradshaw, achieving restoration requires establishing reclaimed mined land conditions that allow balanced restoration of both function and structure, represented by a trajectory within nominal boundaries (large dashed arrow in model). Several alternatives to restoration include neglect (pre-SMCRA), replacement (forest conversions to unused grassland/scrub), and rehabilitation (the forestry reclamation approach) (Fig. 2.5). Although rehabilitation using the FRA provides the best trajectory of these alternatives, it too, when used alone, falls short of a level of ecosystem recovery that would likely meet the expectations of the public as expressed in the final programmatic EIS (US EPA 2005).

An improved approach for Appalachian mines, an *ecosystem reclamation approach* (Fig. 2.6), might begin with a geomorphic landscape design that would accomplish AOC by creating a landscape that mimics stable, natural mountain slopes while

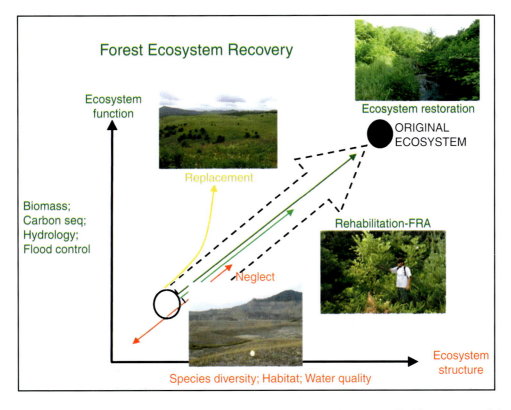

Figure 2.5: Reclamation alternatives after surface mining in the eastern United States coalfield regions (model adapted from Bradshaw 1984).

being cost-effective, attractive and resistant to surficial erosion and mass wasting (Ayres *et al.* 2006; Schor and Gray 2007). Backfill materials would be selected and placed to minimise hydrologic contact with materials with high soluble salt levels to minimise total dissolved solids in stream water (Orndorff *et al.* 2010). Streams would be reconstructed based on pre-mining patterns and capacities using a technique known as natural steam channel design, which seeks to reconstruct the pools, riffles and other habitat features of undisturbed streams, with the goal of restoring the ecological functions that were lost due to the original stream disturbance (Fritz *et al.* 2010; Keystone Stream Team 2007). Soils would be constructed to accommodate native flora and fauna by salvaging topsoil, litter layers, seed pools and coarse woody debris, and mixed with suitable overburden materials to achieve a minimum depth of 1.2 m (Skousen *et al.* 2011). All steps of the FRA would be incorporated to ensure native forest biodiversity, productivity and connectivity that would potentially support native wildlife including interior forest species (McComb *et al.* 1989; Wood *et al.* 2006; Wickham *et al.* 2007) (Fig. 2.6).

Adopting the ecosystem reclamation approach

Compared to traditional reclamation approaches used in the Appalachian region, the ERA may appear complex and difficult to implement. However, the practices associated with the steps above are well established in science and practice but not widely applied together in the Appalachian

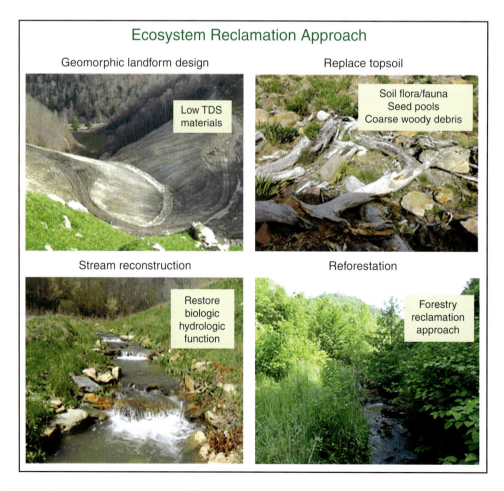

Figure 2.6: Steps implementing an ecosystem reclamation approach to restore pre-mining land capability.

coalfields. An excellent model of an ERA is one developed and used by Alcoa World Alumina Australia on their mines in Western Australia (Gardner and Bell 2007). When and if Appalachian coal operators are compelled or choose to implement an ERA, an iterative process of adaptive management (AM) might be the 'how to' for adopting the ERA. AM is often characterised as 'learning by doing' but that is an oversimplification. It is learning by doing, but in an organised way that combines ongoing operations with monitoring, research, assessment and training. AM was developed by ecologists at the University of British Columbia and initially applied in fishery management. It is now broadly applied in the natural resource fields and is being adopted by various industries. The USA Department of the Interior refined the process for use by all of its agencies (Williams *et al.* 2009). A good AM process is already in place for adopting the FRA by ARRI and its cooperators; its overall structure is an example that could be emulated for adopting an ERA (Fig. 2.7).

AM begins with a clear statement of *objectives and goals* (Step 1); for example, 'return the land to its pre-mining capability and restore all ecosystem services'. With the objectives and goals firmly in mind, the second step is to develop reclamation *guidelines* and procedures by 'reaching back' to the *database* of 'what we know and tools we have'. The database contains all existing knowledge and experience that can be brought to a first version of guidelines. It includes regulations controlling the mining process, practical reclamation experience, baseline information about the site, research results on reclamation processes, and local knowledge of climate, soils, plants and animals and their interactions. Collectively, these guidelines and procedures are called 'best reclamation practices'. Best practices for a forestry reclamation approach based on these criteria are already in place as guideline

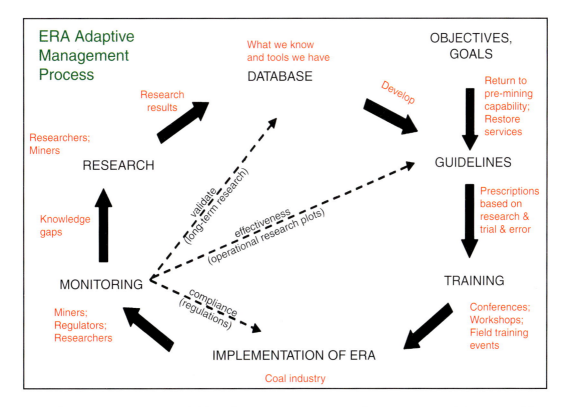

Figure 2.7: An adaptive management model for adopting an ecosystem reclamation approach for sustainable mined land reclamation.

documents for coal operators (http://arri.osmre.gov/Publications/Publications.shtm). These practices are 'best' in the sense that they are based on a thorough analysis of what we know now, but they may not be adequate without further research for implementing a holistic ecosystem reclamation approach. They will be improved as we increase our knowledge database with an applied, operational research plan. Best practices are always subject to change and improvement.

Step 3 of AM is *training* and *implementation* of the best practices. Training should involve all regulatory and mining personnel. Virtually all mine workers in all areas will affect the reclamation process directly or indirectly; it is important that everyone understands the goal and the process for achieving it. Step 4 entails *monitoring* the reclamation process for compliance, effectiveness and validation. Compliance is simply a check to see if the best practices were applied. Federal and state regulators routinely check rule compliance on Appalachian mine operations, but effectiveness monitoring, a short-term check, to see that best practices are working as intended and validation monitoring, long-term statistically-based research, are seldom implemented. Mine operators, regulators and researchers should be involved in monitoring at various levels. Excellent models for monitoring are presented in the treatise by Tongway and Ludwig (2011).

To fill the knowledge gaps in the database from which best practices are formulated, applied research is needed (Step 5). *Research* on several steps of the reclamation process can be done on site in conjunction with operational mining, as well as on adjacent sites and remote locations. Research by independent scientists in cooperation with company personnel, using the scientific method of testing multiple alternative hypotheses, with sound experimental designs and peer review, is essential for science-based mining and reclamation. Cooperative research among agencies, the industry and universities has proven to be cost-effective, operationally feasible, and usually satisfies regulatory and public scrutiny.

AM is much more than simply tracking and changing reclamation direction in the face of a false start. It involves exploring alternative ways to meet reclamation objectives, predicting the outcomes of alternatives based on the current state of knowledge, implementing one or more of these alternatives, monitoring to learn about the impacts of current practices applied, and then using the results to update knowledge and adjust guidelines. Adaptive management focuses on learning and adapting, through partnerships of coal operators, regulators scientists and other stakeholders who learn together how to reclaim mined land in a sustainable way (Williams *et al*. 2009).

Sustainable mining and reclamation

An immediate, short-term reason *why* the Appalachian coal industry might be interested in using an ERA was alluded to above. Adopting an ERA might alleviate litigation and imposition of stricter rule interpretations by the US EPA and other federal agencies. A long-term reason for adopting an ERA is to maintain a 'social license to operate' (Joyce and Thomson 2000; Kurlander 2001). Surface mine reclamation plans contain detailed guidelines for restoring disturbed land to its pre-mining capability and uses. These plans are social contracts with the company's customers and clientele public that ensure the land will be restored in a manner that will allow it to provide pre-mining ecosystem products and services (land capability). The reclamation plan, along with all other company plans of operation, reflects the company's Sustainable Mining and Reclamation Philosophy.

Operating with a sustainable mining and reclamation philosophy has become the rule for corporations involved in extractive and other natural resource industries. In an article in the journal *Resources Policy*, David Humphreys, now chief economist with Rio Tinto, posed the question, 'Can the industry afford sustainable development?' He concluded that environmental and social costs have been more than offset by increased industry productivity. There is no conflict between the interests of mining profitability and sustainable development. He concluded that it is possible to create a confluence of interests between improved industry profitability and the provisions of sustainable development, and that this is not really about costs but about the alignment of mining industry's values with those of the societies in which it operates (Humphreys 2001).

Sustainability is based on the notion that resources be used by current human generations without compromising their availability to future generations (Brundtland 1987). In the context of the minerals sector, the goal should be the integration of economic activity with environmental integrity and social concerns (triple bottom line) (Fig. 2.8). There should be ongoing availability of resources and a productive environment and healthy community at both current and former mining sites (Cowell *et al.* 1999; Mudd 2010). In the past, Appalachian mine operations focused primarily on production with little emphasis on environmental or social concerns except to meet the letter of the law. With increasingly rigorous environmental legislation controlling mining activities and heightened environmental awareness on the part of the general public and local communities, inclusion of mining activities within the sustainable development paradigm has become essential (Humphreys 2001; IIED and WBCSD 2002). In 1999, the program Mining, Minerals and Sustainable Development (MMSD) was initiated globally to develop ways for the mining industry to become sustainable. The International Institute for Environment and Development (IIED) was commissioned to undertake the project working on behalf of the World Business Council for Sustainable Development (WBCSD).

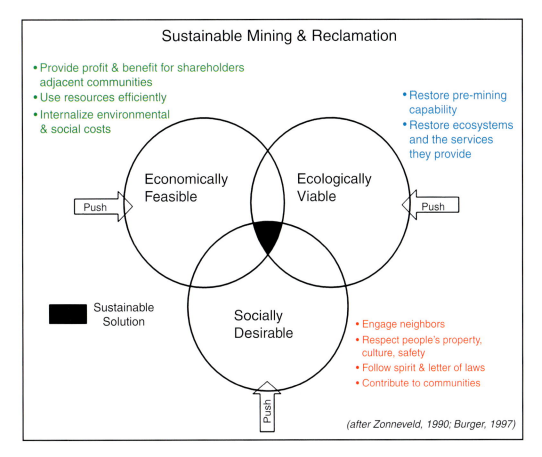

Figure 2.8: The 'triple bottom line' for achieving sustainable mining and reclamation and maintaining a social licence to operate.

In the final report called *Breaking New Ground* (IIED and WBCSD 2002), the MMSD group couched sustainable mining and reclamation within the 'triple bottom line' used by other industries (Fig. 2.8). A mine project is sustainable *economically* if the viability of the project is assured, and if the community will be better off as a result; it is sustainable *ecologically* if pre-mining capability is restored and ecosystem services are recovered; and it is sustainable *socially* if people's wellbeing is maintained or improved. The major organisations representing the mining industry in developed countries have subscribed to sustainability principles and practices. They include the Minerals Council of Australia (www.minerals.org.au/focus/sustainable_development), the European Association of Mining Industries, the Mining Association of Canada, Chile's state-owned CODELCO and the National Mining Association of the United States. And many individual international mining companies adopted sets of company-specific sustainability guidelines: Barrick (www.barrick.com) and BHP Billiton (www.bhpbilliton.com/home/aboutus/sustainability/Pages/default.aspx) are examples and provide models that could be emulated.

Conclusion

The demand for energy throughout the world grows each day, and coal will be needed to meet a large portion of that demand. Coal mining techniques in the Appalachian and Midwestern coalfields of the USA have evolved to mine larger land areas and multiple seams at greater depths. New reclamation approaches, practices and regulations must also evolve to minimise cumulative impacts on aquatic, terrestrial and human resources. Mined land reforestation was used here to illustrate the evolution of reclamation approaches over the past 80 years in the eastern coalfields of the USA. As all stakeholders of the mining and reclamation process appreciate the value of forest ecosystems, there is greater emphasis on ensuring their restoration and proper functioning on reclaimed mined land.

A forestry reclamation approach is supplanting a grassland reclamation approach where forests are the logical post-mining land use. Restoring forestland capability, native species and watershed protection are positive outcomes. However, greater public demand for stream protection, water quality, biodiversity, carbon sequestration, native wildlife habitat and human protection may require a more comprehensive ecosystem reclamation approach. In my view, the components of such an approach already have an established basis in science and could be applied through a process of adaptive management to help the coal industry maintain its social licence to operate.

References

Amichev BY, Burger JA, Rodrigue JA (2008) Carbon sequestration by forests and soils on mined land in the Midwestern and Appalachian coalfields of the U.S. *Forest Ecology and Management* **256**, 1949–1959.

Andrews JA, Johnson JE, Torbert JL, Burger JA, Kelting DL (1998) Minesoil properties associated with early height growth of eastern white pine. *Journal of Environmental Quality* **27**, 192–198. doi:10.2134/jeq1998.00472425002700010027x

Angel PN, Davis VM, Burger JA, Graves D, Zipper CE (2005) The Appalachian Regional Reforestation Initiative. Appalachian Regional Reforestation Initiative, US Office of Surface Mining. Forest Reclamation Advisory No. 1.

Appalachian Regional Reforestation Initiative (ARRI) (2010) Trees for Appalachia's Future – Appalachian Regional Reforestation Initiative. U.S. Office of Surface Mining. Accessed online 12 July 2010. http://arri.osmre.gov/

Ashby W (1990) Factors limiting tree growth in southern Illinois under SMCRA. In: *Proceedings of the 1990 Reclamation Conference and Exhibition*. (Eds J Skousen, J Sencindiver, D Samuel) pp. 287–293. West Virginia University, Morgantown.

Ashby WC (1996) Growth of hardwoods and conifers after 47 years on coal mine soils in southern Illinois. Tree Planters' Notes **47**, 24–29.

Ashby WC (1997) Soil ripping and herbicides enhance tree and shrub restoration on stripmines. *Restoration Ecology* **5**, 169–177. doi:10.1046/j.1526-100X.1997.09720.x

Ayres B, Dobchuk B, Christensen D, O'Kane M, Fawcett M (2006) Incorporation of natural slope features into the design of final landforms for waste rock stockpiles. In: *Proceedings of the 7th International Conference on Acid Rock Drainage*. (Ed. RI Barnhisel) pp. 59–75. American Society of Mining and Reclamation, Lexington, KY.

Baker KL (2009) Costs of reclamation on southern Appalachian coal mines: a cost-effectiveness analysis for reforestation versus hayland/pasture reclamation. M.S. Thesis. Virginia Tech, Blacksburg.

Bolgiano C (1998) *The Appalachian Forest, A Search for Roots and Renewal*. Stackpole Books, Mechanicsburg, Pennsylvania, USA.

Bradshaw AD (1984) Ecological principles and land reclamation practice. *Landscape Planning* **11**, 35–48.

Bragg v. Robertson (1999) The Environmental Law Reporter. 29 ELR 21316 No. No. CivA 2:98-0636, 54 F. Supp. 2d 653/48 ERC 1913/(S.D. W. Va., 06/17/1999). http://elr.info/litigation/%5Bfield_article_volume-raw%5D/21316/bragg-v-robertson

Braun EL (1964) *Deciduous Forests of Eastern North America*. Hafner Publishing Co., New York.

Brundtland GH (Ed.) (1987) *Our Common Future: The World Commission on Environment and Development*. Oxford University Press, Oxford, UK.

Burger JA (1997) Conceptual framework for monitoring the impacts of intensive forest management on sustainable forestry. In: *Forest Management for Bioenergy*. Research Paper 640. (Eds P Hakkila, M Heino, E Puranen) pp. 147–156. The Finnish Forest Research Institute, Vantaa Research Centre.

Burger JA, Evans DM (2010) Ripping compacted mine soils improved tree growth 18 years after planting, In: *27th Annual National Conference of the American Society of Mining and Reclamation*. (Ed. RI Barnhisel) pp. 55–69. Lexington, KY.

Burger JA, Fannon AG (2009) Capability of reclaimed mined land for supporting reforestation with seven Appalachian hardwood species, In: *27th Annual National Conference of the American Society of Mining and Reclamation*. (Ed. RI Barnhisel) pp. 176–191. Lexington, KY.

Burger JA, Zipper CE (2009a) How to Restore Forests on Surface-Mined Land. Virginia Coop. Extension Publ. 460–123.

Burger JA, Zipper CE (2009b) Powell River Project – Restoring the value of forests on reclaimed mined land. Virginia Coop. Extension Publ. 460–138. http://pubs.ext.vt.edu/460/460-138/460-138.html

Burger JA, Graves D, Angel PN, Davis VM, Zipper CE (2005) The Forestry Reclamation Approach. Appalachian Regional Reforestation Initiative, US Office of Surface Mining. Forest Reclamation Advisory No. 2.

Burger JA, Mitchem DO, Zipper CE, Williams R (2008) Hardwood reforestation for Phase III bond release: need for reduced ground cover. In: *Proceedings of the 25th National Meeting of the American Society of Mining and Reclamation*. (Ed. RI Barnhisel) pp. 192–205. Richmond, VA.

Burger J, Davis V, Franklin J, Zipper C, Skousen J, Barton C, Angel P (2009) Tree-compatible ground covers for reforestation and erosion control. Appalachian Regional Reforestation Initiative, Forest Reclamation Advisory Number 6.

Casselman CN, Fox TR, Burger JA (2007) Thinning response of a white pine stand on a reclaimed surface mine in southwest Virginia. *Northern Journal of Applied Forestry* **24**, 9–13.

Chaney WR, Pope PE, Byrnes WR (1995) Tree survival and growth on land reclaimed in accord with Public Law 95–87. *Journal of Environmental Quality* **24**, 630–634. doi:10.2134/jeq1995.00472425002400040013x

Conrad PW, Sweigard RJ, Graves DH, Ringe JM, Pelkki MH (2002) Impacts of spoil conditions on reforestation of surface mine land. *Mining Engineering* **54**, 39–47.

Costanza RR, d'Arge R, de Groot R, Farber S, Grasso M, Hannon B, Limburg K, Naeem S, O'Neill RV, Paruelo J, Raskin RG, Sutton P, VandenBelt M 1997) The value of the World's ecosystem services and natural capital. *Nature* **387**, 253–260. doi:10.1038/387253a0

Cowell SJ, Wehrmeyer W, Argust PW, Robertson GS (1999) Sustainability and the primary extraction industries: theories and practice. *Resources Policy* **25**(4), 277–286. doi:10.1016/S0301-4207(00)00003-9

Czapowskyj MM (1978) Hybrid poplar on two anthracite coal-mine spoils: 10-yr results. USDA Forest Service Northeast For. Exp. Sta. Res. Note. NE-267. USDA Forest Service, Broomall, PA.

Daniels WL, Amos DF (1985) Generating productive topsoil substitutes from hard rock overburden in the southern Appalachians. *Environmental Geochemistry and Health* **7**, 8–15. doi:10.1007/BF01875045

Davidson WH, Hutnik RJ, Parr DE 1984) Reforestation of mined land in the northeastern and north-central U.S. *Northern Journal of Applied Forestry* **1**, 7–11.

Davis G (Ed.) (1965) A guide for revegetating bituminous strip-mine spoils in Pennsylvania. USDA Forest Service.

Doolittle WT (1962) Southern Appalachian white pine plantations: site, volume, and yield. United States Department of Agriculture, Forest Service Station Paper SE-149.

Emerson P, Skousen J, Ziemkiewicz P (2009) Survival and growth of hardwoods in brown versus gray sandstone on a surface mine in West Virginia. *Journal of Environmental Quality* **38**, 1821–1829. doi:10.2134/jeq2008.0479

Fields-Johnson C, Zipper CE, Burger JA, Evans DM (2012) Forest restoration on steep slopes after coal surface mining in Appalachian USA: Soil grading and seedling effects. *Forest Ecology and Management* **270**, 126–134.

Franklin JA, Zipper CE, Burger JA, Skousen JG, Jacobs DF (2012) Influence of herbaceous ground cover on forest restoration of eastern US coal surface mines. *New Forests* doi:10.1007/s11056-012-9342-8

Fritz KM, Fulton S, Johnson BR, Barton CC, Jack JD, Word DA, Burke RA (2010) Structural and functional characteristics of natural and constructed channels draining a reclaimed mountaintop removal and valley fill coal mine. *Journal of the North American Benthological Society* **29**, 673–689. doi:10.1899/09-060.1

GAO (2009) Surface coal mining: characteristics of mining in mountainous areas of Kentucky and West Virginia. United States Government Accountability Office. GAO-10–21.

Gardner JH, Bell DT (2007) Rauxite mining restoration by Alcoa World Alumina Australia in Western Australia: Social, political, historical, and environmental contexts. *Restoration Ecology* **15**, S3–S10. doi:10.1111/j.1526-100X.2007.00287.x

Groninger JW, Fillmore SD, Rathfon RA (2006) Stand characteristics and productivity potential of Indiana surface mines reclaimed under SMCRA. *Northern Journal of Applied Forestry* **23**, 94–99.

Haering KC, Daniels WL, Galbraith JM (2004) Appalachian mine soil morphology and properties: effects of weathering and mining method. *Soil Science Society of America Journal* **68**, 1315–1325. doi:10.2136/sssaj2004.1315

Heal G (2000) *Nature and the Marketplace: Capturing the Value of Ecosystem Services*. Island Press. Washington, DC.

Holl KD (2002) Long-term vegetation recovery on reclaimed coal surface mines in the eastern USA. *Journal of Applied Ecology* **39**, 960–970. doi:10.1046/j.1365-2664.2002.00767.x

Humphreys D (2001) Sustainable development: Can the mining industry afford it? *Resources Policy* **27**, 1–7. doi:10.1016/S0301-4207(01)00003-4

IIED and WBCSD (2002) *Breaking New Ground*. International Institute for Environment and Development and World Business Council for Sustainable Development. Earthscan Publications Ltd. London, UK, and Sterling, VA.

Jeldes IA, Hoomehr S, Wright WC, Schwartz JS, Lane DE, Drumm EC (2010) Stability and erosion on steep slopes constructed by the forest reclamation approach in the southern Appalachian region. In: *26th Annual National Conference of the*

American Society of Mining and Reclamation. (Ed. RI Barnhisel) pp. 470–488. Lexington, KY.

Jones AT, Galbraith JM, Burger JA (2005) A forest site quality classification model for mapping reforestation potential of mine soils in the Appalachian coalfield region. In: *22th Annual National Conference of the American Society of Mining and Reclamation*. (Ed. RI Barnhisel) pp. 523–539. Lexington, KY.

Joyce S, Thomson I (2000) Earning a social license to operate: Social acceptability and resource development in Latin America. *Canadian Mining and Metallurgical Bulletin* **93**, 1037

Kelting DL, Siegel CM, Burger JA (1997) Value of commercial forestry as a post-mining land use. In: *Proceedings of the 14th National Meeting of the American Society for Surface Mining and Reclamation*. (Ed. JE Brandt) pp. 344–348. Princeton, WV.

Keystone Stream Team (2007) Guidelines for natural stream channel design for Pennsylvania waterways. Alliance for the Chesapeake Bay.

Kurlander LT (2001) Newmont Mining: the social license to operate. Global Executive Forum. University of Colorado, Institute for International Business and Center for International Business Education and Research.

McComb WC, McGarigal K, Fraser JD, Davis WH (1989) Planning for basin-level cumulative effects in the Appalachian coal field. *Transactions of the 54th North American Wildland and Natural Resources Conference*. **54**, 102–112.

Medvick C (1980) Tree planting experiences in the eastern interior coal province. In: *Symposium on Trees for Reclamation in the Eastern U.S.* pp. 85–91. Interstate Mining Compact Commission and USDA Forest Service.

Mudd GM (2010) The environmental sustainability of mining in Australia: Key mega-trends and looming constraints. *Resources Policy* **35**, 98–115. doi:10.1016/j.resourpol.2009.12.001

Orndorff Z, Daniels WL, Beck M, Eick M (2010) Leaching potentials of coal spoil and refuse: acid-base interactions and electrical conductivity. In: *27th National Meeting of the American Society for Surface Mining and Reclamation*. (Ed. RI Barnhisel) pp. 736–766. Lexington, KY.

Palmer MA, Bernhard ES, Schlesinger WH, Eshelman KN, Foufoula-Georgiou E, Hendryx MS, Lemly AD, Likens GE, Loucks OL, Power ME, White PS, Wilcock PR (2010) Mountaintop mining consequences. *Science* **327**, 148–149. doi:10.1126/science.1180543

Putz FE, Canham CD (1992) Mechanisms of arrested succession in shrublands: root and shoot competition between shrubs and tree seedlings. *Forest Ecology and Management* **49**, 267–275. doi:10.1016/0378-1127(92)90140-5

Ricketts TH, Dinerstein E, Olson DM, Loucks CJ, Eichbaum W, DellaSalla D, Kavanagh K, Hedao P, Hurley P, Carney K, Abell R, Walters S (1999) *Terrestrial Ecoregions of North America: A Conservation Assessment*. Island Press, Washington DC.

Rodrigue JA, Burger JA (2004) Forest soil productivity of mined land in the mid-western and eastern coalfield regions. *Soil Science Society of America Journal* **68**, 833–844.

Schor HJ, Gray DH (2007) *Landforming: An Environmental Approach to Hillside Development, Mine Reclamation and Watershed Protection*. John Wiley & Sons, Inc.

Showalter J, Burger JA, Zipper CE, Galbraith JM, Donovan P (2007) Physical, chemical, and biological mine soil properties influence white oak seedling growth: a proposed mine soil classification model. *Southern Journal of Applied Forestry* **31**, 99–107.

Skousen J, Ziemkiewicz P, Venable C (2006) Tree recruitment and growth on 20-year-old, unreclaimed surface mined lands in West Virginia. *International Journal of Mining, Reclamation and Environment* **20**, 142–154. doi:10.1080/17480930600589833

Skousen J, Gorman J, Pena-Yewtukhiw E, King J, Stewart J, Emerson P, DeLong C (2009) Hardwood tree survival in heavy ground cover on reclaimed land in West Virginia: mowing and ripping effects. *Journal of Environmental Quality* **38**, 1400–1409. doi:10.2134/jeq2008.0297

Skousen J, Zipper C, Burger J, Angel P, Barton C (2011) Selecting topsoil substitutes for forestry mine soils. In: *Proceedings of the American Society for Mining and Reclamation*. (Ed. RI Barnhiesel). pp. 591–609. Lexington, KY.

SMCRA (2006) *Public Law 95–87: Surface Mining Control and Reclamation Act of 1977*. US Code, Title 30, Chapter 25. Washington, DC.

Sweigard R, Burger J, Graves D, Zipper C, Barton C, Skousen J, Angel P (2007a) Loosening compacted soils on mined sites. Appalachian Regional Reforestation Initiative, Forest Reclamation Advisory Number 4.

Sweigard R, Burger J, Zipper C, Skousen J, Barton C, Angel P (2007b) Low compaction grading to enhance reforestation success on coal surface mines. Appalachian Regional Reforestation Initiative, US Office of Surface Mining. Forest Reclamation Advisory Number 3.

Taylor TJ, Agouridis CT, Warner RC, Barton CD (2009) Runoff curve numbers for loose-dumped spoil in the Cumberland Plateau of Eastern Kentucky. *International Journal of Mining, Reclamation and Environment* **23**(2), 103–120. doi:10.1080/17480930802176389

Tongway DJ, Ludwig JA (2011) *Restoring Disturbed Landscapes: Putting Principles into Practice*. Island Press, Washington, DC.

Torbert JL, Burger JA (1990) Tree survival and growth on graded and ungraded minesoil. *Tree Planters' Notes* **41**, 3–5.

Torbert JL, Tuladhar AR, Burger JA, Bell JC (1988) Minesoil property effects on the height of ten-year-old white pine. *Journal of Environmental Quality* **17**, 189–192. doi:10.2134/jeq1988.00472425001700020004x

Torbert JL, Schoenholtz SH, Burger JA, Kreh RE (2000) Growth of three pine species on pre- and post-SMCRA land in Virginia. *Northern Journal of Applied Forestry* **17**, 95–99.

US EPA (US Environmental Protection Agency) (2005) Mountaintop mining/valley fills in Appalachia: final programmatic environmental impact statement (FPEIS). Philadelphia, PA Report no. EPA 9–03-R-05002. [Publication EPA 9–03-R-05002 contains only the comments on the Draft Programmatic Environmental Impact Statement (DPEIS) and ancillary information. Technical information is contained in the DPEIS [USEPA] US Environmental Protection Agency Mid-Atlantic Mountaintop Mining (4 August 2006; http://www.epa.gov/region3/mtntop/eis.htm)]

US EPA (US Environmental Protection Agency) (2009) Memorandum of understanding implementing the interagency action plan on Appalachian surface coal mining. Office of Research and Development, National Center for Environmental Assessment, Washington, DC.

US Office of Surface Mining (OSM) (2010) Annual Evaluation Reports for States and Tribes. US Dept. of the Interior, Washington, DC. http://www.osmre.gov/Reports/EvalInfo/EvalInfo.shtm. Accessed 13 December 2010.

Utley JH (2008) Early strip mining in Hopkins County, Kentucky. Unpublished paper.

Vimmerstedt JP (1962) Southern Appalachian white pine plantations: site, volume, and yield. USDA Forest Service. Southeastern For. Exp. Sta. Pap. No. 149.

Vogel WG (1981) A guide for revegetating coal minesoils in the Eastern United States. Gen. Tech. Rep. NE-68. Broomall, PA: USDA Forest Service, NE For. Exp. Sta.

Wickham JD, Riitters K, Wade T, Coan M, Homer C (2007) The effect of Appalachian mountaintop mining on interior forest. *Landscape Ecology* **22**, 179–187. doi:10.1007/s10980-006-9040-z

Williams BK, Szaro RC, Shapiro CD (2009) Adaptive Management. US Dept. of the Interior Technical Guide. Adaptive Management Working Group, US DOI, Washington, DC.

Williamson DL, Gray RB (1996) Evaluation of natural succession on reclaimed coal mine land in western Kentucky. In: *Proceedings of the 13th National Meeting of the American Society for Surface Mining and Reclamation*. (Eds WL Daniels, JA Burger, CE Zipper) pp. 629–636. Princeton, WV.

Wood PB, Bosworth SBB, Dettmers R (2006) Cerulean warbler abundance and occurrence relative to large-scale edge and habitat characteristics. *The Condor* **108**, 154–165. doi:10.1650/0010-5422(2006)108[0154:CWAAOR]2.0.CO;2

Zipper CE (1987) Mined land reclamation and development potentials in southwestern Virginia's coal-mining region. Virginia Coop. Exten. Serv. Publ. No. 69.

Zipper CE, Burger JA, Skousen JG, Angel P, Barton CD, Davis V, Franklin J (2011) Restoring forests and associated ecosystem services on Appalachian coal surface mines. *Environmental Management* **47**, 751–765. doi:10.1007/s00267-011-9670-z

Zonnveld IS (1990) Scope and concepts of landscape ecology as an emerging science. In: *Changing Landscapes: An Ecological Perspective*. (Eds IS Zonneveld, RTT Forman) pp. 2–20. Springer-Verlag, New York.

3

Hidden costs of mining in ecologically sensitive areas

Howard D Smith

Summary

As resources become scarcer, the mining industry will begin to place greater focus upon mineral deposits that are currently not subject to development because they are physically larger but of lower quality than existing major deposits or are contained within areas protected by virtue of their ecological value. Exploitation of these resources will create an increased risk of immediate environmental damage as well as long-term impacts on ecological and socio-cultural sustainability of the affected region. This can be interpreted as a potential future cost that is not generally considered adequately by the mining company or its regulators during mining operations and one that has historically been passed onto the public following mine closure.

Using the Ranger uranium mine as an example, some of the hidden socio-cultural and ecological costs relevant to Australia's ecologically sensitive Kakadu National Park are described. The principles used to describe these are then applied to other mining situations in Tanzania and Namibia, and discussed in terms of those countries' regulatory systems. While there is a special focus here on uranium, many of the hidden costs that are identified are clearly applicable to other commodities although the quantum values of those costs may differ considerably.

Introduction

As mineral resources become depleted and scarcer their value as commodities increases, thereby invigorating exploration and instigating technological change required for the processing of ores having challenging metallurgies or low mineral grades. Other deposits that are currently inaccessible by virtue of their location in ecologically sensitive areas become attractive and the target of intense and extended political and legislative agendas aimed at their exploitation. One of the best examples of this rests with the uranium mining industry which has shown a massive upsurge in interest since mid-2005 in the wake of upward spiralling spot prices and increased government support. In Australia, this is reflected not only by the abolition of the Federal Government's three uranium mines policy in 2007 but also in development of a pro-uranium mining policy in Western Australia in 2008.

Modern society has a need for minerals, but this must be balanced by the need to maintain a sustainable environment that has a high degree of biodiversity. In Australia, ecologically sensitive areas such as Kakadu National Park are offered protection under the *Environment Protection and Biodiversity Conservation Act 1999* – developed in response to the United Nation's Convention on Biological Diversity (amongst other drivers such as streamlined federal involvement in environmental impact assessment). While similar legislation designed to protect environmental values is found in other countries, many of these countries are beleaguered by poverty and the desire for an increased standard of living means that legislation is often not as well enforced as in Australia – and this places the integrity of protected areas at risk. This appears to be a market-driven issue and the continued efforts to pursue mining at Koongarra and Jabiluka (both of which are currently excised from Kakadu) suggest that it is not confined solely to developing countries.

Irrespective of whether a location is protected by legislation or not, where industrial activity such as mining is in conflict with the need for environmental protection, best practice and strong legislative and regulatory frameworks assist in ensuring that ecological integrity and sustainable management of the environment are maintained. The framework that exists in Australia is strict and has led to a perception within some mining companies that Australia's regulatory and political regimes and the subsequent project delays that they create makes mining too difficult and costly – and this is why many companies now seek profitability overseas (Perry 2008).

Consequently many Australian exploration companies are now targeting African countries well endowed with large but low-grade uranium deposits. Most of these countries have only basic mining laws and mining regulation and often the legislation supporting environmental protection in the face of their developing nuclear industries is weak. Application of voluntary standards applied to the acquisition of international finance for major projects (Equator Principles Association 2006; International Finance Corporation 2006) would help to enhance environmental and social performance of the mining industry in Africa, provided that the environmental and social management systems and processes developed in Australia by the industry over the past few decades are also implemented, continuously improved and adapted to the African situation.

Exploitation of deposits where regulation is poor will pose a significant threat to the environment and may lead to increases in associated long-term ecological and socio-cultural costs through damage to or deterioration of biotic communities in the post-mining environment that provide many critical ecosystem services to the environment and society. For example, soils are involved with nutrient storage and cycling while a series of complex interactions between microbes and invertebrates that break down organic matter and pollutants naturally purifies fresh water. Serious damage to or loss of these functions may have severe impacts on human health and socio-economics, through contamination of water supplies and the associated loss of natural resources or agricultural output on which local communities rely. Unfortunately, the current level of understanding of how these types of activities interact in many sensitive ecosystems is often poor and leads to a situation where the true ecological and socio-cultural costs that might be related to the loss of these sensitive systems either in part or as a whole cannot be fully assessed. Efforts to understand the costs associated with their loss are complicated further by the absence of a standard and meaningful way of expressing them in purely financial terms.

Globally, arguments over which areas should continue to be protected due to their ecological sensitivity will intensify as the mining industry seeks to expand and meet demand in the face of shrinking major resources. Currently, there are no well structured or universally accepted definitions of what constitutes an ecologically sensitive area, so legal definitions will undoubtedly develop as the mining industry and environmentalists seek to defend their positions. In this paper, ecologically sensitive areas are considered to be fragile areas where a small change to any part of the ecosystem produces a measurable response upon that ecosystem, surrounding ecosystems and/or any social and cultural systems that draw upon them. Baseline measures of the area's vegetation and wildlife biodiversity values, its natural resources and assessments of how they are used by society are required if a credible and accurate measure of the area's net worth is to be established. To achieve this, biodiversity must be measured not only against this baseline, but also expressed in terms of intergenerational equality (i.e. the ecosystem's worth to the environment and the inhabitants of that environment both now and well into the future).

By necessity the definition provided here is broad, but not unreasonably so because until recently the long-term cost to socio-economic and cultural systems was rarely, if ever, included as part of the overall cost-benefit analysis of mining projects. The failure to account for these costs exists partly because there is currently no accounting system suitable for quantifying the net worth of an ecosystem in monetary terms and partly because the symbiotic functions that exist between society, culture and ecosystem are poorly defined. Residual liability that comes from long-term disruption of ecosystems during mining operations and rehabilitation is therefore passed directly onto the landowners or the government following closure of the mining venture and lease relinquishment. Except perhaps where obligations to undertake perpetual care exist, landowners are generally ill-prepared and under-resourced and therefore incapable of dealing with the legacies of poorly handled rehabilitation and closure processes.

Even though it is not yet practicable to quantify the net worth of ecological and cultural systems in monetary terms, qualitative assessments may still be made. Baseline species and functions of ecosystems can be identified and interpreted in terms of cultural and socio-economic function, and from these the amount of loss can be estimated. It is the absence of market values and measures of productivity for many ecosystem services and goods that makes the outcomes qualitative, rather than quantitative, and difficult to apply as part of a cost-benefit analysis to mining developments.

With each mining venture, there are also attendant technological and capital costs related to the mine's operations that must be considered. These costs can be determined across the life of the mining venture, are generally easy to quantify in monetary terms and are readily reflected in the company's annual balance or profit and loss sheets. However, the costs associated with damage to environmental, cultural and socio-economic systems are less tangible and therefore difficult to quantify – and it is these which are considered in this paper to constitute the principal hidden costs of mining. Although these costs may be partly mitigated by improvements to processing technology, rehabilitation procedures and development of high quality regulatory or monitoring systems, the overall value cost of environmental protection generally increases with mounting ecological sensitivity. This is because greater effort and input is required to achieve acceptable social and cultural outcomes at all phases of the project.

There has been considerable effort placed into defining ways to calculate the net worth of ecosystems (Boyd and Banzhaf 2006; Field 1997). Rather than focus on accounting and financial theories that might be applicable, this paper seeks to examine in more detail the links between society, culture and ecosystem and use them to evaluate the long-term and as yet unquantifiable cost to socio-economic and cultural systems caused by mining activities. Three examples of ecologically sensitive areas affected by uranium mining (Australia's Kakadu National Park, Tanzania's Bahi wetland ecosystems and the Namib-Naukluft National Park desert ecosystem in Namibia) are used to describe the impacts and risks posed by uranium mining. It is acknowledged that this approach will not provide quantifiable values or definitions for true socio-economic and cultural costs, but nevertheless has the potential to provide guidance on how they might be factored into upfront and ongoing evaluation of mining projects.

Determining the value of an ecosystem

The first step in identifying hidden costs of mining impacts is an attempt to apply financial theory and accounting systems to determine the worth of each unimpacted ecosystem, which is no trivial matter. In the absence of an ecosystem value, present and future values of a mining operation can be readily calculated using standard accounting and valuation techniques based upon current and predicted commodity prices (although there is some degree of uncertainty involved). Using this approach is generally unsuitable for valuing an entire ecosystem because commodity prices do not exist for many species. Complete ecological baseline values can only be calculated using complete species inventories and a detailed knowledge of how they interact with each other and this becomes more difficult as ecosystem complexity increases. If this can be undertaken with a sufficient degree of reliability and validity, it may then be possible to make a quantitative estimate of any ecological change in terms of species abundance and distribution patterns. This would only be applicable if 'commodity' values are first established for all species in a manner consistent with a market-based economy. There are three major problems with this approach.

First, it is difficult to obtain a complete inventory of species for complex or unique ecosystems because of the high degree of biodiversity that exists. Second, it is difficult to completely identify and model how large numbers of species interact and how the loss of a single species might affect the overall ecosystem. Third, while many species have cultural or a broader economic value in a subsistence society, they cannot be ascribed a commodity price because markets do not exist and because their true perceived values are deeply rooted in the different world views particular to cultural groups, thus making values less tangible in standard accounting terms. Differences in world view leads to development of different non-financial valuation systems across cultures, which makes a complete cost-benefit analysis over the life of the mining project (and into the future) expressed in monetary terms almost impossible.

Other methods, such as non-market approaches (also known as contingent valuation) have been applied, but these generally result in an underestimation of the true environmental value. Introduction of specific indices based around the contribution of ecosystem to human welfare (Boyd and Banzhaf 2006) has expanded this approach, but ecological valuations still remain limited because the economic accounting principles used are concerned with ecological end products only and do not necessarily account for loss of specific intermediate processes and/or elements that make up nature.

A dynamic valuation process (Hertzler 2009) which combines non-market and bio-economic models to estimate social price for natural capital in terms of a 'scarcity rent' describing total value of ecosystem services has also been proposed. Scarcity rent is defined as the cost of completely consuming a finite resource and making benefits of the extracted resource unavailable to future generations. This approach is problematic because it provides an estimation of what the public perceives is the worth of the ecosystem impacted, and interprets it in terms of human welfare, or a willingness to pay for its loss. Like the market-based approach, there are no inbuilt mechanisms to account for the effect of partial or total loss of any single species on the overall ecosystem and, as a result, the most significant environmental issue identified – an intangible notion of the overall integrity of an ecologically sensitive area – remains undescribed in financial terms.

Collectively, these accounting-based approaches suggest that there are two hidden aspects of costs that must be considered in greater detail: the potentially calculable cost based around the monetary value of species damaged or lost to the ecosystem and the potentially incalculable cost ascribed to intangible damage to cultural, social and spiritual systems. If these incalculable components are to be understood fully, the link between social, cultural and spiritual activities and the ecosystem specific to the mining project must first be established.

A further significant limitation common to these approaches is that none of them consider alternative economic systems based upon the social, cultural and spiritual views of indigenous or other subsistence-based cultural groups who also make claim upon the resources the ecosystem has to offer. Indigenous economic systems typically do not value natural resources according to scarcity or in monetary terms, but are often more concerned with expressing value according to well developed cultural and spiritual requirements. Although this bears some superficial similarities to the human welfare approach, there is no direct financial analogue that can be calculated. In an indigenous society, the close affinity between human activities and ecosystems means that loss of a single species from the ecosystem may be far more damaging to spirit, culture and health than a simple monetary evaluation might otherwise indicate.

The three examples studied here (Kakadu National Park, Bahi wetlands and the Namib-Naukluft National Park) all sustain some form of subsistence economy based primarily around traditional or indigenous values, but also which has been influenced by various non-indigenous cultural activities. The fundamental differences between them relate primarily to their geographical locations, the amount of rainfall received by each and the social and cultural structures that have developed in response. Two are wetlands and one is a desert environment and this results in significant differences in the ecosystems and patterns of human use that are observed at each. The three ecosystems also have something in common, because all are now at risk from the impacts of uranium mining. This risk is explained later in greater detail and evaluated in terms of potential contamination of the ecosystem, threat to human health and change in cultural activities. However, each area will first be described in greater detail in order to better understand its specific ecosystem and cultural use patterns and appreciate how they might be valued.

Wetland ecosystems

Water and wetlands are vitally important for human wellbeing and maintenance of high biodiversity as they connect people to landscapes and provide a wealth of ecosystem services for support of livelihoods and sustenance of culture, especially in the context of indigenous culture. Kakadu National Park is a tropical monsoon wetland ecosystem characterised by a rich diversity and abundance of flora and fauna linked to a strong cultural framework (Finlayson *et al.* 2006). The operating Ranger uranium mine now constitutes part of and impacts upon the contiguous cultural and environmental landscapes of the park, and the principal cultural and economic activities revolve around tourism, scientific research and traditional Aboriginal practices. With an annual rainfall typically in excess of 1500 mm, Kakadu has a high capacity for dilution of water-borne contaminants to levels below which toxic effects might be observed, but there will be an accumulation of radionuclides and other toxicants in sediments and soils that will not undergo dilution and will instead continue to increase over time.

In comparison, the Bahi lakes in Tanzania are wetland ecosystems located in a semi-arid environment with annual rainfall around 500 mm. Compared to Kakadu, species abundance and diversity is lower because of the high degree of salinity. There is a greatly reduced capacity for dilution of contaminants and as a result of lower rainfall these precipitate as salts that mix with sediments as the lakes dry out. Based upon fishing, agriculture and production of salt, the cultural and socio-economic framework of Bahi is also significantly different to the traditional subsistence use of Kakadu and, like Kakadu, Bahi is also at risk from the impacts of uranium mining. To better understand the risks to each, their respective ecosystems, cultural activities and socio-economic relationships will now be briefly described.

Kakadu National Park – Australia

Located in Australia's north, Kakadu National Park contains two wetlands (Woolwonga and Magela, Fig. 3.1) both of which are listed under the Ramsar Convention and both of which support a strong indigenous cultural heritage spanning over 50 000 years. Subsequent definition of their ecological and cultural criteria as part of the World Heritage nomination process has marked Kakadu as an important culturally and ecologically sensitive area which is now under constant pressure from tourism, feral animals and industrial development associated with mining activities.

At landscape scale, Kakadu is an example of a complex of unique and sensitive ecosystems, providing a habitat for over 1600 plant and 500 animal and bird species (United Nations Education Programme 2005), many of which are believed to be endemic, rare or threatened by extinction. For example, of the seven species of freshwater turtle found in Kakadu, the pig-nosed turtle (*Carettochelys insculpta*) is of significant scientific interest as it is the sole surviving member of a family that has undergone some 40 million years of independent adaptation to its environment. Kakadu also contains a significant portion of the total range of four species of fish: the exquisite rainbowfish (*Melanotaenia exquisita*), Mariana's or the Magela hardyhead (*Craterocephalus marianae*), the sharp-nosed grunter (*Syncomistes butleri*) and Midgley's grunter (*Pingalla*

Figure 3.1: The Magela wetland facing north from Ubirr rock where groups of Aboriginal people camped in rock shelters and took advantage of the enormous variety of foods available from the East Alligator River, the woodlands, and the surrounding stone country. (*Photograph courtesy HD Smith*).

midgleyi) that have been recorded only from the northern part of Australia's Northern Territory.

In terms of impact assessment, the Ramsar listed wetlands would ordinarily provide an interesting point for comparative study as both have suffered the impact of uranium mining: Woolwonga in the 1950s and 1960s during mining in the Upper South Alligator Valley (Mudd 2002); and Magela since the onset of mining at Ranger in 1981 (Senate Environment, Communications, Information Technology and the Arts References Committee 2003). There are also significant uranium deposits located at Jabiluka and Koongarra, which if mined would contribute further to those impacts. Mining is currently vetoed at both of these deposits by the Mirarr Aboriginal land owners of the Park – in the case of Koongarra under the *Aboriginal Land Rights* (*Northern Territory*) *Act 1976* and at Jabiluka by agreement with the mining company involved.

Unfortunately, comprehensive baseline studies and full species inventories for Magela and Woolwonga were never completed, so it is impossible to estimate a realistic 'economic value' of the wetlands' ecological sensitivity even if appropriate accounting tools did exist. Neither is it possible to document the full extent of ecological change that has occurred as a result of mining. This situation is further confounded as there have been ongoing impacts by feral animals such as the pig (*Sus scrofa*) and water buffalo (*Bubalus bubalis*) and several invasive weed species.

A recent ecological risk analysis has implied that at a landscape scale, risk to the Magela wetland's ecosystem from feral animals and weeds is far more significant than that posed by mining (Bayliss *et al.* 2006). However, this analysis is confined to a single area downstream of the mine and makes no consideration of cultural ramifications such as the perception that food resources have become contaminated and are therefore unusable. Further, it does not consider the wider impacts that the Ranger mine has had on the environment due to the complete destruction of the Djalkmarra billabong's ecosystem, the loss of access to cultural resources due to present fencing and the long-term contamination of the operational areas.

Greater account should also be taken of the long-term risk that radionuclide and chemical contamination poses to ecosystem services and through them, on human health and cultural activity. Ecosystem services are considered here to be components of nature, directly enjoyed, consumed or used to yield human wellbeing. They include resource components (e.g. water and plant species) and are tied together by physical, biological and chemical interactions or ecosystem functions and processes (Boyd and Banzhaf 2006). As a result, weeds and feral animals may upset the ecological balance. Many ecosystem services will initially remain fundamentally intact (as the role of one species has been substituted by another); however, they are expected to become less resilient to perturbation and this could result in complete collapse of the ecosystem, with attendant and significant socio-economic and cultural impacts. Chemical poisoning and radioactive contamination has a more dramatic effect on the level of human interaction with ecosystem services as even the species that survive are perceived to be contaminated and no longer suitable for traditional use.

Although there have been major changes to the local social structure and cultural behaviour since the arrival of Europeans, Mirarr still place a high value on the natural environment and periodically undertake traditional activities. These include visitation to important ceremonial grounds and art sites and the collection of subsistence foods and culturally useful natural resources. It is noted that these activities have undergone significant decline in recent years in conjunction with an increasing perception that the environment surrounding the Ranger uranium mine has been and continues to be contaminated with radioactive mine waste. It is a testament to the strength of Mirarr culture that the value they place on the environment has endured throughout the life of the mine.

To date, there have been few attempts to determine an economic value for Kakadu National Park. In the 1990s, the contingent valuation approach was used to determine the 'value' of the Kakadu Conservation Zone; an area of mineralisation in the southern end of the park including Coronation Hill and El Sherana, thought to contain significant deposits of gold, palladium and platinum. At AU$435 million, it was calculated to be more than four times the net present value of the proposed mining venture (Carson *et al.* 1994), but indigenous and not economic values were cited as the main reason why the mining projects did not proceed (Hamilton 1996). The low number derived suggests that the valuation was incomplete (in that cultural values may not have been included), but more significantly that the potential political cost to the Australian Government through poor public relations far outweighed the value of the mine at that time. As a consequence, plans for mining in the Conservation Zone were scrapped.

A similar application of this approach to the wider Kakadu National Park undertaken a few years earlier suggested that the economic value of preserving the park was as high as AU$647 million per annum (Stratton and Whitten 2009). This is considerably higher than that earlier expressed for the Conservation Zone, principally because it focuses on a much larger area of land that contains the two Ramsar listed wetlands and considers a wider array of cultural sites that are of importance to the tourism industry. However, it too appears to be an underestimation and contains no consideration of the values that ecosystem services and components provide to Mirarr socio-economics and culture.

Bahi wetlands – Tanzania

Fed by the Bubu River originating in the Mbulu highlands, the Bahi wetlands (refer Fig. 3.2) are located in Singida District of central Tanzania, near the nation's capital, Dodoma. They support a high level of biodiversity, and have been identified as one of the most important wetlands habitats in Tanzania that meets the Ramsar 1% criteria for nomination – which requires the site to regularly support 1% of the individuals in a population of one species or subspecies of waterbird (Ramsar Convention Secretariat 2006). Bahi is also of vital socio-economic importance to the Tanzanian national economy and the communities that live in the area as it supports a range of livelihood activities including cultivation, fisheries, livestock grazing, salt production and the harvesting of natural products.

The large mudflats surrounding the wetlands are generally considered to be saline deserts and devoid of macrophytic vegetation, although smaller halophytic plant species have been found growing on the saline soils fringing the main lake. The dominant halophytes identified are *Cyperus laevigatus*, *Dactyloctenium* spp., *Juncus maritimus*, *Salvadora persica*, *Sporobolus spicatus*, *Sporobolus robustus*, *Suaeda monoica* and *Triplocephalum holstii* (McGinley 2007). The wetlands are inhospitable to many plants and animals because of the water's heat and high salinity, but continue to be home to huge flocks of lesser (*Phoenicopterus minor*) and greater flamingo (*Phoenicopterus ruber roseus*) and serve as an important breeding ground for the African Spoonbill (*Platalea alba*) (Dinesen 2004).

These birds serve important ecological functions by filtering blue-green algae from the saline water and feeding on copepod larvae that thrive in the wetlands. There are efforts underway to protect these species through the development of a single species action plan (Childress *et al.* 2008) under the 1995 Den Hague Agreement on the Conservation of African–Eurasian Migratory Waterbirds, but the principal focus of this plan is on nearby Lake Natron and not Bahi. However, the Bahi wetlands is now known to be one of eight alkaline lakes over a 940 km north–south transect within the Rift Valley to which these birds appear to be confined (Childress *et al.* 2007) and moves are now underway to afford it some level of protection.

The white-lipped or alkaline tilapia (*Oreochromis alcalica*) is endemic to the saline lakes of the Rift Valley (Lott 1992) and because of its abundance, is an important contributor to the socio-economics of the surrounding villages. It has been noted that sustainability of current practices has begun to decrease and that detailed ecological studies are

Figure 3.2: The Bahi wetlands in Tanzania, an area used for rice growing and salt harvesting. The surrounding villages also use the nearby grasslands for animal husbandry. This area is known to hold widespread but low-level uranium mineralisation that several foreign companies are seeking to exploit. (*Photograph courtesy A Lyamunda and M Kurz*).

needed to allow recommendations regarding specific conservation needs of the area's resources to be developed. Since the interaction of these factors directly and indirectly affects the ecological resilience of the surrounding dryland areas, understanding of the values and potentials of the wetlands in relation to people's livelihoods is also important for the development of sustainable dryland management strategies (Shemdoe *et al*. 2007).

A major constraint to the sustainable management of this area rests with the lack of knowledge held by government planners, natural resource managers and local communities of the diverse benefits that Bahi provides and the nature of the anthropogenic threats to their ecological character (McCartney and Rebelo 2010). For example, at the Bahi Swamps, the wildlife is under the control of the Director of Wildlife but the wetland is managed by the villagers (Mpemba 1993).

The Bahi wetlands have become of interest to uranium exploration companies following the first recognition of uranium accumulations near the centre of Bahi Swamp in 1953. Since then other large, but low-grade areas of mineralisation have been identified within the smaller playa lakes inside the catchment area to the north-west and these are now the subject of pre-feasibility mining studies. This has resulted in control and management of the ecosystem and its resources moving away from the villagers and coming under greater influence of the mining companies involved. Additional pressure is placed on local people with uranium mining being promoted as a means for generating local wealth, despite the risk it poses to the ecological integrity of the wetland and the livelihoods of the people who depend on it for agriculture.

As a result of the proposed mining activities, local communities are now faced with a choice between alleviating poverty in the short term by accepting the risks, or continuing with their existing agrarian systems. This is a difficult choice, because while the presence of a uranium industry might inject a large amount of financial income over the short-term, long-term damage to Bahi's ecosystem will result in a loss of sustainability and decrease in socio-economic value well beyond the life of mining. For example, should the delicate chemical balance of the lakes change as a result of mining activities, then the tilapia fishing industry might be lost forever.

Attempts at estimating the socio-economic value of the Bahi wetlands to nearby villages have been undertaken, but it has not been fully defined (Bakobi 1993; Mbogoro and Smith 2011). Annual income from the wetlands has been estimated to be as much as AU$1000 per household (Mwakaje 2009), but these values relate only to human derived income and do not consider the contribution of multifarious ecosystem services and how they would affect this value should they be lost. It could also be argued that these values have been underestimated because Tanzania has a depressed economy and the values will be significantly lower than what might be expected for a comparable situation in a well-developed country.

Desert ecosystems

Like wetland ecosystems, water is the key determinant of biodiversity and patterns of cultural use within a desert. Biodiversity is significantly lower than in a wetland, but the ecosystem services contained may be just as complex. Patterns of cultural use for the desert ecosystem will differ significantly from those of the wetlands because of the limited availability of resources and because human populations in deserts tend to be smaller, potentially leading to a decrease in cultural activity. Net cultural value is therefore expected to be lower than for a wetland, but due to the scarcity of resources, the intergenerational impact may well be more pronounced.

Contamination within a desert ecosystem is expected to be more highly localised than in a wetland, because there is limited opportunity for water-borne contaminants to be transferred across the landscape. One of the more significant impacts in the desert can be expected through drawdown of groundwater from subterranean aquifers; where even small changes to the water balance or chemistry have the capacity to produce large and long-term impacts leading to a significant loss in the diversity of species of ecological and cultural interest.

Namib-Naukluft National Park – Namibia

Namib-Naukluft National Park, located on the western coast of Namibia, is part of the world's oldest desert and contains the ecologically sensitive World Heritage Listed Welwitschia Plains. Receiving only ~100 mm of rainfall annually, the park's principal source of water is fog blown in from the Atlantic Ocean that helps sustain the welwitschia and the permanent underground aquifers of dry river valleys from where the culturally important nara melons draw their sustenance. *Welwitschia mirabilis* (refer Fig. 3.3a) is a long-lived, taxonomically distinct plant endemic to the Namib Desert and semi-desert areas. It provides shelter for most of the smaller endemic and semi-endemic resident fauna of the gravel plains while its leaves trap litter, allowing organic material to accumulate in otherwise infertile desert soils.

The Welwitschia Plains also support a vast diversity of endemic lichens and succulent and other species of flowering plants, which display an array of adaptations to cope with the harsh conditions. This creates an extremely sensitive ecological

Figure 3.3: *Welwitschia habilis* (a) and the !Nara melon (*Acanthosicyos horrida*) (b): both important contributors to the ecosystem and socio-economics of the Namib desert.

habitat and network of food webs which include many endemic insects and arachnids. The nara melon (*Acanthosicyos horrida*) (refer Fig. 3.3b), although relatively scarce remains culturally important, being a popular and important part of the native Topnaar people's diet, with other parts of the plant being used as medicines or a skin moisturiser. Archaeological evidence indicates that it has been a staple food and was transported and perhaps even traded for at least 8000 years (Wilkins-Ellert 2004).

The ecological integrity of parts of the Namib-Naukluft Park is now at greater risk because of a planned increase in uranium mining in the area near Swakopmund. Five uranium mining licences have so far been issued and a further 11 are anticipated, meaning that a greater focus on the capacity of this ecologically sensitive area to carry a massive cumulative impact is required. Information contained within Environmental Impact Assessments for new mines near Swakopmund (Digby Wells and Associates 2008; Turgis Consulting 2008) tends to indicate that direct and indirect impacts of uranium mining taking place in and around the region have the potential to profoundly alter a unique ecology and the way people perceive and use this section of the national park (Juergens 2008). Extinctions of species, loss of sense of place for residents and visitors and a fracturing of the integrity of the Namib Desert may all occur should these projects proceed (Ministry of Mines and Energy 2010).

It is a concern that the rate at which progress to mining has occurred has already outpaced implementation of the high standard of environmental, social and health and safety precautions that are needed to ensure sustainable and responsible development of the desert ecosystem. In addition, the positive and negative cumulative socio-economic and ecological impacts associated with the development of multiple uranium companies within the Namib Desert do not appear to be well understood. A strategic environmental assessment foreshadowed in 2008 was completed in 2010 (Ministry of Mines and Energy 2010), several years after approval of the newer mines at Langer-Heinrich and Trekkopje; and it is only recently that baseline values of the area have been qualitatively described (Henschel 2008) and the socio-economic aspects of biophysical degradation of the environment discussed (Hoadley and Limpitlaw 2008). No attempt has yet been made to formally evaluate either in monetary terms.

Risk to ecosystems

Complete prediction and evaluation of the impact of any mining activity on ecosystems is difficult because the synergies between various components of those systems are not well known. Ecological models based on conceptual understanding of environmental interactions are often incomplete in the face of this missing synergistic information, making it difficult to appreciate how the ecosystem might adapt until one component of the ecosystem has been reduced significantly or effectively eradicated – and of course, it is then too late. Understanding of ecosystem services and function remains limited and as a compromise, most environmental impact assessment procedures utilise risk assessment systems to evaluate the potential of impact. This approach means that the default position becomes one of 'a manageable impact' which implies that some damage to ecosystems is considered to be acceptable by the industry, if not the community.

Uranium is to be recovered at all three example sites, which means that there is no significant difference in the actual type of risk to ecological sustainability and biodiversity observed in Kakadu and the Bahi wetlands and the Namib-Naukluft. The principal risk is radioactive contamination and the long-term problems associated with toxicity and chemistry of the toxic progenies of the uranium sequence (refer Fig. 3.4). How they translate into direct ecological impacts will be a function of the ease with which they are transported across the respective environment, how well they are diluted and the species-specific sensitivity to each of the radionuclides and chemical toxicants.

The United Nations Scientific Committee on the Effects of Atomic Radiation (UNSCEAR) has reviewed the generally accepted position that assessment of risk from radionuclides should consider protection of human health as a priority rather than one that recognises the more immediate risk to plants, animals and other ecological organisms (Linsley 1997). However, it is now accepted that a greater focus on the natural environment is warranted as radiation doses necessary to produce significant deleterious effects are very difficult to establish given the large number of long-term confounding factors and levels of radiosensitivity

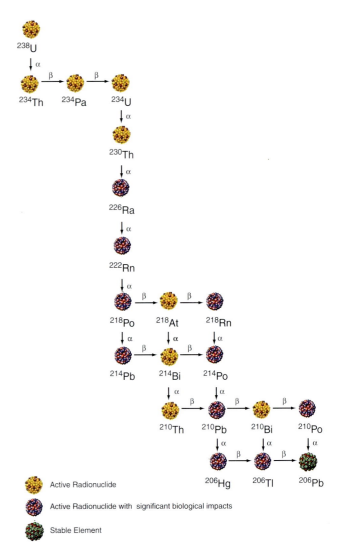

Figure 3.4: Natural decay series for ^{238}U – the principal route of radionuclide and chemical contamination (after Argonne National Laboratory 2005).

present in natural plant and animal communities in both terrestrial and aquatic environments.

To better understand the exact nature of the risk posed by radionuclides, an understanding of how they are generated is required. During the natural radioactive decay of ^{235,238}U there are two types of radiological particles (α and β) and several toxic chemical risks that must be considered. After processing, ^{230,234}Th are the controlling isotopes and, together with their progeny, contribute ~70% of the radioactivity of the original ore to the tailings. In consequence, tailings from uranium mines will continue to emit radiation for close to 90 000 years as the decay chain passes first through ^{230}Th and then ^{226}Ra, the two longest lived isotopes after ^{234}U. The decay chain ultimately leads to long-term increase in lead contamination when the isotopes decay to their stable end product ^{206}Pb, but not before transition through two highly toxic chemical species: polonium (^{210}Po) and thallium (^{206}Tl). In the context of protecting the environment from the effects of ionising radiation associated with uranium mining, the α-emitter ^{226}Ra is the uranium progeny of most interest as it is known to bioaccumulate, confers a high effective dose to biota (Sauerland *et al.* 2005) and may substitute for calcium in animal and human bones (Neuzil and Dysart 1984; Rosenthal *et al.* 1963).

Mining also releases large volumes of more chemically conservative by-products: manganese (Mn), magnesium (Mg) and sulphate (SO_4) and trace amounts of metalloids and heavy metals such as arsenic (As) and cadmium (Cd). The amount of each is dependent solely upon the geochemical nature of the base ore being mined. There is little information available from mines located in the Namib-Naukluft National Park and mining has yet to commence at Bahi, but it is possible to discuss the extent of chemical and radiological risk to the ecosystems of Magela.

Ranger uranium mine has been in continuous operation since 1981 but monitoring has not always followed strict regulatory requirements (Senate Environment, Communications, Information Technology and the Arts References Committee 2003), meaning it is possible that a large volume of contaminated water has passed undetected into the Magela system and that significant long-term ecological impacts may have already occurred. Over the course of a year, assuming mine site water discharges occur and allowing for tailings seepage rates, more than 1000 kg of uranium could enter the Ramsar listed wetlands compared to an estimated 25 kg that existed before the mine was developed (Rogers and Mudd 2011). Direct monitoring of impacts at the Magela wetlands is not undertaken and complete ecological baselines were never established, so it is not possible to definitively and directly determine the damage done to ecological services from the contaminants that have come from the mine. Given the large volumes of water and high river flows during the wet season, a high degree of dilution exists and this results in decreased probability of directly measuring an impact.

The Commonwealth's Supervising Scientist routinely concludes that the downstream environment (including the Magela wetlands) has remained protected from the effects of uranium mining at Ranger (Supervising Scientist 2009), but this is based on a limited amount of non-representative grab sampling and monitoring taken from single downstream and upstream statutory monitoring points on a weekly basis. During a high rainfall and subsequent flooding event in 2007 (Fig. 3.5), where much of the monitoring equipment was damaged, lost or remained offline for several days, the claim that the environment remained protected was still maintained (Supervising Scientist 2007). In reality, it is impossible to state with any degree of certainty what increased load of contaminants was washed into the wetlands or how widely it was dispersed, so the full extent of contamination that occurred during the period of flooding remains unknown.

Current understanding of the ecological integrity of the Magela system is based upon a limited amount of species monitoring in Magela Creek and at some small billabongs close to the mine site. Assessment of toxicity impacts is limited to studies of freshwater snail egg production (*Amerianna cumingi*) and bioaccumulation of ^{235}U and ^{226}Ra in mussels. Specific monitoring of black-banded rainbowfish (*Melanotaenia nigrans*) larvae commenced in 1992 but ceased in 2006, so information on changes to biodiversity is now restricted to data from monitoring of benthic macroinvertebrate communities in streams and fish communities in billabongs (Marchant 1982).

This system of monitoring is supported by routine measurement of chemical values in Magela Creek and by some site-specific ecotoxicology

Figure 3.5: Floodwater on the South Alligator floodplain near the Ranger uranium mine in late February 2007. The mine is located towards the elevated area in the distance. Although bunds and other contingencies were in place, parts of the mine flooded and protection of the Magela wetlands from water contaminated by the mine could not be fully guaranteed.

studies (Hogan *et al.* 2008; Holdway 1992), but the basic premise is that the ecosystem remains protected if toxicity guidelines based upon a few species are not breached. Given the nature of the risk, the number of species that contribute to overall biodiversity and the value of Kakadu National Park to the public and to the Mirarr custodians of the land, a more comprehensive assessment of impacts on the ecosystem of the Ramsar listed Magela wetland appears to be warranted. It would also be of interest and of value to compare species abundance and biodiversity of the Magela wetlands after 30 years of mining activities to that which existed before development of the mine. However, as previously noted, comprehensive baseline values were never established, and without this information, it is impossible to assess the total cost that has been incurred by the environment and Mirarr socio-cultural systems as a result of mining activities.

Risk to cultural systems

Indigenous cultures have a close affinity with their environments and view them as essential resources of economic, physical, cultural and spiritual goods. At Kakadu, where the Mirarr custodians still maintain direct links with their environment, anything that impacts upon the environment, subsequently impacts upon Mirarr socio-cultural systems, their economics and their spiritual beliefs. The Ranger Project Area remains a part of the contiguous ecological and cultural landscape that is Kakadu

National Park (Smith 2009), so it becomes necessary to consider the effect that mining has upon the wider landscape if costs to cultural systems are to be fully assessed.

Following discussions with the Mirarr and others, the pre-mining cultural landscape of the Ranger Project Area was defined and placed into the context of the wider Kakadu region (refer Fig. 3.6a) and by comparing this with current mining activities, a projected post-mining cultural environment has been inferred (refer Fig. 3.6b). Comparison of the two environments reveals that the Mirarr plan to avoid the area immediately affected by mining because of long-term contamination by radionuclides, and will therefore lose access to resources, camping sites and walking trails. In addition, several ceremonial sites and a burial site may also be avoided.

Loss of resources is a matter of particular importance to any subsistence society that draws heavily on its immediate environment for food. Despite work that has been carried out to date, there remains relatively little available information on radionuclides in terrestrial animals and plants and continued study of radionuclide concentrations in Aboriginal bush foods collected from Kakadu will be required for the foreseeable future, in order to provide Aboriginal people with the assurance that their food resources remain protected from mining operations.

Recent studies have shown that ^{210}Po, ^{210}Pb and ^{226}Ra are more easily transferred to the edible portions of native fruit and vegetables than ^{234,238}U and ^{230}Th, but are present in low concentrations (Ryan *et al.* 2005). Wide variation in radionuclide concentration is observed both across and within species, making it difficult to determine from the presented data what might constitute acceptable values for background and human consumption and what might be considered contaminated. Few data are available for bioaccumulation of radionuclides in other classes of aquatic and semi-aquatic fauna, but

Figure 3.6: (a) Cultural Landscape of the Ranger Project Area, 2009. Ceremonial sites are shown in red, camping sites in white, burial sites in blue and culturally significant sites in yellow. (b) Cultural landscape as envisaged upon rehabilitation and surrender of the lease. Anticipated avoidance areas are shown as concentric brown circles with a probable 'no-go' area defined by the opaque circle (Smith 2009).

the measured concentrations of ^{210}Po are found to be very high in liver, kidney and intestine samples from buffalo and turtle, and in freshwater mussel and pig flesh (Martin *et al.* 1995; Martin and Ryan 2004). Conversely, waterfowl species used as traditional food sources (e.g. pygmy goose and magpie goose) were found to have concentrations of heavy metals below recommended health levels (Brennan *et al.* 1992) and ^{226}Ra in mussels collected from Mudjinberri billabong although consistent with regional background levels (Bolhoffer *et al.* 2011) are high enough to be of serious concern (McLaughlin 1985).

Indigenous people are concerned that an additional load of uranium entering Magela Creek from mining activities may push ^{226}Ra levels in mussels beyond that which is considered safe for regular human consumption. The ease with which specific radionuclides may pass into the food chain and the conflicting interpretations of results obtained from monitoring creates a mixed response to claims that the Ranger uranium mine has had minimal environmental impact on natural food systems. This suggests that additional monitoring would assist in providing greater clarity, and since the current monitoring regime makes no direct measures on human health, measures including high confidence level studies designed to check for the presence of radioactivity in humans would be useful. These are discussed in more detail below.

The risk to cultural and social systems at Bahi may be significantly greater than that for Kakadu because Bahi is a major contributor to national rice and fish production in Tanzania and a means by which poverty in the local villages is alleviated. Sensitivity of tilapia and rice to radionuclides and other forms of chemical contamination has not been established so it is not possible to predict what might occur should uranium mining commence. Bahi is already susceptible to serious degradation and loss of sustainability (Yanda *et al.* 2005) and current land management issues and conflicts between local land owners that have arisen through overuse will inevitably increase should mining of the wetlands occur under substandard legislative and regulatory regimes.

Other than for agricultural purposes, the patterns of cultural use of the Bahi wetlands appear to be poorly reported and limited to one or two archaeological notes (Culwick 1931; Masao 1976). These suggest that a subsistence economy existed in the past and that some parts of the wider wetland region may have had particular spiritual importance, but there does not appear to be sufficient information extant to produce a visual estimate of the changes to the full cultural landscape as presented here for Kakadu. A similar scenario is found for the Namib-Naukluft National Park, where the sensitivity of key species such as *Welwitschia habilis* and the nara melon to radionuclide and chemical contamination is yet to be determined and a full cultural landscape recorded.

Cultural landscaping is an important part of the rehabilitation process and could possibly be used to estimate how patterns of land use have changed and what might be required to rectify or compensate for them, even if that change cannot be quantified in monetary terms. As a part of the closure and post-mining stewardship process, criteria based on cultural considerations are essential for Kakadu (Australian Government 1999) are now being devised (Smith 2008). It will be possible to place a post-mining value for partial restoration of the ecological and cultural environment because once closure is attained, it will be feasible to calculate a monetary cost for the work completed. However, the amount required will be dependent upon the precise nature and location of the work being undertaken. For example, it is likely that costs associated with attempting to reconstruct an ecosystem for Kakadu will differ significantly from those required for Bahi and Namibia-Naukluft National Park because of differences in mine dimensions, local employment costs and the amount of research required for re-establishment of key species.

Risk to human health

The direct links between human health and damage to ecosystems caused by mining is another important cultural factor that has historically been poorly dealt with by industry and its regulators, even though health is one of the few 'hidden' costs for which a monetary value might be determined through the welfare system. Difficulties arise because many health matters are easily confounded by life choices, making it difficult to ascribe any particular effect to a specific cause. For example, ^{222}Rn has been identified as the most likely cause of lung cancer and other diseases amongst uranium mine workers in Nevada (Schubauer-Berigan *et al.* 2009), while a putative link between the increase in uranium mining activities and the incidence of cancer amongst Aboriginal people in the Kakadu region has recently been identified (Tatz *et al.* 2006), but it is well known that cancer can be caused by several other sources such as cigarette smoking.

Unfortunately, the type of data collected by the health authorities is variable and not collated in any systematic way (Taylor 1999) and has led to criticism of Tatz *et al.*'s (2006) work. Despite this absence of measurable scientific data, the UMPNER inquiry (Commonwealth of Australia 2006) still concluded that the reported increase in cancer incidence in Aboriginal people of the Kakadu region, if it was verified, was not attributable to radiation exposure arising from the mining of uranium in the region. UMPNER's conclusions are based solely upon assumptions drawn from monitoring the environment near the Ranger uranium mine and do not include a direct measure of the health of the affected people.

There are methods by which impacts on the health of people using ecosystems potentially contaminated by radionuclides can be directly measured. ^{226}Ra has similar chemistry to Ca, and can replace it in the human skeletal system, so it has been suggested that if a human population has experienced chronic exposure to Sr or Ra radionuclides (Neuzil and Dysart 1984; Rosenthal *et al.* 1963) and/or heavy metal contamination then traceable evidence of this will be found in the milk-teeth of children (Council for Scientific and Industrial Research 2008). Unfortunately, this potentially valuable study has never been progressed and until it and others of a similar nature are completed, the direct impact of contamination on the health of people who utilise a radiologically contaminated area and the costs imposed upon the wider health system will never be truly known. Good scientific practice would ensure the social and physical health of Aboriginal peoples living in the communities surrounding current or proposed uranium mines in Kakadu, Bahi and Namib-Naukluft is measured directly instead of concluding that human health is protected based on results from a limited amount of analysis of local species. A similar situation is encountered in Namibia, where valid questions have been raised about the risk to human health from uranium mining and radiation (Shindondola-Mote 2009) and maintenance and access to relevant records appears to be poor.

During radioactive decay two types of particles, each of which have different epidemiological and environmental effects, are emitted. Alpha particles cannot penetrate most matter they encounter as they travel relatively slowly and therefore present a risk only over short distances in response to their electric charge and large mass. External exposure is not a major concern because α particles lack the energy to penetrate the outer dead layer of skin, but if α-emitters (e.g. ^{226}Ra or ^{222}Rn) have been inhaled, ingested or absorbed into the blood stream, sensitive living tissue can be exposed to α radiation. The resulting biological damage may increase the probability of cancer and particularly lung cancer in humans following inhalation. Other α-emitters such as ^{210}Po are soluble and circulate through the body to every tissue and cell in levels much higher than ^{222}Rn and, as previously mentioned, can be easily transferred into edible parts of plants. Consequently, ^{210}Po has a combined radiological and chemical effect causing genetic damage and early death from diseases such as liver and bladder cancer, stomach ulcer, leukaemia, liver cirrhosis, cardiovascular diseases and lung cancer (Carvalho and Oliveira 2006; Rego 2009).

Direct exposure to β particles is more hazardous because emissions from strong sources can penetrate the skin. If ingested, particles released directly to living tissue from isotopes such as ^{206}Tl or ^{214,210}Pb create damage at the molecular level, which can disrupt cell function. β particles are much smaller and have less charge than α particles, so they generally travel further into tissues leading to widely dispersed cellular damage and cancer.

Chemical toxicity varies according to the nature and concentration of the isotopic element generated. For example, the effects of lead are well documented (Juberg *et al.* 1997) and it is now known that ^{218,214,210}Pb exhibit a toxic chemical effect following accumulation in the bodies of water organisms and soil organisms even at very low concentration. Consumption of lead contaminated foods can lead to a range of epidemiological effects including damage to the kidneys, brain and central nervous system.

The toxicology of thallium (^{206}Tl), bismuth (^{214}Bi) and mercury (^{208}Hg) have all been extensively reviewed (Fowler and Sexton 2007; Peter and Viraraghavan 2005; Zahir *et al.* 2005) and although present in very low concentration are highly soluble. Each is readily transported into the food chain and is toxic to humans and some fish and plant species. However, further research is required to clearly define transport mechanisms, bioaccumulation and longevity of these radionuclides in ecosystems.

Conclusions

Wherever mining is undertaken, risks to ecosystem, cultural and social structures and human health exist. These risks vary according to the mineral commodity, extent of human reliance upon the impacted ecosystem and the ability of companies and governments to work within appropriate environmental monitoring, legislative and regulatory frameworks. With each of these risks comes a cost, if not to the company then to the government or the general public during and after mining operations. However, it is difficult to place a monetary value on many of these risks because they are based upon different perceptions of ecological values and appropriate accounting tools therefore do not exist. Evaluation is further complicated in many cases because adequate baseline information is lacking, and monitoring systems from which the extent of impact can be determined are not comprehensive. Consequently, many of the risk-related costs of mining remain hidden.

The perceived value of an ecosystem increases with mounting sensitivity (mainly because the perceived loss of function is greater), but it is very difficult to consistently express this value in mathematical or financial terms mainly because socio-cultural reliance on the ecosystem cannot be defined by standard accounting theory. It is even more difficult to develop a ranking or comparative system of valuation because the wide disparity in social structures, cultural norms and environmental types does not allow a standard point against which ecological sensitivity can be referenced. In short, different ecosystems are valued differently by different cultural groups and, as a result, there are no common grounds for comparison of value across those groups.

These differences in values are reflected in a lack of consistent and systematic means of accounting by which we can evaluate (i) the 'worth' of a species within an ecological structure; and (ii) the cost of damage to a cultural system that relies upon that species. Without a consistent approach, it becomes difficult to balance the long-term cost of mining upon ecological and socio-cultural values against the shorter-term income returned from exploiting a commodity. Consequently, a mining project's total value cannot be fully assessed in the absence of ecological and cultural 'worth', which makes it difficult to determine whether or not exploitation of a resource is of net positive benefit or negative impact to society. However, in the absence of suitable accounting procedures, a risk-based approach may be useful for exploring the qualitative assessment of mining impacts on socio-economic, cultural and health systems.

Uranium mining creates some of the highest risks to ecologically and culturally sensitive areas and represents one of the most contentious and emotive activities undertaken by the mining industry. The principal risks are twofold – chemical and radiological – each of which will have immediate and long-term effects. The examples used here indicate that the risk to Kakadu National Park and the Bahi wetlands will be similar, while those of Namib-Naukluft National Park will be significantly different because of its desert ecosystem and subsequent diminished capacity for water borne transport and distribution of radionuclides and other chemical species throughout the local environment. Social and cultural responses to ecological change (real or otherwise perceived) brought on by uranium mining are likely to be significant and different for each of these areas. Again, the costs associated with this are difficult to assess because appropriate descriptions of the cultural landscapes of each of the affected areas are still being developed.

History shows that attempts to mine uranium in ecologically sensitive areas create challenges and complex political, social and technical problems that require innovative solutions. Powerful legislation and regulation, linked with comprehensive monitoring programs, are one means by which the risk and therefore the hidden cultural and social costs can be mitigated. Legislation and regulation in Australia's Northern Territory is strong and sophisticated, but even there the monitoring programs in place are not comprehensive and need review and improvement if the full impacts on ecosystems are to be assessed.

The situation in Tanzania and Namibia is far more problematic as legislation is still being developed. Environmental and health monitoring is hardly comprehensive and regulation is effectively non-existent. Both countries look towards uranium mining as one means of alleviating an unacceptably low standard of living, but instead may inadvertently create long-term financial, environmental and social problems if the political will to implement adequate regulation of uranium mining does not exist. Development of legislation, regulation and monitoring systems comes at significant cost to governments and to mining companies and, even when these systems are in place, environmental impacts are never completely mitigated during and after the operational phase of mining. As a result, attendant risks to local socio-economic, cultural and ecosystems that are ultimately borne by the general public or land owners, remain without recompense and represent a potential source of future litigation should a means to calculate their worth ever be determined.

This paper has placed a strong focus on the biophysical aspects of ecological sensitivity and how they relate to society. It has not considered other technical costs that will be required to ensure an adequate level of environmental protection is maintained during operations and in the post-mining environment, neither does it consider costs to reputation brought about by political activities and poor public relations that appear to be common to uranium mining projects. These are matters that still need to be raised and discussed at length as ecologists and the mining industry seek to develop an understanding of exactly what are considered to be ecologically sensitive areas and how they should be valued.

One relevant concern that still needs to be raised is just how far is society willing to go when making decisions about potential damage to the world's ecological heritage. Protecting known deposits that are located in ecologically sensitive areas by legislation is only a temporary solution because it is only a matter of time before political pressure is brought to bear to enable their exploitation. Should this occur, governments and local communities must be in a position to undertake a comprehensive cost-benefit analysis of the project that considers not only the net return to society from mineral exploitation, but also the net loss to society through increased health costs and loss of vital ecosystem services, socio-economic and cultural materials. It is therefore essential that a systematic means of evaluating these be developed and applied worldwide during planning of the mine to improve cost-benefit analyses and assist in protecting the world's most sensitive ecosystems from any type of development.

There are clearly considerable gaps in the mining industry's ability to fully address the socio-economic and cultural costs attributed to mining ventures in ecologically sensitive areas. It is therefore recommended that to help address these gaps, the following steps are taken before commencement of mining activities:

- Full inventories of species are recorded.
- Levels of production and use of natural resources are measured.
- Relevant systems of directly monitoring human health are developed and baselines measured.
- More comprehensive and targeted systems of monitoring of key species are applied.
- A means of either calculating a monetary value for non-marketable natural resources is determined. If this cannot be done, then a means of equating qualitative estimates in terms applicable to cost-benefit analyses must be devised.

These are significant tasks, but if they cannot be achieved in the face of 'sophisticated' technical, environmental and social management systems that exist in countries like Australia, then what hope is there for protecting the ecologically sensitive areas of less sophisticated and developing countries of Africa? For it is in these situations in particular, where there is limited legislation and regulation, that the true cost of mining in ecologically sensitive areas may remain hidden and not become fully visible until well into the future.

Acknowledgements

The author gratefully acknowledges the indigenous Mirarr landowners, without whose kind assistance and permission, the cultural mapping of the Ranger uranium mine project area could not have been undertaken, and the Northern Land Council for encouraging this work and permitting its publication.

References

Argonne National Laboratory (2005) Human Health Fact Sheet – Natural decay series: Uranium, Radium and Thorium. Argonne National Laboratories. http://www.ead.anl.gov/pub/doc/natural-decay-series.pdf

Australian Government (1999) Environmental Requirements of the Commonwealth of Australia for the Operation of Ranger Uranium Mine. Department of the Environment, Water, Heritage and the Arts, Canberra.

Bakobi BLM (1993) Conservation of wetlands of Tanzania. In: *Wetlands of Tanzania. Proceedings of a Seminar on Wetlands of Tanzania*. 27–29 November 1991, Morogoro, Tanzania. (Eds GL Kamukala, SA Crafter) pp. 15–25.

Bayliss P, van Dam R, Boyden J, Walden D (2006) Ecological risk assessment of Magela floodplain to differentiate mining and non-mining impact. Supervising Scientist Report 189. Supervising Scientist, Darwin NT.

Bolhoffer A, Brazier J, Humphrey C, Ryan B, Esparon A (2011) A study of radium bioaccumulation in freshwater mussels, *Velesunio angasi*, in the Magela Creek catchment, Northern Territory, Australia. *Journal of Environmental Radioactivity* **102**, 964–974.

Boyd J, Banzhaf S (2006) What are ecosystem services? The need for standardised environmental accounting units. Discussion Paper RFF–02–06. Resources for the Future, Washington.

Brennan KG, Noller BN, LeGras C, Morton SR, Dostine PL (1992) Heavy metals in waterbirds from the Magela Creek Flood Plain, Alligator Rivers Region, Northern Territory, Australia. Supervising Scientist Technical Memorandum 36. Australian Government Publishing Service.

Carson RT, Wilks L, Imber D (1994) Valuing the preservation of Australia's Kakadu Conservation Zone. *Oxford Economic Papers (New Series) – special issue on environmental economics* **46**, 727–749.

Carvalho FP, Oliveira JM (2006) Polonium in cigarette smoke and radiation exposure of lungs. *Czechoslovak Journal of Physics* **56**(4), D697–D703. doi:10.1007/s10582-006-0568-6

Childress B, Hughes B, Harper D, van den Bossche W (2007) East African flyway and key site network of the Lesser Flamingo (*Phoenicopterus minor*) documented through satellite tracking. *The Ostrich* **78**(2), 463–468. doi:10.2989/OSTRICH.2007.78.2.55.135

Childress B, Nagy S, Hughes B (2008) International Single Species Action Plan for the Conservation of the Lesser Flamingo (*Phoenicopterus minor*). AEWA Technical Series No. 34. Bonn, Germany.

Commonwealth of Australia (2006) Uranium Mining, Processing and Nuclear Energy – Opportunities for Australia? Report to the Prime Minister by the Uranium Mining, Processing and Nuclear Energy Review Taskforce, December 2006.

Council for Scientific and Industrial Research (2008) High Confidence Study of Children Potentially Affected by Radionuclide and Heavy Metal Contamination Arising from the Legacy of Mine Water Management Practices on the Far West Rand of South Africa. Project Concept Note: CSIR, South Africa.

Culwick AT (1931) Ritual use of rock paintings at Bahi, Tanganyika Territory *MAN* **31**(41–59), 33–36.

Digby Wells and Associates (2008) Environmental Impact Assessment and Environmental Management Plan for the Valencia Uranium Project. Report to Valencia Uranium Pty Ltd, April 2008, Digbey Wells and Associates, Randburg, South Africa.

Dinesen L (2004) African Spoonbill *Platalea alba* breeding in Bahi Swamp, Tanzania. *Bulletin of the African Bird Club* **11**, 44–45.

Equator Principles Association (2006) The Equator Principles – a financial industry benchmark for determining, assessing and managing social and environmental risk in project financing http://www.equator-principles.com/documents/Equator_Principles.pdf

Field BC (1997) *Environmental Economics – An Introduction*. 2nd edition. McGraw-Hill, New York.

Finlayson CM, Lowry J, Bellio MG, Nou S, Pidgeon R, Walden D, Humphrey C, Fox G (2006) Biodiversity of the wetlands of the Kakadu Region, northern Australia. *Aquatic Sciences* **68**, 374–399. doi:10.1007/s00027-006-0852-3

Fowler BA, Sexton MJ (2007) Bismuth. In: *Handbook of the Toxicology of Metals*. 3rd edition. (Eds GF Nordberg, BA Fowler, M Nordberg, L Friberg) pp. 433–443. Academic Press, Burlington, USA.

Hamilton C (1996) Mining in Kakadu – Lessons from Coronation Hill: A lecture to the Parliamentary Library 'Vital Issues' Seminar Series. The Australia Institute, Discussion Paper 9, July 1996.

Henschel J (2008) The value of the Namib. *ROAN News* **3**, 34–36.

Hertzler G (2009) Dynamic non-market valuation of ecosystem services. Final Report to Land & Water Australia, Land and Water Australia.

Hoadley M, Limpitlaw D (2008) Managing the socio-economic impacts of uranium mining in Namibia. *ROAN News* **3**, 28–30.

Hogan A, van Dam R, Houston M, Lee N (2008) Toxicity of Ranger mine RP2 and Pit 3 waters to native freshwater species: 2007 wet season. Supervising Scientist Report 197, Supervising Scientist, Darwin NT.

Holdway DA (1992) Uranium mining in relation to toxicological impacts on inland waters. *Ecotoxicology (London, England)* **1**(2), 75–88. doi:10.1007/BF00831889

International Finance Corporation (2006) International Finance Corporation's Performance Standards on Social & Environmental Sustainability. World Bank, Washington, USA. http://www.ifc.org/ifcext/sustainability.nsf/Content/PerformanceStandards

Juberg DR, Kleiman CF, Kwon SC (1997) Position paper of the American Council on Science and Health: lead and human health. *Ecotoxicology and Environmental Safety* **38**(3), 162–180. doi:10.1006/eesa.1997.1591

Juergens N (2008) Commentary on the EIA for Trekkopje Uranium Project. *ROAN News* **3**, 15–18.

Linsley G (1997) Radiation and the environment: assessing effects on plants and animals. *International Atomic Energy Agency Bulletin* **39**(1), 17–20.

Lott C (Ed.) (1992) *The Spectrum Guide to Tanzania*. Camperapix Publishers International, Nairobi, Kenya.

Marchant R (1982) The macroinvertebrates of Magela Creek, Northern Territory. Supervising Scientist for the Alligator Rivers Region, Research Report 1, Australian Government Publishing Service.

Martin P, Hancock GJ, Johnston A, Murray AS (1995) Bioaccumulation of radionuclides in traditional Aboriginal foods from the Magela and Cooper Creek systems. Supervising Scientist for the Alligator Rivers Region, Research Report 11, Australian Government Publishing Service.

Martin P, Ryan B (2004) Natural-series radionuclides in traditional Aboriginal foods in tropical northern Australia: a review. *The Scientific World Journal* **4**, 77–95. doi:10.1100/tsw.2004.6

Masao FT (1976) The later stone age and the rock paintings of central Tanzania. PhD Thesis, Simon Fraser University, Toronto.

Mbogoro DK, Smith HD (2011) Economic and Ecological Research of the Bahi Swamp. Report to CESOPE, Dodoma, Tanzania, November 2010.

McCartney MP, Rebelo LM (2010) The value of the Bahi wetland, Tanzania, and threats to its ecological character; *presentation to the International Symposium on Wetlands in a Flood Pulsing Environment. 1–5 February 2010. Maun, Botswana.*

McGinley M (2007) East African Halophytics. In: *Encyclopedia of Earth*. http://www.eoearth.org/article/East_African_halophytics

McLaughlin D (1985) Problems of communicating radiation doses to Aboriginal members of the general public in the Alligator Rivers region. *Radiation Protection in Australasia* **3**(1), 67–70.

Ministry of Mines and Energy (2010) Strategic Environmental Assessment for the central Namib Uranium Rush. Ministry of Mines and Energy, Windhoek, Republic of Namibia.

Mpemba EB (1993) Wildlife resources and tourism in wetlands of Tanzania. In: *Wetlands of Tanzania. Proceedings of a Seminar on Wetlands of Tanzania.* (Eds GL Kamukala, SA Crafter) pp. 61–66. 27–29 November 1991, Morogoro, Tanzania.

Mudd GM (2002) Uranium mill tailings in the Pine Creek Geosyncline, northern Australia: past, present and future hydrogeological impacts. In: *Proceedings of the Uranium Mining & Hydrogeology III–3rd International Conference*, Freiberg, Germany, Sept. 2002, pp. 83–840.

Mwakaje AG (2009) Wetlands, livelihoods and sustainability in Tanzania. *African Journal of Ecology* **47**(supplement 1), 179–184. doi:10.1111/j.1365-2028.2008.01067.x

Neuzil EF, Dysart ME (1984) The determination of ^{90}Sr and ^{228}Ra in human teeth by age groups and in other substances. *The International Journal of Applied Radiation and Isotopes* **35**(12), 1113–1115. doi:10.1016/0020-708X(84)90143-1

Perry J (2008) Australian companies at forefront of African uranium 'revival'. *Australian Journal of Mining Online*. http://www.theajmonline.com.au/mining_news/news/2008/october/13

Peter ALJ, Viraraghavan Y (2005) Thallium: a review of public health and environmental concerns. *Environment International* **31**, 493–501. doi:10.1016/j.envint.2004.09.003

Ramsar Convention Secretariat (2006) The Ramsar Convention Manual: a guide to the Convention on Wetlands (Ramsar, Iran, 1971) 4th edition, Ramsar Convention Secretariat, Gland, Switzerland.

Rego B (2009) The Polonium Brief: a hidden history of cancer, radiation and the tobacco industry. *Isis* **100**, 453–484. doi:10.1086/644613

Rogers B, Mudd GM (2011) Water quality, water management and the Ranger Uranium Project: guidelines, trends and issues. *Water, Air, and Soil Pollution* **217**, 347–363.

Rosenthal HL, Gilster J, Bird JT (1963) Strontium-90 content of deciduous human incisors. *Science* **140**, 176–177. doi:10.1126/science.140.3563.176

Ryan B, Martin P, Iles M (2005) Uranium-series radionuclides in native fruits and vegetables of northern Australia. *Journal of Radioanalytical and Nuclear Chemistry* **264**(92), 407–412.

Sauerland S, Martin P, Humphrey C (2005) Radium-226 in Magela Creek, northern Australia: application of protection limits from radiation for humans and biota. *Radioprotection* **40**(Supplement 1), S451–S456.

Schubauer-Berigan MK, Daniels RD, Pinkerton LE (2009) Radon exposure and mortality among white and American Indian uranium miners: an update of the Colorado Plateau Cohort. *American Journal of Epidemiology* **169**(6), 718–730. doi:10.1093/aje/kwn406

Senate Environment, Communications, Information Technology and the Arts References Committee (2003) Regulating the Ranger, Jabiluka, Beverly and Honeymoon Uranium Mines. Senate Environment, Communications, Information Technology and the Arts References Committee (SECITARC), Commonwealth of Australia, October 2003.

Shemdoe RS, Kingazi SP, Kitula R, Chaggu EJ (2007) Reducing stresses on wetland resources in dryland ecosystems of Mpwapwa District, central Tanzania: where do we start? *Journal of Sustainable Development in Africa* **9**(1), 197–206.

Shindondola-Mote H (2009) *Uranium Mining in Namibia – the mystery behind low-level radiation*. Labour Resource and Research Institute, Windhoek, Namibia.

Smith HD (2008) Using traditional ecological knowledge to develop closure criteria in tropical Australia. In: *Mine Closure 2008 – Proceedings of the Third International Seminar on Mine Closure*, 14–17 October 2008, Johannesburg, South Africa. (Eds A Fourie, M Tibbett, I Weiersbye, P Dye) pp. 47–56.

Smith HD (2009) Strangers in a foreign land – developing cultural closure criteria for mines in Australia's Northern Territory. In: *Mine Closure 2009 – Proceedings of the Fourth International Seminar on Mine Closure*, 9–11 September 2009, Perth, Australia. (Eds A Fourie, M Tibbett) pp. 3–12.

Stratton A, Whitten S (2009) Putting a value on ecosystem goods and services. *Ecos* **148**(2), 17.

Supervising Scientist (2007) Annual Report 2006–2007, Supervising Scientist, Darwin.

Supervising Scientist (2009) Annual Report 2008–2009, Supervising Scientist, Darwin.

Tatz C, Cass A, Condon J, Tippett G (2006) Aborigines and Uranium: Monitoring the Health Hazards, AIATSIS Research Discussion Paper 20, Australian Institute of Aboriginal and Torres Strait Islander Studies, Canberra.

Taylor J (1999) Aboriginal people in the Kakadu region: social indicators for impact assessment. CAEPR Working Paper 4/1999. Centre for Aboriginal Economic Policy Research, Australian National University.

Turgis Consulting (2008) Trekkopje Uranium Project: Final ESIA Report.

United Nations Education Programme (2005) Protected areas and world heritage: Kakadu National Park, Northern Territory, Australia, United Nations Education Programme – World Conservation Monitoring Centre, Geneva.

Wilkins-Ellert MH (2004) *Acanthosicyos horridus* – record from Protabase. In *PROTA* (*Plant Resources of Tropical Africa / Ressources végétales de l'Afrique tropicale*). (Eds GJH Grubben, OA Denton) Wageningen, Netherlands.

Yanda PZ, Majule AE, Mwakaje AG (2005) Wetland utilisation, poverty alleviation and environmental conservation in semi arid areas of Tanzania – the case of Singida Region; presented at 10th Annual Workshop, Research on Poverty Alleviation, Dar-es-Salaam, Tanzania.

Zahir F, Rizwi SJ, Haq SK, Khan RH (2005) Low dose mercury toxicity and human health. *Environmental Toxicology and Pharmacology* **20**(2), 351–360. doi:10.1016/j.etap.2005.03.007

ENDEMISM IN IRONSTONE GEOSYSTEMS

4

Iron geosystems: priority areas for conservation in Brazil

*Claudia M Jacobi, Flávio F do Carmo,
Felipe F do Carmo and Iara C de Campos*

Summary

Brazil includes some of the most important metal-rich regions in the world. In particular, its ferriferous formations represent almost 20% of global iron ore reserves. Besides its huge mineral potential, the geoecologic systems constituted by ferriferous formations and their mountaintop outcrops – locally known as cangas – also provide environmental services, such as water reserves, that are vital to society. Natural communities characterised by high local and regional diversities, together with speleological and paleoenvironmental sites, confer a high conservation value to these systems. Many components of the local flora are endemic to cangas, and their associated caves are characterised by peculiar trophic resources and a rich invertebrate fauna. In spite of their significant environmental heritage, iron-rich areas are among the most threatened ecological systems in Brazil, due to their restricted distribution in association with important iron ore deposits, and their poor representation in the National Protected Areas System. Their vulnerability is inherently high, since mineral rights to mining companies (from exploration licences to extraction permits) covers almost 100% of all known canga sites, threatening 166 rare plants. Furthermore, according to the National Mining Plan, the majority of iron ore reserves are scheduled to be mined in the next 29 years. In addition, there is paucity of research, public information and legal instruments to promote their protection. This is probably one of the most challenging scenarios for environmental conservation in Brazil.

Introduction

Ironstone outcrops constitute, together with banded iron formations (BIFs), remarkable Brazilian geoecological systems. Because they are associated with mega-deposits of iron ore they are among the most threatened landscapes in the country (Jacobi and Carmo 2008a). Most of these formations are located on mountaintops, and host natural communities characterised by high local and regional diversity, with many examples of rare, endemic and threatened species. The high conservation value of these metal-rich areas is enhanced by the presence of significant speleological and paleoenvironmental sites and important aquifers.

In Brazil, there are still no public policies that specifically address the environmental and geoeconomic peculiarities related to ironstone outcrops and other ferriferous (iron-rich) formations, and multidisciplinary studies at a regional scale are lacking. At the same time, public and private demands for geological studies are greater than ever, at increasingly refined scales, with the objective of optimising the knowledge of iron ore deposits. The unprecedented growth of the Brazilian mineral production in response to the boost in global demand is a sign of the current and projected economic pressure on iron ecosystems which put at risk their biological, hydrological, paleoenvironmental and archaeological heritage.

Our aim is to present cangas and BIFs geosystems as key Brazilian conservation targets, based on the anthropogenic pressures, high diversity both locally and regionally, and ecosystem and cultural values. This is a necessary step to contribute to an effective land use plan aimed at the rational use of natural resources of Brazil.

Iron geosystems

Geology and geographical setting

With a continental area of 8.51 million km², Brazil harbours a geological diversity that places it among the largest world reserves of metallic ores. Currently, six of these minerals account for the main investments and mineral production in the country: iron, bauxite, manganese, chromium, nickel and niobium (Jacobi et al. 2011). Brazil has one of the largest world occurrences of ferriferous formations (Klein 2005), in which large high-quality iron ore mega-deposits occur, almost 20% of the world reserves. This key metallic mineral commodity represents more than 90% of the production and national reserves in Brazil (Table 4.1).

The main ferriferous formations – a generic denomination for sedimentary rocks that exhibit iron contents above 15% (Kimberley 1978) – are composed mostly by the banded iron formations (BIFs), namely sedimentary deposits with laminated rocks formed by alternating layers of silica and hematite-magnetite, as well as carbonates and iron silicates (Klein 2005; Spier et al. 2007). These BIFs date from the Late Archean to the Paleoproterozoic (Rosière and Chemale 2000) and are considered key features for the understanding of the evolution of the Earth surface (Spier et al. 2007). In Brazil the main BIF type is itabirite.

Genetically associated with these formations, and covering them, are Cretacean limonite-cemented iron-caps formed *in situ* over the iron-enriched host rocks (Dorr and Barbosa 1963; Vasconcelos et al. 2008) known in Brazil as 'cangas'. These structures may reach 30 m deep and tens of square kilometres. They usually occur in the highest portions of the landscape, constituting extensive plateaus separated by valleys and scarps (Fig. 4.1). The main iron geosystems (Fig. 4.2) and deposits associated with iron ore are located in East (Quadrilátero Ferrífero, Morro do Pilar, Peixe Bravo and Caetité), North (Carajás), and South-west Brazil (Urucum) (Dorr 1969; Walde 1986). They overlap with several outstanding world conservation sites (Table 4.2), namely the Atlantic Rainforest and Cerrado (Brazilian savannah), which are two

Table 4.1. Geo-economic summary of the main Brazilian metallic minerals (year 2008)

	Iron	Bauxite	Manganese	Chromium	Nickel	Niobium	Total
Reserves (million t)	40 000	3600	59	13.9	10	5.2	43 688.1
World reserves rank	5°	3°	7°	5°	7°	1°	
Production (million t/yr)	360	28.1	3.1	0.7	0.08	0.09	392.07
World production rank	2°	3°	1°	#1°	10°	1°	
Exports FOB (million US$)	16 538	293	616	16.5	376	1626	19 465.5

\# Americas
Source: IBRAM (2009), DNPM (2009, 2010a, 2010b).

Figure 4.1: Ironstone formations in the Quadrilátero Ferrífero. (a) close view of a canga; and of (b) banded iron formations; (c) Itatiaiuçu Peak, a huge BIF block standing 100 m above the surrounding cangas; (d) typical plateaus formed by cangas (1400 m a.s.l.), with the city of Belo Horizonte in the background.

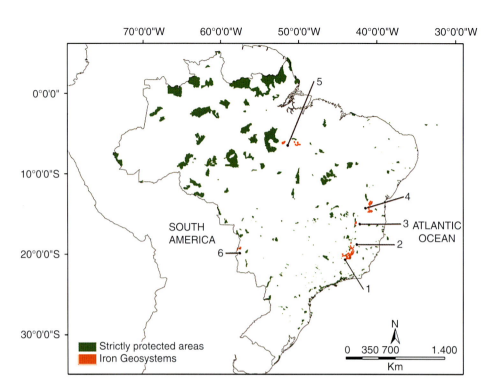

Figure 4.2: Location of main iron geosystems and all strictly protected areas in Brazil. 1) Quadrilátero Ferrífero; 2) Morro do Pilar; 3) Peixe Bravo; 4) Caetité; 5) Carajás; 6) Urucum.

of the 34 world biodiversity hotspots (Mittermeier *et al.* 2004); three of the 37 earth's last wild places (Mittermeier *et al.* 2002): Amazon Rainforest, Pantanal (world's largest floodplain) and Caatinga (the Brazilian semi-arid biome); and one of the world centres of plant diversity (Giulietti *et al.* 1997), the Espinhaço Range.

Ferruginous karst systems

Ferriferous formations are home to around 20% of the almost 8000 caves officially registered in Brazil (CECAV 2011; Piló and Auler 2009). In spite of their abundance, few scientific works have been produced regarding key aspects of their genesis, geology and ecology. Basic speleological surveys are more common, because they are required by the authorities to release mining permits.

Iron caves were ignored until recently in the Brazilian speleological scenario. The first descriptions were by Noël Aimé Pissis in a work published in 1842 (Piló and Auler 2005). But it wasn't until the end of 1950 that the first systematic surveys on iron geosystems were performed, following an accord signed between the Departamento Nacional de Produção Mineral (National Department of Mineral Production) (DNPM) and the United States Geological Survey (USGS). The aim of this agreement was to refine the geological mapping of the Quadrilátero Ferrífero (or Iron Quadrangle). Among the geologists working on the project was George C. Simmons, who produced a seminal paper on caves (Auler 2005; Simmons 1963). In the 1980s, Brazilian speleological groups initiated the inventory of iron caves in Carajás (Moreira and Paiva 1988; Pinheiro and Maurity 1988), describing the first archaeological sites in ferruginous caves.

Aside from caves, other macroforms such as canyons, dolines, ruiniform cliffs and a significant subterranean or partially subterranean network drainage, typical of karstic systems (Ford and Williams 1989), are also recurrent in iron geosystems. There are dissolution cupules in the roof and walls, and rock dissolution by drainage, among other microforms that are also characteristic of

Table 4.2. Geographical characteristics of the main Brazilian iron geosystems

Site	No. of mines[1]	Iron formation types[2]	Max. altitude (m)	Conservation interest	Mean rainfall (mm/yr)[3]
Quadrilátero Ferrífero	46	Metamorphic itabirites Contain iron ore oxide facies, such as haematites, magnetites and martites Age: Paleoproterozoic (2.6–2.1 Ga.)	1850	Atlantic Rainforest, Cerrado, Espinhaço	1650
Urucum	5	Anchimetamorphic jaspilites associated with glacial sediments Age: Proterozoic (0.6 Ga.)	950	Pantanal	1000
Carajás	1	Anchimetamorphic jaspilites Contain hematite and martite, followed by magnetite and goethite Age: Archean (2.7–2.6 Ga.)	800	Amazon Rainforest	2100
Peixe Bravo	licensing stage	Hematite diamictites and banded hematite quartzites, probably associated with glacial sediments Age: Proterozoic (0.6 Ga.?)	1000	Atlantic Rainforest, Caatinga, Cerrado, Espinhaço	800
Morro do Pilar	licensing stage	Itabirites with oxide facies Contain hematite, magnetite, siderite and ankerite Age: Archean/Proterozoic.	1100	Atlantic Rainforest, Cerrado, Espinhaço	1400
Caetité	licensing stage	Metamorphic itabirites with oxide facies Age: Archean/Proterozoic.	1100	Caatinga, Cerrado, Espinhaço	850

1. (DNPM 2007)
2. (Alecrim 1982; Rosière and Chemale 2000; Schobbenhaus and Coelho 1986)
3. (IBGE 2011)

karstic systems (Maurity and Kotschoubey 2005; Silva and Salgado 2009). It is possible then to hypothesise that geochemical dissolution had a pronounced role in structuring the regional topography, with elements typical of karstic landscapes (Carmo *et al.* 2011).

One of the most important elements of this system is its association with karstic-like aquifers, in which water is stored in cavities, channels and pores, as opposed to non-karstic sites, where water is stored in water tables (Hardt and Pinto 2009). In the Quadrilátero Ferrífero, BIFs and cangas constitute a system of aquifers with very large hydrogeological potential, with reserves estimated in 4×10^9 m^3. In these systems of itabiritic aquifers, the median outflow measured in springs is 400 m^3/day, reaching 12 000 m^3/day in some situations (CPRM 2004). Cangas, in particular, represent important recharge areas to subjacent aquifers due to their porosity and permeability (CPRM 2004). Thousands of springs and several sources associated with itabiritic aquifers provide water to Belo Horizonte, fifth in GNP and the sixth largest city in the country.

An important target area for future studies is a recently discovered speleological site located in the Peixe Bravo valley (Carmo *et al.* 2010), where tens of ferruginous caves were found, among which some with probably the highest halls (almost 8 m) for this lithotype, considering that the usual height of these cave halls is less than 3 m.

Paleoenvironmental and archaeological heritage

Four sites associated with ironstone formations were selected as candidates by the Comissão Brasileira dos Sítios Geológicos e Paleobiológicos (Brazilian Commission of Geologic and Palaeobiologic Sites)

(SIGEP) to integrate the UNESCO World Heritage list, due to their relevant geologic testimony of the evolution on earth. Three of these are located in the Quadrilátero Ferrífero. The first, Serra da Piedade, is an important geological heritage site, due to its excellent exposure of Paleoproterozoic itabirite, indicating atmospheric changes. It is also intimately related to the history of gold exploration in the Brazilian hinterland during the 17th century (Ruchkys et al. 2007). The second, Pico do Itabirito, is the type locality chosen by German geologist Wilhelm von Eschweg to describe the predominant BIF type in Brazil, itabirite, in 1822. In the third site, Fonseca, there is an important evidence of the tertiary flora in the tropical southern hemisphere. This fossil deposit, topped by cangas, is characterised by a great variety of angiosperm families, among which Melastomataceae and Mimosaceae are the most abundant (Berry 1935; Mello et al. 2000). The most notable fossil is that of *Eriotheca prima* (Malvaceae), and represents the first record of the Brazilian Caenozoic (Mello et al. 2000). The fourth site is in Urucum, with late-Precambrian iron formations evidencing traces of Neoproterozoic glaciation (SIGEP 2011).

Ceramic and stone artefacts located in Carajás caves and in the Quadrilátero Ferrífero, and petroglyphs in canga surfaces in Urucum are archaeological sites that indicate human occupation between 1500 to 8000 years B.P. (Baeta and Piló 2005; Girelli 1994; Oliver 2008). Recently, geometric scratches distributed over large areas inside some iron caves were discovered in the Peixe Bravo valley. These parallel V-shaped traces are similar to other archaeological records of rupestral art (Carmo et al. 2011).

Biodiversity
Plant communities

The mineralogical make up of the outcrops can reach 90% of iron oxides and hydroxides, and the soil – when present – is very acid, shallow and with low percentage of organic matter. The ample spectrum of habitats described for this system (Jacobi et al. 2007) such as caves, ponds, rock pools, bare rock and fissures (Fig. 4.3) creates ecological conditions that generally differ from the adjacent landscape – or matrix – in which they are imbedded, which are usually rainforests or savannahs. Two main types of physiognomies may be distinguished, depending on topographical soil accumulation capacity: open areas, where typical rupicolous herbs and shrubs are established; and forest islands, where depressions favour the concentration of shrubs and trees. These tree associations show milder microclimatic conditions (Fig. 4.4). Among the dominant genera are *Eremanthus* Less (Asteraceae), *Clusia* L. (Clusiaceae), *Mimosa* L., and *Senna* Mill. (Fabaceae), *Leandra* Raddi and *Miconia* Ruiz and Pav. (Melastomataceae), *Eugenia* L. and *Myrcia* DC. *ex* Guill. (Myrtaceae), *Psychotria* L. (Rubiaceae), *Callisthene* Mart., and *Vochysia* Aubl. (Vochysiaceae).

In open areas, temperatures that reach almost 70°C and relative humidity of less than 10% have been measured close to exposed rock substratum (Fig. 4.4). Great temperature and air humidity amplitudes are found both in time (annually and daily) and space (in microhabitats separated by a few centimetres). These microclimatic conditions, common to other rock outcrop types, induce high levels of evapotranspiration in plants that indirectly inhibit photosynthesis due to reduced stomatal opening. This extreme ecophysiological situation is reflected in adaptations such as succulency, xeromorphism and desiccation tolerance, i.e. the ability of survive cycles of dehydration and subsequent rehydration (Kluge and Brulfert 2000; Porembski and Barthlott 2000; Szarzynski 2000). Among the genera dominant that present species with these adaptations are the succulents *Discocactus* Pfeiff., *Pilosocereus* Byles and Rowley and *Tacinga* Britton and Rose (Cactaceae, a Brazilian endemic genus), *Acianthera* Scheidw. and *Hoffmannseggella* H.G.Jones, endemic to East Brazil (Orchidaceae), xeromorphic species of *Aspilia* Thouars, *Baccharis* L., *Lychnophora* Mart. (a Brazilian endemic genus), and *Symphyopappus* Turcz. (Asteraceae), *Chamaecrista* Moench and *Mimosa* L. (Fabaceae), *Croton* L. (Euphorbiaceae), *Byrsonima* Rich. *ex* Kunth (Malpighiaceae), *Tibouchina* Aubl.

Figure 4.3: Habitat heterogeneity in cangas. (a) Column cacti (*Pilosocereus pachyclados*) over bare rock; (b) forest islands are found in cangas above 1500 m, where humidity supports a diversity of epiphytes; (c) ferns and *Blastocaulon* sp. (Eriocaulaceae) are exclusive of penumbral habitats, such as cave entrances; (d) patches of *Vellozia graminea* (Velloziaceae); (e) pond on doliniform depression in one of the cangas with highest altitude. Pictures a and c are from Peixe Bravo and the others from the Quadrilátero Ferrífero.

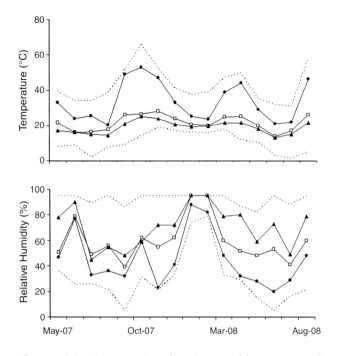

Figure 4.4: Mean microclimatic variables measured monthly at 15:00 hours in a Quadrilátero Ferrífero canga (1450 m a.s.l.). ● open vegetation/bare rock; ▢ forest island; ▲ cave. Max–min (dotted lines) in 24 hours.

(Melastomataceae), *Lippia* L. and *Stachytarpheta* Vahl (Verbenaceae). Desiccation-tolerant species of *Vellozia* Vand. (Velloziaceae) and *Trilepis* Nees (Cyperaceae) are also common.

A considerable within-site (α) diversity correlated with microhabitat heterogeneity seems to characterise plant communities in cangas. In only 10 ironstone outcrops located throughout the Quadrilátero Ferrífero, with a maximum distance of 90 km among them and with a total area of less than 450 ha, almost 1100 species of vascular plants have been inventoried (unpublished data). A previous list (Jacobi and Carmo 2008a) on the vegetation of six cangas, which totalled 458 species of vascular plants, noted that less than 5% of species were common to all outcrops. This high regional (β) diversity probably resulted from the combined effect of landscape isolation, climatic amplitude and mineral variation of the substrate, particularly weathering.

There are great knowledge gaps regarding plant communities in Brazilian iron geosystems. In the Quadrilátero Ferrífero, Carajás and Urucum, floristic and community structure studies have been performed in small areas, of which few have been published (Jacobi *et al.* 2007; Mourão and Stehmann 2007; Secco and Mesquita 1983; Silva *et al.* 1996; Viana and Lombardi 2007). There are still no floristic or plant ecology publications on the Caetité, Morro do Pilar and Peixe Bravo sites.

Rare plants

In Brazil, the geographic distribution of 166 rare plant species (*sensu* Giulietti *et al.* 2009) are superimposed totally or partially with the main iron geosystems (Jacobi *et al.* 2011). These species represent roughly 7% of all known rare plant species in the country. Plants are considered rare if they had a known distribution restricted to 1° latitude and 1° longitude, corresponding to 10 000 km^2 (Giulietti *et al.* 2009). This criterion is more restrictive than the one proposed by Eken *et al.* (2004) for the determination of Key Biodiversity Areas (KBA). Moreover, plants with linear distribution, such as those on mountain ranges, certainly occupy a much smaller area and most of these narrow endemics are known to only one location. These 166 species are distributed in 45 families and 91 genera of vascular plants (see Appendix 4.1). Among the families with most number of species are Melastomataceae (17 spp. = 10.2%), Eriocaulaceae (15 spp. = 9%), Fabaceae (14 spp. = 8.4%), Velloziaceae (13 spp. = 7.8%), Cactaceae (11 spp. = 6.6%), Asteraceae and Verbenaceae (7 spp. = 4.2% each), Apocynaceae and Bromeliaceae (6 spp. = 3.6% each). The richest genera are *Chamaecrista* (12 spp.), *Vellozia* (9 spp.), *Paepalanthus* (7 spp.), *Microlicia* (5 spp.), *Actinocephalus* (endemic to Brazil), *Barbacenia, Lavoisiera, Luxemburgia* (endemic to Brazil), *Stachytarpheta* and *Symplocos* (4 spp. each).

The Quadrilátero Ferrífero harbours 40% of these rare plants, followed by Peixe Bravo (28%), Morro do Pilar (16%), Caetité and Carajás (6.5% each), and Urucum (3%). Probably, the peculiar environmental setting together with geographic isolation and the age of cangas and BIFs have contributed to the evolution of species with distribution restricted to one or few sites, such as the cactus *Arthrocereus glaziovii* and the milkweed *Minaria monocoronata*, both known to only a few ironstone outcrops in the Quadrilátero Ferrífero.

Biospeleological richness

The few biospeleological studies point to an invertebrate community with extreme structural complexity, high richness of troglomorphic organisms (i.e. bearing morphological, physiological or behavioural adaptations to cave environments) and high β diversity, when compared to communities of other lithotypes. Ferreira (2005) found 334 invertebrate morphospecies from 90 families in 13 ferruginous caves. Frequent genera are *Endecous* (Ensifera: Phalangopsidae), *Zelurus* (Heteroptera: Reduviidae), *Mesabolivar* (Araneida: Pholcidae), *Ctenus* (Araneida: Ctenidae), *Plato* (Araneida: Theridiosomatidae), and *Pseudonannolene* (Spirostreptida: Pseudonannolenidae). He concluded that in iron caves the most abundant trophic resource is the roots growing from plants established on the duricrust, that reach the cavities via microchannels and fissures, which are very common in cangas, but are an unusual resource in other Brazilian caves. In some dripping spots, columnar structures are formed from the anastomosed root growth. These are rare structures only found in some Czech caves (Ferreira 2005).

In ferruginous geosystems, the occurrence of sub-surface interstitial space among compartments probably favours the migration of fauna through the cavities, enhancing the richness mediated by the balance between colonization and extinction (Silva 2008). This condition, associated to the presence of microchannels, makes possible the frequent transit of fauna between the epigeal (aboveground) and hypogeal (caves) environments, enhancing its diversity (Ferreira 2005).

Caves in ferruginous rocks have the lowest similarity and largest mean richness of troglomorphic invertebrates (5.33 spp./cave) when compared with carbonatic, siliciclastic and magmatic caves (Silva 2008). A great variety of troglomorphic groups have evolved in this system, with several occurrences of invertebrates new to science, such as the harvestman *Canga renatae* (Da Silva *et al.* 2010) from Carajás. The evolutionary processes that have led to the formation of these relict groups were attributed to the superficiality of the caverns, the net of microchannels and the aboveground environmental harshness by Ferreira (2005) and Silva (2008). Ferreira (2008) stressed that the loss of extensive areas of cangas due to iron ore mining has caused large disconnection of troglomorphic species populations. Possibly in the past the cave systems must have exhibited subterranean connectivity, maintaining genetic variability, since individuals of the same troglomorphic species have been found in caves separated by distances of 30 km.

Environmental liability and public policies on natural resources use

Surface mining is always characterised by high environmental impact to evolutionary exceptional ecosystems as well as threats to human health (Finger *et al.* 1999; Gibson *et al.* 2010; Jacobi and Carmo 2008b; Palmer *et al.* 2010). The impacts resulting from the extraction of iron ore are classified as direct, of high magnitude, and irreversible, according to the current Brazilian legislation. This classification is mainly due to the way in which iron ore is extracted and processed, resulting in the complete destruction of the landscape in the region where the deposits are located. The cangas on top of these deposits are discarded and the ore is usually obtained using thousands of tons of explosives. The opencast pits may reach diameters of 8 km and a perimeter of 30 km. In some cases, the pits may be around 0.5 km deep and expose the water table. Nowadays, the foreseen operational life for the largest mines is between 12 and 18 years. During this period, it is necessary to reduce the water table level by pumping water from the pit. After concluding the operation, artificial lakes are formed in which evaporation exceeds groundwater inflow, constituting large groundwater sinks (*sensu* Johnson and Wright 2001).

The extraction process generates a material with low iron concentration, called spoil tips and stored near the pits. In Brazil, for each ton of iron ore, another ton of spoil material is produced. If the ore enrichment is done via wet milling, it generates spoils in the form of pulp. The most usual method for disposing of this toxic material is in tailings dams, frequently located in natural valleys. In most cases the vegetation of these valleys, usually forest communities with surface hydrology, is lost.

Besides the environmental impacts inherent to opencast mining, the indirect pollution potential must also be considered. The deposits, and consequently the main mining sites, are frequently located on the top or sides of mountains that shape the catchment. The iron system itself constitutes an aquifer with high capacity of water recharge and storage. Therefore, the pollution potential throughout the system is high when alterations of great magnitude occur in the highest parts of the relief. A study performed in the Velhas basin, in the Quadrilátero Ferrífero, showed that water pollution from heavy metals, notably arsenic and antimony, generated by mining activities may dissipate for over 400 km from the pollution source (Veado *et al.* 2006).

Currently Brazil is the second largest world provider or iron ore. In 2010, around 370 million tons were produced, most of which extracted from the Quadrilátero Ferrífero geosystems. This region alone concentrates almost 80% of the mines in activity today, followed by Urucum and Carajás (Table 4.2). The main buyer of this commodity is China, where 60% of all the exported iron ore is sent. Considering that the Chinese deficit of this raw material is ~1 billion tons/year, continued growth of this sector is predicted (IBRAM 2010). The government foresees that the annual production might reach 1.1 billion tons by 2030 (MME 2010). This corresponds to the Brazilian production of the 1990 decade, or the world production of 2003. To attain these goals, about US$40 billion will be invested during the 2010–2015 period (IBRAM 2011). According to a survey on Brazilian transnational companies in South America (Novoa 2009), a considerable portion of this investment comes from public funds via the *Banco Nacional de Desenvolvimento Econômico e Social* (National Development Bank) (BNDES).

Nowadays, the areas granted by the National Department of Mineral Production to the iron mining companies, ranging from geological surveys to mining operation permits, reach ~300 000 km^2 of continental Brazil (Jacobi *et al.* 2011). In these areas are located 99% of the ironstone outcrops. According to the National Mining Plan, a program that sets the mineral sector parameters, all the known profitable reserves in the country will have been mined in the next 29 years (MME 2010).

There are still no adequate public policies for the geoeconomic and environmental specificity inherent to iron geosystems. At the same time, the financial resources are not sufficient to effectively protect this unique natural asset (Jacobi *et al.* 2011). In 2008, the Sistema Nacional de Unidades de Conservação (National Protected Areas System) (SNUC) received federal resources of about US$200 million (MMA 2009). Considering that the strictly protected areas in Brazil correspond to roughly 500 000 km^2, the 2008 budget was of meagre US$400/$km^2$.

The great increase in environmental impact assessment studies that the government now requires from polluting industries has generated positive changes in the environmental legislation, particularly related to the speleological patrimony. In the 1988 Federal Constitution, natural subterranean caves were considered patrimony of the State. In 1990, a decree established that all cavities should be preserved, together with their immediate surroundings (*a priori* 250 m). This caused the delay or cessation of several business enterprises, including mining projects. Recently, more flexible laws were passed so as to reduce the alleged antagonism between the country's economic growth and the preservation of speleological sites. In 2008, a new decree classified caves in four relevance types according to attributes based on their geology, biology, hydrology, palaeontology, scenery, and historic, cultural and socio-economic aspects. With this change, every area with speleological potential that is a target of impacting activities must be the object of environmental impact studies to determine the degree of relevance of its caves. This said, only caves considered of maximum relevance are granted full protection, with no degradation allowed. Cavities of high, medium or low relevance may be impacted or destroyed, provided that the company takes conservation compensatory measures, not necessarily – very rarely, indeed – associated with the target geosystem. Although this new decree establishes the conservation of cavities of maximum relevance to substitute those marked for destruction, the Sociedade Brasileira de Espeleologia (Brazilian Speleological Society) (SBE) fears that this new change in the legislation will come to reduce the number of known caves by 70% (SBE 2011).

Perspectives for the conservation of ferruginous geosystems

Surface mining is one of the most powerful activities capable of altering the landscape associated with iron geosystems in Brazil. The same situation may be found in other iron ore-producing countries, among them Australia. Floristic surveys of BIF ranges (DEC 2009; Gibson *et al.* 2007) and reviews of groundwater resources (Johnson and Wright 2001) already propose that biodiversity conservation issues be taken into account in environmental assessments of any mining proposal. These initiatives and others, such as reinforcing taxation of polluting industries and mandatory ecosystem economic value studies, could be adapted to the Brazilian reality.

A significant contribution to the conservation of iron geosystems should be the revision of the current Brazilian list of threatened species. Only 13% of the rare plants associated to these systems (see Appendix 4.1) are cited in the official List of Threatened Flora Species (MMA 2008b), in part because species with insufficient data are not included. Given that these 'red lists' represent a legal instrument for the protection of biodiversity (Rodrigues *et al.* 2006), this lack of representation needs to be reassessed by the Ministry of the Environment, because it jeopardises the conservation of species and their environment. In support of this opinion, Rapini *et al.* (2009) state that the species cited in the catalogue of Brazilian Rare Plants (Giulietti *et al.* 2009) could be classified as threatened by extinction, because they fit the criterion of restricted distribution (B1 *sensu* Standards IUCN and Petitions Subcommittee 2010). Besides, in the case of the 166 rare plants associated with iron geosystems, their habitats are practically 100% superimposed with the allotment of mining titles in various stages, from geological surveys to operation permits and, therefore, directly threatened by mining activities (Jacobi *et al.* 2011). Inadequate planning could lead to the loss of rare plant species, unique to the biological Brazilian patrimony, if their sites of occurrence are not acknowledged by the licensing agencies as extremely vulnerable areas (Rapini *et al.* 2009).

The Convention on Biological Diversity declared that among the essential actions to promote the conservation of biological diversity is the establishment of a system of protected areas, where special measures need to be taken for the *in situ* conservation of ecosystems and natural habitats. The identification of priority areas in Brazil constitutes a relevant initiative to subsidise the location of new protected areas (Rylands and Brandon 2005). The government has recognised the extreme environmental importance of regions where the main iron geosystems are located and indicated them as priority areas for conservation, sustainable and common use of the Brazilian biodiversity. In most cases, the priority action has been the creation of protected areas (MMA 2008a). In spite of this, iron geosystems are still not directly contemplated by conservation and management actions. Even under direct immediate threat, cangas are not yet well represented in the SNUC. Currently, less than 1% of these ecosystems are contained in strictly protected areas, such as national and state parks. In the Peixe Bravo, Caetité and Morro do Pilar geosystems, there are still large well preserved areas, not yet subjected to the impacts of opencast mining. However, the first site environmental impact studies, necessary for obtaining operating licences, are in progress, while the other two of these licences have already been granted.

Considering this scenario, the recent proposal of the Chico Mendes Biodiversity Conservation Institute (ICMBio) of the first national park with the objective of specifically preserving iron geosystems is a great step forward. The target area for this new protected area is Serra do Gandarela (Quadrilátero Ferrífero), home to thousands of hectares of canga immersed in one of the last remnants of Atlantic Rainforest in the region. At the same time, this site represents a target area for iron ore exploitation. Indeed, the environmental licences are already in their final stage, which will allow the beginning of operations of one of the largest iron ore mines in the country. This situation is probably one of the most challenging scenarios for conservation in Brazil. Beyond the short-term economic benefits generated by mining, the conservation of iron geosystems would bring benefits not only to the biota but to society, granting the maintenance of ecosystem services, not to mention the perpetuation of cultural and environmental heritage. In this perspective, the incorporation of economic value studies of goods and environmental services – such as water recharge and storage – would represent a relevant factor in the rational approach of the use of natural resources. The economic value would grant that the natural capital be adequately incorporated in the cost-benefit analyses of mining activities and in the compensatory measures related to the licensing of activities that cause significative environmental impacts (Camphora and May 2006; May *et al*. 2000).

Acknowledgements

This research was partially supported by the Foundation for Research Support of Minas Gerais (FAPEMIG, grant APQ387/08) and the Brazilian National Research Council (CNPq, grant 479834/2008-3). The contents of this chapter, however, do not necessarily reflect the views of these agencies. The Peixe Bravo geosystems research is a joint collaboration with the Minas Gerais Public Prosecutors of São Francisco River, Rio Pardo and Rio Verde Grande basins (PJSF) and the Interinstitutional Group of Studies and Environmental Actions of Northern Minas (NIEA-NM). We thank the Military Environmental Police of Taiobeiras (MG) for logistic support. We thank the Brazilian National Research Council (CNPq) and the Coordination for the Improvement of Higher-education Personnel (CAPES) for scholarships to all the authors. The valuable suggestions of two anonymous reviewers are duly acknowledged.

References

Alecrim JD (1982) *Recursos minerais do estado de Minas Gerais*. METAMIG, Belo Horizonte.

Auler AS (2005) O legado espeleológico de George Simmons. *O Carste* 17, 73.

Baeta A, Piló H (2005) Arqueologia do Quadrilátero Ferrífero: aspectos preliminares sobre sua ocupação. Estudo de caso do Capão Xavier. *O Carste* 17, 116–131.

Berry EW (1935) Tertiary plants from Brazil. *Proceedings of the American Philosophical Society* 75, 565–590.

Camphora AL, May PH (2006) A valoração ambiental como ferramenta de gestão em unidades de conservação: há convergência de valores para o bioma Mata Atlântica? *Megadiversidade* 2, 24–38.

Carmo FF, Carmo FF, Leles BP, Jacobi CM (2010) *Primeiros registros de cavernas ferruginosas no norte de Minas Gerais*. Boletim Eletrônico da Sociedade Espeleológica do Brasil 163, Campinas. http://www.sbe.com.br

Carmo FF, Carmo FF, Salgado AAR, Jacobi CM (2011) New speleological site in ferruginous systems, in Rio do Peixe Bravo valley, Northern Minas Gerais. *Espeleo-Tema* 22, 79–93.

CECAV – Centro Nacional de Estudo, Proteção e Manejo de Cavernas (2011) *Mapa de Potencialidades de Ocorrência de Cavernas Baseada na Litologia*. Brasília. http://www.icmbio.gov.br/cecav/index.php

CPRM – Companhia de Pesquisas e Recursos Minerais (2004) *Projeto APA Sul RMBH: hidrogeologia, mapa hidrogeológico escala 1:50.000* **8**, SEMAD/CPRM, Belo Horizonte.

Da Silva MB, Pinto-da-Rocha R, Giribet G (2010) *Canga renatae*, a new genus and species of Cyphophthalmi from Brazilian Amazon caves (Opiliones: Neogoveidae). *Zootaxa* **2508**, 45–55.

DEC – Department of Environment and Conservation2009) Floristic surveys of the Banded Ironstone Ranges. *Information Sheet* **23**, 1–2.

DNPM – Departamento Nacional de Produção Mineral (2007) *Universo da Mineração Brasileira*. Brasília. http://www.dnpm.gov.br

DNPM – Departamento Nacional de Produção Mineral (2009) *Economia Mineral do Brasil*. DNPM, Brasília.

DNPM – Departamento Nacional de Produção Mineral (2010a) *SIGMINE: Informações Geográficas da Mineração*. http//sigmine.dnpm.gov.br

DNPM – Departamento Nacional de Produção Mineral (2010b) *Sumário Mineral* **29**, Brasília. http://www.dnpm.gov.br

Dorr JVN (1969) Physiographic, stratigraphic and structural development of the Quadrilátero Ferrífero, Minas Gerais. *U.S. Geological Survey Professional Paper* **641-A**, 110

Dorr JVN, Barbosa ALM (1963) Geology and ore deposits uf the Itabira District, Minas Gerais, Brazil. *U.S. Geological Survey Professional Paper* **341-C**, 110

Eken G, Bennum L, Brooks TM, Darwall W, Fishpool LDC, Foster M, Knox D, Langhammer P, Matiku P, Radford E, Salaman P, Sechrest W, Smith ML, Spector S, Tordoff A (2004) Key biodiversity areas as site conservation targets. *Bioscience* **54**, 1110–1118. doi:10.1641/0006-3568 (2004)054[1110:KBAASC]2.0.CO;2

Ferreira RL (2005) A vida subterrânea nos campos ferruginosos. *O Carste* **17**, 106–115.

Ferreira RL (2008) Ecossistemas subterrâneos nos campos ferruginosos. In *Simpósio Afloramentos Ferruginosos no Quadrilátero Ferrífero: biodiversidade, conservação e perspectivas de sustentabilidade*. Belo Horizonte, Minas Gerais, CD, pp. 52–65.

Finger A, Stolton S, Dudley N (Eds) (1999) *Metals from the forests: mining and forest degradation, Arborvitae, special issue*. World Wide Fund for Nature & International Union for Conservation of Nature, Gland, Switzerland.

Ford D, Williams P (1989) *Karst geomorphology and hydrology*. Unwin Hyman, London.

Gibson N, Coates DJ, Thiele KR (2007) Taxonomic research and the conservation status of flora in the Yilgarn Banded Iron Formation ranges. *Nuytsia* **17**, 1–12.

Gibson N, Yates CJ, Dillon R (2010) Plant communities of the ironstone ranges of South Western Australia: hotspots for plant diversity and mineral deposits. *Biodiversity and Conservation* **19**, 3951–3962. doi:10.1007/s10531-010-9939-1

Girelli M (1994) *Lajeados com gravuras na região de Corumbá, MS*. PhD dissertation. Universidade do Vale do Rio dos Sinos, Brazil.

Giulietti AM, Pirani JR, Harley RM (1997) *Espinhaço Range Region, Eastern Brazil, Site SA20. Centres of plant diversity: South America*. http://botany.si.edu/projects/cpd/sa/sa20.htm

Giulietti AM, Rapini A, Andrade MJG, Queiroz LP, Silva JMC (2009) *Plantas Raras do Brasil*. Conservation International, Belo Horizonte.

Hardt R, Pinto SAF (2009) Carste em litologias não carbonáticas. *Revista Brasileira de Geomorfologia* **10**, 99–105.

IBGE – Instituto Brasileiro de Geografia e Estatística (2011). http://www.ibge.gov.br

IBRAM – Instituto Brasileiro de Mineração (2009) *Informações e Análises da Economia Mineral Brasileira (The Brazilian Mining Association)*. IBRAM, Brasília. http://www.ibram.org.br

IBRAM – Instituto Brasileiro de Mineração (2010) *Informativo do Instituto Brasileiro da Mineração, Ano V* **37**. IBRAM, Brasília. http://www.ibram.org.br

IBRAM – Instituto Brasileiro de Mineração (2011) *Informativo do Instituto Brasileiro da Mineração. Ano VI* **41**. IBRAM, Brasília. http://www.ibram.org.br

Jacobi CM, Carmo FF (2008a) The contribution of ironstone outcrops to plant diversity in the Iron Quadrangle, a threatened Brazilian landscape.

Ambio **37**, 324–326. doi:10.1579/0044-7447(2008) 37[324:TCOIOT]2.0.CO;2

Jacobi CM, Carmo FF (2008b) Diversidade dos campos rupestres ferruginosos no Quadrilátero Ferrífero, MG. *Megadiversidade* **4**, 24–32.

Jacobi CM, Carmo FF, Vincent RC, Stehmann JR (2007) Plant communities on ironstone outcrops – a diverse and endangered Brazilian ecosystem. *Biodiversity and Conservation* **16**, 2185–2200. doi:10.1007/s10531-007-9156-8

Jacobi CM, Carmo FF, Campos IC (2011) Soaring Extinction Threats to Endemic Plants in Brazilian Metal-Rich Regions. *Ambio* **40**, 540–543.

Johnson SL, Wright AH (2001) Central Pilbara Groundwater Study, Water and Rivers Commission, Hydrogeological Record Series, Report HG 8.

Kimberley MM (1978) Paleoenvironmental Classification of Iron Formations. *Economic Geology and the Bulletin of the Society of Economic Geologists* **73**, 215–229. doi:10.2113/gsecongeo.73.2.215

Klein C (2005) Some Precambrian banded iron formations (BIFs) from around the world: Their age, geologic setting, mineralogy, metamorphism, geochemistry, and origin. *The American Mineralogist* **90**, 1473–1499. doi:10.2138/am.2005.1871

Kluge M, Brulfert J (2000) Ecophysiology of vascular plants on Inselbergs. In *Inselbergs – biotic diversity of isolated rock outcrops in tropical and temperate regions*. (Eds S Porembski and W Barthlott) pp. 37–48. Springer-Verlag, Berlin.

Maurity CW, Kotschoubey B (2005) Evolução recente da cobertura de alteração no Platô N1 Serra dos Carajás-PA. Degradação, pseudo-carstificação, espeleotemas. *O Carste* **17**, 78–91.

May PH, Veiga Neto F, Chévez O (2000) *Valoração econômica da biodiversidade: estudos de caso no Brasil*. PROBIO, Ministério do Meio Ambiente, Secretaria de Biodiversidade e Florestas, Brasília.

Mello CL, Sant' Anna LG, Bergqvist LP (2000) The palaeontological site of Fonseca, Minas Gerais state, Brazil (Fossil plants of the Tertiary of Brazil), In: *Sítios Geológicos e Paleontológicos do Brasil*. (Eds C Schobbenhaus, DA Campos, ET Queiroz, M Winge & M Berbert-Born). http://www.unb.br/ig/sigep/sitio086/sitio086.htm

Mittermeier RA, Mittermeier CG, Gil PR, Pilgrim J, Fonseca G, Konstant WR, Brooks T (2002) *Wilderness: Earth's Last Wild Places*. Conservation International & CEMEX, Mexico.

Mittermeier RA, Gil PR, Hoffman M, Pilgrim J, Brooks T, Mittermeier CG, Lamoreux J, Fonseca GAB (2004) *Hotspots revisited: earth's biologically richest and most endangered terrestrial ecoregions*. CEMEX & Agrupacion Sierra Madre, Mexico.

MMA – Ministério do Meio Ambiente (2008a) Áreas Prioritárias para Conservação, Uso Sustentável e Repartição de Benefícios da Biodiversidade Brasileira: Atualização - Portaria MMA n°9, de 23 de janeiro de 2007. *Ministério do Meio Ambiente* (*Série Biodiversidade*) **31**, MMA Secretaria de Biodiversidade e Florestas, Brasília.

MMA – Ministério do Meio Ambiente (2008b) Instrução Normativa n° 6, de 23 de setembro de 2008. *Lista Oficial das Espécies da Flora Brasileira Ameaçadas de Extinção*.

MMA – Ministério do Meio Ambiente (2009) *Pillars for the Financial Sustainability of the National Protected Areas System*. MMA, Brasília.

MME – Brazilian Ministry of Mines and Energy (2010) *National Mining Plan, 2030*. MME, Brasília.

Moreira JRA, Paiva RS (1988) Levantamento bioespeleológico preliminar da Serra dos Carajás-PA-Brasil. In *Congresso de Espeleologia da América Latina e do Caribe 1*. Belo Horizonte, Brazil, Anais, 142–150.

Mourão A, Stehmann JR (2007) Levantamento da flora do campo rupestre sobre canga hematítica couraçada remanescente na Mina do Brucutu, Barão de Cocais, Minas Gerais, Brasil. *Rodriguésia* **58**, 775–786.

Novoa LF (2009) O Brasil e seu "desbordamento": o papel central do BNDES na expansão das empresas transnacionais brasileiras na América do Sul. In *Empresas transnacionais brasileiras na América Latina: um debate necessário*. (Eds Instituto Rosa Luxemburgo Stiftung, Movimento dos atingidos por barragem (MAB), Rede Social Justiça e Direitos Humanos, Campanha Justiça nos Trilhos, Instituto Políticas alternativas para o Cone Sul (PACS), Rede Brasil sobre Instituições Financeiras Multilaterais and Rede Brasileira de Justiça Ambiental) pp. 187–204. Expressão Popular, São Paulo.

Oliver JR (2008) The Archaeology of Agriculture in Ancient Amazônia. In *The Handbook of South American Archaeology*. (Eds H Silvermann and WH Isbell) pp. 185–216. Springer, New York.

Palmer MA, Bernhardt ES, Schlesinger WH, Eshleman KN, Foufoula-Georgiou E, Hendryx MS, Lemly AD, Likens GE, Loucks OL, Power ME, White PS, Wilcock PR (2010) Mountaintop mining consequences. *Science* **327**, 148–149. doi:10.1126/science.1180543

Piló LB, Auler A (2005) Cavernas em minério de ferro e canga de Capão Xavier, Quadrilátero Ferrífero, MG. *O Carste* **17**, 92–105.

Piló LB, Auler A (2009) Geoespeleologia das cavernas em rochas ferríferas da região de Carajás, PA. In *Congresso Brasileiro de Espeleologia*. Montes Claros, Belo Horizonte, Brazil, Anais, 181–186.

Pinheiro RVL, Maurity CW (1988) As cavernas em rochas intempéricas da Serra dos Carajás Brasil. In *Congresso de Espeleologia da América Latina e do Caribe*. Belo Horizonte, Brazil, Anais, 179–186.

Porembski S, Barthlott W (2000) Granitic and gneissic outcrop (Inselbergs) as centers of diversity for desiccation-tolerant vascular plants. *Plant Ecology* **151**, 19–28. doi:10.1023/A:1026565817218

Rapini A, Andrade MJG, Giulietti AM, Queiroz LP, Silva JMC (2009) Introdução. In *Plantas Raras do Brasil*. (Eds AMGiulietti, A Rapini, MJG Andrade, LP Queiroz and JMC Silva) pp. 23–28. Conservation International, Belo Horizonte.

Rodrigues ASL, Pilgrim JD, Lamoreaux JF, Hoffmann M, Brooks TM (2006) The value of the IUCN Red List for conservation. *Trends in Ecology & Evolution* **21**, 71–76. doi:10.1016/j.tree.2005.10.010

Rosière CA, Chemale F, Jr (2000) Brazilian Iron Formations and their Geological Setting. *Revista Brasileira de Geociencias* **30**, 274–278.

Ruchkys UA, Renger FE, Noce CM, Machado MMM (2007) Serra da Piedade, Iron Quadrangle, state of Minas Gerais: from the myth of Sabarabuçu towards the historical, geological, natural and religious heritage. In *Geological and Palaeontological Sites of Brazil*. (Eds M Winge, C Schobbenhaus, M Berbert-Born, ET Queiroz, DA Campos, CRG Souza and ACS Fernandes). http://www.unb.br/ig/sigep/sitio129/sitio129english.pdf

Rylands AB, Brandon K (2005) Brazilian protected areas. *Conservation Biology* **19**, 612–618. doi:10.1111/j.1523-1739.2005.00711.x

SBE - Brazilian Society of Speleology (2011) Intervenção em sítios espeleológicos só com anuência do IBAMA e do ICMBIO. *Boletim Eletrônico da Sociedade Brasileira de Espeleologia* **185**. http://www.cavernas.org.br

Schobbenhaus C, Coelho CES (Eds) (1986) *Principais depósitos minerais do Brasil: Ferro e metais da indústria do aço*. 2nd edn. DNPM/CVRD/CPRM, Brasília – DF.

Secco RS, Mesquita AL (1983) Nota sobre a vegetação de canga da Serra Norte. *I Boletim do Museu Paraense Emílio Goeldi Série Botânica* **59**, 1–13.

SIGEP - Brazilian Comission of Geologic and Paleobiologic Sites (2011) Morraria do Urucum, Corumbá, Ladário, MS. http://sigep.cprm.gov.br/propostas/morraria_do_urucum_MS.htm

Silva M (2008) Ecologia e conservação das comunidades de invertebrados cavernícolas na Mata Atlântica Brasileira. PhD dissertation. Instituto de Ciências Biológicas, Universidade Federal de Minas Gerais, Brazil.

Silva JR, Salgado AAR (2009) Mapeamento das unidades de relevo da Serra do Gandarela –Quadrilátero Ferrífero/MG. *Geografias* **5**, 107–125 [UFMG].

Silva MFF, Secco RS, Lobo MGA (1996) Aspectos ecológicos da vegetação rupestre da Serra dos Carajás, estado do Pará, Brasil. *Acta Amazonica* **26**, 17–44.

Simmons GC (1963) Canga caves in the Quadrilátero Ferrífero, Minas Gerais, Brazil. *The National Speleological Society Bulletin* **25**, 66–72.

Spier CA, Oliveira SMB, Sial AN, Rios FJ (2007) Geochemistry and genesis of the banded iron formations of the Cauê Formation, Quadrilátero Ferrífero, Minas Gerais, Brazil. *Precambrian Research* **152**, 170–206. doi:10.1016/j.precamres.2006.10.003

Standards IUCN and Petitions Subcommittee (2010) Guidelines for Using the IUCN Red List Categories and Criteria. Version 8.1. Prepared by the Standards and Petitions Subcommittee in March 2010. http://intranet.iucn.org/webfiles/doc/SSC/RedList/RedListGuidelines.pdf

Szarzynski J (2000) Xeric islands: environmental conditions on inselbergs. In *Inselbergs – biotic diversity of isolated rock outcrops in tropical and temperate regions.* (Eds S Porembski and W Barthlott) pp. 37–48. Springer-Verlag, Berlin.

Vasconcelos P, Monteiro H, Farley K, Spier K (2008) Cangas: Self-healing protective covers that inhibit the erosion of deeply weathered banded iron-formations. In *VII Simpósio Nacional de Geomorfologia: I Encontro Latino-Americano de Dinamica e Diversidade de Paisagens e Geomorfologia*. Belo Horizonte, Brazil, Anais.

Veado MARV, Arantes IA, Oliveira AH, Almeida MRMG, Miguel RA, Severo MI, Cabaleiro HL (2006) Metal pollution in the environment of Minas Gerais State – Brazil. *Environmental Monitoring and Assessment* **117**, 157–172. doi:10.1007/s10661-006-8716-9

Viana PL, Lombardi JA (2007) Florística e caracterização dos campos rupestres sobre canga na Serra da Calçada, Minas Gerais, Brasil. *Rodriguésia* **58**, 159–177.

Walde DHG (1986) Geologia do Ferro. In *Principais Depósitos Minerais do Brasil: Ferro e Metais da Indústria do Aço*. 2nd edn. (Eds C Schobbenhaus and CES Coelho) pp. 3–6. DNPM/CVRD/CPRM, Brasília – DF.

Appendix 4.1. Brazilian rare plants (*sensu* Giulietti *et al.* 2009) whose known distribution overlaps either totally or partially with the main ferruginous geosystems √: threatened species according the official Brazilian list (MMA 2008b)

Species	Iron geosystem
Acanthaceae	
Staurogyne minarum (Nees) Kuntze.	Quadrilátero Ferrífero
Staurogyne vauthieriana (Nees) Kuntze.	Quadrilátero Ferrífero
Staurogyne warmingiana (Hiern) Leonard √	Quadrilátero Ferrífero
Stenandrium hatschbachii Wassh.	Peixe Bravo
Stenandrium stenophyllum Kameyama √	Peixe Bravo
Amaranthaceae	
Gomphrena centrota Holzh.	Urucum
Gomphrena serturneroides Suess.	Caetité
Pfaffia argyrea Pedersen √	Peixe Bravo
Annonaceae	
Annona saffordiana R.E. Fries.	Quadrilátero Ferrífero
Apocynaceae	
Ditassa longisepala (Hua) Fontella and E.A. Schwarz.	Quadrilátero Ferrífero
Mandevilla semirii M.F.Sales, Kin.-Gouv. and A.O. Simões.	Peixe Bravo
Marsdenia bergii Morillo.	Carajás
Matelea morilloana Fontella.	Caetité
Minaria monocoronata (Rapini) T.U.P. Konno and Rapini.	Quadrilátero Ferrífero
Minaria refractifolia (K.Schum.) T.U.P. Konno and Rapini.	Morro do Pilar
Aquifoliaceae	
Ilex loranthoides Mart. ex Reissek.	Quadrilátero Ferrífero
Araceae	
Philodendron biribirense Sakur. and Mayo.	Morro do Pilar
Araliaceae	
Schefflera lucumoides (Decne. and Planch. ex Marchal) Frodin and Fiaschi.	Quadrilátero Ferrífero
Arecaceae	
Syagrus werdermannii Burret.	Caetité
Asteraceae	
Antheremanthus hatschbachii H.Rob. √	Peixe Bravo
Aspilia grazielae J.U.Santos √	Urucum
Chionolaena lychnophorioides Sch. Bip.	Quadrilátero Ferrífero

(Continued)

Appendix 4.1. (Continued)

Species	Iron geosystem
Dasyphyllum trychophyllum (Baker) Cabrera.	Quadrilátero Ferrífero
Senecio gertii Zardini.	Peixe Bravo
Stevia resinosa Gardner.	Quadrilátero Ferrífero
Viguiera corumbensis Malme √	Urucum
Bignoniaceae	
Amphilophium perbracteatum A.H. Gentry.	Caetité
Lundia damazii DC.	Quadrilátero Ferrífero
Bromeliaceae	
Dyckia bracteata (Wittm.) Mez.	Quadrilátero Ferrífero
Dyckia densiflora Schult.f.	Quadrilátero Ferrífero
Encholirium irwinii L.B. Sm.	Peixe Bravo
Encholirium reflexum Forzza and Wand.	Peixe Bravo
Orthophytum humile L.B. Sm.	Peixe Bravo
Orthophytum supthutii E.Gross and Barthlott.	Morro do Pilar
Cactaceae	
Arrojadoa albicoronata (Van Heek et al.) P.J. Braun and Esteves.	Peixe Bravo
Arrojadoa multiflora F. Ritter.	Caetité
Arthrocereus glaziovii (K. Schum.) N.P. Taylor and Zappi.	Quadrilátero Ferrífero
Brasilicereus markgrafii Backeb. and Voll √	Peixe Bravo
Cipocereus laniflorus N.P. Taylor and Zappi √	Quadrilátero Ferrífero
Discocactus horstii Buining and Brederoo √	Peixe Bravo
Discocactus pseudoinsignis N.P. Taylor and Zappi.	Peixe Bravo
Micranthocereus auriazureus Buining and Brederoo √	Peixe Bravo
Micranthocereus polyanthus (Werderm.) Backeb. √	Caetité
Micranthocereus violaciflorus Buining.	Peixe Bravo
Pilosocereus fulvilanatus (Buining and Brederoo) F. Ritter.	Peixe Bravo
Caryophyllaceae	
Paronychia fasciculata Chaudhri.	Quadrilátero Ferrífero
Celastraceae	
Maytenus radlkoferiana Loes.	Quadrilátero Ferrífero
Convolvulaceae	
Bonamia corumbaensis Hoehne.	Urucum

Appendix 4.1. (Continued)

Species	Iron geosystem
Ipomoea carajensis D. Austin √	Carajás
Ipomoea cavalcantei D.F. Austin √	Carajás
Ipomoea marabensis D. Austin and R. Secco.	Carajás
Cyperaceae	
Hypolytrum paraense M. Alves and W.W. Thomas.	Carajás
Droseraceae	
Drosera graomogolensis T. Silva.	Peixe Bravo
Eriocaulaceae	
Actinocephalus callophyllus (Silveira) Sano.	Morro do Pilar
Actinocephalus ciliatus (Bong.) Sano.	Morro do Pilar
Actinocephalus falcifolius (Koern.) Sano.	Quadrilátero Ferrífero
Actinocephalus glabrescens (Silveira) Sano.	Morro do Pilar
Eriocaulon aquatile Koern.	Morro do Pilar
Eriocaulon carajense Moldenke.	Carajás
Leiothrix gomesii Silveira.	Quadrilátero Ferrífero
Leiothrix milho-verdensis Silveira.	Morro do Pilar
Paepalanthus conduplicatus Koern.	Quadrilátero Ferrífero
Paepalanthus digitiformis Hensold.	Morro do Pilar
Paepalanthus flaviceps Koern.	Quadrilátero Ferrífero
Paepalanthus garimpensis Silveira.	Quadrilátero Ferrífero
Paepalanthus grão-mogolensis Silveira.	Peixe Bravo
Paepalanthus langsdorffii (Bong.) Koern.	Quadrilátero Ferrífero
Paepalanthus xiphophyllus Ruhland.	Quadrilátero Ferrífero
Erythroxylaceae	
Erythroxylum nelson-rosae Plowman.	Carajás
Fabaceae	
Chamaecrista aristata (Benth.) H.S. Irwin and Barneby.	Peixe Bravo
Chamaecrista caracencis (H.S. Irwin and Barneby) H.S. Irwin and Barneby.	Quadrilátero Ferrífero
Chamaecrista fodinarum H.S. Irwin and Barneby.	Peixe Bravo
Chamaecrista fuscescens (Benth.) H.S. Irwin and Barneby.	Morro do Pilar
Chamaecrista hatschbachii H.S. Irwin and Barneby.	Peixe Bravo
Chamaecrista itabiritoana (H.S. Irwin and Barneby) H.S. Irwin and Barneby.	Quadrilátero Ferrífero

(Continued)

Appendix 4.1. (Continued)

Species	Iron geosystem
Chamaecrista phyllostachya (Benth.) H.S. Irwin and Barneby.	Peixe Bravo
Chamaecrista pilicarpa (Harms) H.S. Irwin and Barneby.	Quadrilátero Ferrífero
Chamaecrista stillifera (H.S. Irwin and Barneby) H.S. Irwin and Barneby.	Peixe Bravo
Chamaecrista strictifolia (Benth.) H.S. Irwin and Barneby.	Peixe Bravo
Chamaecrista tephrosiifolia (Benth.) H.S. Irwin and Barneby.	Peixe Bravo
Chamaecrista ulmea H.S. Irwin and Barneby.	Peixe Bravo
Harpalyce parvifolia H.S. Irwin and Arroyo.	Peixe Bravo
Lonchocarpus variabilis R.R. Silva and A.M.G. Azevedo.	Urucum
Gentianaceae	
Deianira damazioi E.F. Guim.	Quadrilátero Ferrífero
Lamiaceae	
Hyptidendron claussenii (Benth.) Harley √	Quadrilátero Ferrífero
Hyptis piranii Harley.	Peixe Bravo
Hyptis rhypidiophylla Briq. √	Quadrilátero Ferrífero
Hyptis tricephala A. St-Hil. ex. Benth.	Quadrilátero Ferrífero
Lauraceae	
Cryptocarya sellowiana P.L.R. de Moraes.	Quadrilátero Ferrífero
Nectandra venulosa Meisn.	Morro do Pilar
Persea pedunculosa Meisn.	Quadrilátero Ferrífero
Lentibulariaceae	
Utricularia physoceras P. Taylor.	Carajás
Lythraceae	
Cuphea disperma Koehne.	Morro do Pilar
Cuphea rubro-virens T.B. Cavalc.	Peixe Bravo
Lythraceae	
Cuphea teleandra Lourteig √	Peixe Bravo
Diplusodon aggregatifolius T.B. Cavalc.	Peixe Bravo
Diplusodon saxatilis Lourteig.	Morro do Pilar
Malpighiaceae	
Banisteriopsis salicifolia (DC.) B. Gates.	Quadrilátero Ferrífero
Melastomataceae	
Cambessedesia pityrophylla (Mart. ex DC.) A.B. Martins.	Quadrilátero Ferrífero

Appendix 4.1. (Continued)

Species	Iron geosystem
Cambessedesia salviifolia (Cham.) A.B. Martins.	Morro do Pilar
Eriocnema acaulis Triana √	Quadrilátero Ferrífero
Eriocnema fulva Naudin √	Quadrilátero Ferrífero
Lavoisiera bradeana Barreto.	Morro do Pilar
Lavoisiera mellobarretoi Markgr.	Peixe Bravo
Lavoisiera punctata Mart. and Schrank ex DC.	Quadrilátero Ferrífero
Lavoisiera tetragona Mart. and Schrank ex DC.	Morro do Pilar
Marcetia hatschbachii A.B. Martins.	Peixe Bravo
Microlicia cuspidifolia Mart.	Quadrilátero Ferrífero
Microlicia glazioviana Cogn.	Quadrilátero Ferrífero
Microlicia microphylla (Naudin) Cogn.	Quadrilátero Ferrífero
Microlicia suborbicularifolia Hoehne.	Quadrilátero Ferrífero
Microlicia trichocalycina DC.	Morro do Pilar
Trembleya calycina Cham.	Quadrilátero Ferrífero
Trembleya hatschbachii Wurdack and E. Martins.	Peixe Bravo
Trembleya rosmarinoides DC.	Quadrilátero Ferrífero
Monimiaceae	
Macropeplus schwackeana.	Quadrilátero Ferrífero
Myrtaceae	
Accara elegans (DC.) Landrum.	Quadrilátero Ferrífero
Campomanesia prosthecepala Kiaersk.	Quadrilátero Ferrífero
Eugenia blanda Sobral.	Peixe Bravo
Ochnaceae	
Luxemburgia ciliatibracteata Sastre.	Morro do Pilar
Luxemburgia corymbosa A.St.-Hil.	Quadrilátero Ferrífero
Luxemburgia flexuosa Sastre.	Morro do Pilar
Luxemburgia speciosa A.St.-Hil.	Morro do Pilar
Ouratea hatschbachii K. Yamamoto.	Peixe Bravo
Olecaceae	
Chionanthus greenii Lombardi.	Quadrilátero Ferrífero
Orchidaceae	
Habenaria itacolumia Garay √	Quadrilátero Ferrífero

(Continued)

Appendix 4.1. (Continued)

Species	Iron geosystem
Orobanchaceae	
Agalinis schwackeana (Diels) V.C. Souza and Giul.	Quadrilátero Ferrífero
Esterhazya nanuzae V.C. Souza.	Morro do Pilar
Passifloraceae	
Passiflora hypoglauca Harms.	Quadrilátero Ferrífero
Picramniaceae	
Picramnia ferrea Pirani and W.W. Thomas.	Carajás
Plantaginaceae	
Stemodia lobata J.A. Schmidt.	Quadrilátero Ferrífero
Poaceae	
Axonopus carajasensis M.N.C. Bastos √	Carajás
Colanthelia distans (Trin.) McClure.	Quadrilátero Ferrífero
Paspalum brachytrichum Hack.	Quadrilátero Ferrífero
Polygalaceae	
Securidaca acuminata A.St.-Hil. and Moq.	Quadrilátero Ferrífero
Rubiaceae	
Borreria semiamplexicaule E.L. Cabral.	Carajás
Simaroubaceae	
Simaba suaveolens A.St.-Hil.	Quadrilátero Ferrífero
Symplocaceae	
Symplocos angulata Brand.	Quadrilátero Ferrífero
Symplocaceae	
Symplocos insolita Aranha, P.W. Fritsch and Almeida.	Morro do Pilar
Symplocos microstyla Aranha, P.W. Fritsch and Almeida.	Quadrilátero Ferrífero
Symplocos saxatilis Aranha, P.W. Fritsch and Almeida.	Morro do Pilar
Turneraceae	
Turnera caatingana Arbo.	Caetité
Turnera coccinea Arbo.	Peixe Bravo
Turnera involucrata Arbo.	Caetité
Velloziaceae	
Barbacenia reflexa L.B. Sm. and Ayensu.	Peixe Bravo
Barbacenia riparia (N.L. Menezes and Mello-Silva) Mello-Silva.	Peixe Bravo

Appendix 4.1. (Continued)

Species	Iron geosystem
Barbacenia umbrosa L.B. Sm. and Ayensu.	Peixe Bravo
Barbacenia williamsii L.B. Sm.	Quadrilátero Ferrífero
Vellozia bradei Schulze-Menz.	Peixe Bravo
Vellozia costata L.B. Sm. and Ayensu.	Morro do Pilar
Vellozia gigantea N.L. Menezes and Mello-Silva.	Quadrilátero Ferrífero
Vellozia graomogolensis L.B. Sm.	Peixe Bravo
Vellozia luteola Mello-Silva and N.L. Menezes.	Peixe Bravo
Vellozia maxillarioides L.B. Sm.	Peixe Bravo
Vellozia prolifera Mello-Silva.	Peixe Bravo
Vellozia sellowii Seub.	Quadrilátero Ferrífero
Vellozia spiralis L.B. Sm.	Peixe Bravo
Verbenaceae	
Bolchea chascanoides Moldenke.	Peixe Bravo
Lantana gracilis T. Silva.	Caetité
Lippia rhodocnemis Mart. and Schauer.	Morro do Pilar
Stachytarpheta ajugifolia Schauer.	Quadrilátero Ferrífero
Stachytarpheta discolor Cham.	Morro do Pilar
Stachytarpheta galactea S. Atkins.	Caetité
Stachytarpheta glandulosa S. Atkins.	Caetité
Xyridaceae	
Xyris nigricans L.A. Nilsson √	Quadrilátero Ferrífero
Xyris villosicarinata Kral and Wand.	Quadrilátero Ferrífero

5

Flora and vegetation surveys of the ironstone ranges of the Yilgarn Craton

*Neil Gibson, Rachel Meissner,
Adrienne S Markey and Wendy A Thompson*

Summary

A series of ancient subdued ironstone ranges stretch over 700 km from the edge of the species-rich south-west of Western Australia into the arid interior. This area, well to the south of the major iron ore producing region of the Pilbara, has had little systematic flora survey. Limited surveys in early 1990s indicated that some of these ironstone ranges were a repository for localised endemic plant species and restricted vegetation communities. Following the rapid growth in the price of iron ore and the expansion of iron ore exploration and mining in these ranges in the early 2000s, a major survey program was undertaken to document the biodiversity values of these ranges in order to provide a regional basis for the assessment of development proposals. Between 2005 and 2008 some 24 ranges were surveyed. These surveys initially concentrated in areas of high mineral prospectively but now cover a wide cross-section of the ironstone ranges across the Yilgarn Craton. The surveys have highlighted the unique nature of the vegetation on each range; the occurrence of species with distributions restricted to, or centred on these ranges; and have identified over 20 new species. These data illustrate the high biodiversity values of the ironstone ranges of the Yilgarn and will be integral in the assessment of future resource development proposals over the ranges.

Introduction

The Yilgarn Craton is one of the most ancient land surfaces on the planet (Fig. 5.1). It comprises granite basement rocks 2600–3000 million years old, most of which has been above sea level and has not been glaciated for the last 260 million years (Anand and Butt 2010). On this ancient landscape are found a series of ironstone ranges. The rocks of these ranges, called Banded Iron Formation (BIF), are of similar age to the granites and have undergone extensive transformation and weathering over millions of years. The ranges occur from the boundary of the species-rich south-west (roughly coincident with the 300 mm rainfall isohyet) into the arid zone for over 700 km (Fig. 5.1). The shrublands that dominate the ranges contrast strongly with the vegetation of the adjacent plains, which gradually change from eucalypt woodlands and sand plains to hummock grasslands (*Triodia*) and mugla woodlands (*Acacia aneura*) further inland (Fig. 5.2). These

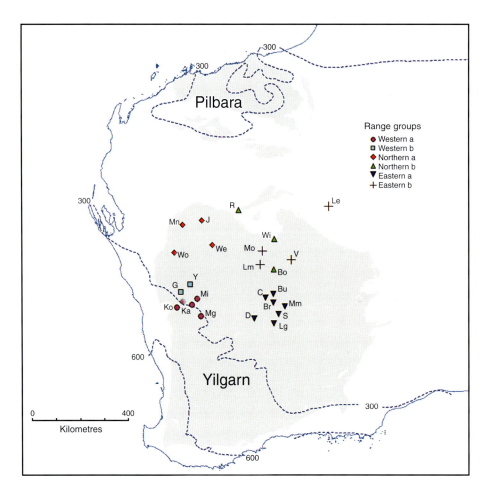

Figure 5.1: Distribution of the Pilbara and Yilgarn Cratons with the location of the 24 ironstone ranges studied and their broad biogeographical groupings. Rainfall isohyets for 300 mm (arid zone boundary) and 600 mm shown as dashed lines. Details of the individual ranges can be found in Appendix 5.1.

ranges rarely rise more than 200 m above the surrounding plain and until recently have been little studied. The pioneering mapping work of John Beard in the late 1970s identified a unique series of vegetation communities occurring along topographic gradients on some of these ranges near the boundary of the south-west (Beard 1981). These patterns and the occurrence of endemic species restricted to these ranges were later confirmed by surveys undertaken in the 1990s (Gibson *et al.* 1997; Gibson 2004 and references therein).

Modernisation and industrial development in China has been one of the main drivers for the recent study of the patterns of the flora and vegetation of these ranges in southern Western Australia. Historically, most of iron ore production in Australia has come from the extensive BIF ranges in the Pilbara Craton of north-western Australia (Fig. 5.1). The five-fold increase in the price of iron ore between 2000 and 2008 (ABARE 2010) has lead to the development of mines in the small ranges of the Yilgarn Craton located well south of the Pilbara. Even at full production, mines in these ranges will only produce a small fraction of the ore being extracted in the Pilbara.

This most recent wave of exploration and mining in the ironstone ranges of the Yilgarn was lead by the recommissioning and expansion of the Koolyanobbing mine by Portman Pty Ltd in early 2000. Their studies in the Mt Jackson Range–Windarling Ridge–Helena and Aurora Range area documented several highly localised endemic

Figure 5.2: (a) Ironstone ridge of Jasper Hill in the western ranges. Shrublands on the range give way to eucalypt woodlands on the surrounding flats (Photo: S Dillon); (b) Close up of the banding typical of ironstones (Photo: S Dillon); (c) Ridge on Mt Ida in the eastern ranges. Shrublands again dominate the range with mulga (*Acacia aneura*) woodlands on the flats (Photo: R Meissner).

species and identified restricted vegetation types specific to individual ranges. The projected impacts on the conservation values of these ranges were so high that initially it was recommended that environmental approval be refused (EPA 2002).

By 2004, the Western Australian Department of Industry and Resources had identified over 30 ironstone ranges across the Yilgarn that they considered prospective for iron ore. Given the almost total lack of knowledge of the flora and vegetation of these ranges and the high conservation values identified in the ranges studied to date, it was considered that detailed surveys were urgently needed to provide a regional context for future resource development proposals. To that end the Western Australian Department of Environment and Conservation (DEC) funded a survey program covering 24 of the ironstone ranges over the next four years (2005–2008). In addition a review of the known biological and economic values of the ranges was commissioned (Government of Western Australia 2007). Twenty-three papers have been published describing the vegetation and flora of the individual ranges based on these surveys (Appendix 5.1). A meta-analysis of the patterns across these 24 ranges has recently been completed (Gibson *et al.* 2012).

Recent analysis of the patterns of the vegetation and flora of four other ironstone ranges (further south) found species' richness and composition to be strongly correlated with broad scale spatial pattern and local environmental factors (Gibson *et al.* 2010). It was suggested that the concentration of species with distributions largely restricted to ironstone substrates (referred to as ironstone specialists) could be related to soil chemistry but this was not investigated in that study.

In this paper we provide an overview of the patterns in the distribution of the flora and vegetation described to date, we investigate the broad scale biographical pattern found across the 24 ranges and further investigate the correlation between soil chemistry and distribution of ironstone specialist species.

Methods

On each range a minimum of 50 20-m by 20-m permanently located plots were established to sample geographic, topographic and geological substrates as far as practical given access restrictions (Fig. 5.1; Appendix 5.1). Ironstone substrates were the primary target for sampling but some associated geologies were included to examine faithfulness of vegetation patterning to particular substrates. Most of the ranges occur in the arid zone (< 300 mm annual rainfall) with the major rainfall gradient decreasing towards the north-east. Associated with this gradient was a gradual change in the seasonality of rainfall, with the south-western ranges receiving most of their rainfall in winter and the north-eastern ranges being more influenced by summer rainfall events. Full details of sampling procedures can be found in papers listed in Appendix 5.1. They broadly involved compiling lists of all vascular plants, and recording a standard set of site variables (topography, slope, aspect, percentage bare ground, percentage litter cover, etc.) from each plot. For simplicity, all taxa at species level and below are referred to as species. Plant nomenclature follows current usage at Western Australian Herbarium (1998–).

In an analysis of biogeographic patterns across the ranges a Sørensen dissimilarity matrix was calculated from the 1217 site by 906 species matrix. The centroids of the plots from each range were calculated and these were further analysed using non-metric multidimensional scaling ordination and group average cluster analysis. To examine degree of congruence of the floras between ranges, an analysis of similarity (ANOSIM) was undertaken on the Sørensen dissimilarity matrix using the range as a factor. All analyses were undertaken in the Primer v 6 and Permanova+ extensions (Anderson *et al.* 2008; Clarke and Gorley 2006).

The correlation of edaphic factors and the distribution of ironstone specialist species were investigated for a subset of six ranges (western groups in Fig. 5.1) where high numbers of specialist ironstone species were found on some ranges. If, as previously suggested, the distribution of ironstone specialist species was related to specific edaphic factors then there should be little difference between the soils of the ranges with high numbers of ironstone specialists. Edaphic differences between ranges were examined using an ANOSIM on a Euclidean distance matrix calculated from pH, electrical conductivity, organic carbon, nitrogen, phosphorous and potassium values using range as a factor. Between 10 and 21 plots occurred on the ridges of these ranges and soil chemistry was analysed from a bulked sample of 20 subsamples gridded across the plot and taken from the top 10 cm. Soil sampling and analysis methods are detailed in the papers listed in the Appendix 5.1. All soil variables except pH were log-transformed and normalised before analysis. A similar analysis using data from all plots on these six ranges was also undertaken and produced essentially similar results.

Results

Flora

The database from these surveys contains over 29 000 records and these are supported by 6800 voucher specimens that have been lodged at the Western Australian Herbarium. Some 906 species were recorded from the 1217 plots established across the ranges, representing 269 genera in 74 families. The most common genera were *Acacia* (80 taxa), *Eremophila* (47), *Senna* (31), *Eucalyptus* (28), *Rhodanthe* (18), *Calandrinia* (17), *Ptilotus* (17), *Maireana* (16), and *Grevillea* (15). The richest families were Fabaceae (140 taxa), Asteraceae (103), Myrtaceae (75), Poaceae (64), Chenopodiaceae (57), Scrophulariaceae (44), Malvaceae (32), Proteaceae (24), Goodeniaceae (22) and Lamiaceae (22). This taxonomic spectrum is typical of the transition from the south-west to the arid zone. The flora on the ranges varied from 93 to 321 species with a general decrease in perennial species per plot with decreasing rainfall ($r^2 = 0.32$, $P < 0.0001$: Fig. 5.3) but with the total species recorded per plot being somewhat confounded by varying patterns in annual occurrence as a result of seasonal rainfall differences.

Threatened species in Western Australia are protected under both State and Commonwealth legislation and are generally referred to as Declared Rare Flora (DRF). New populations of four DRF species were found during the surveys on four of

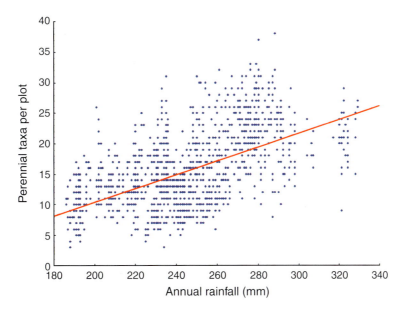

Figure 5.3: The number of perennial species recorded per plot was strongly correlated with mean annual rainfall ($r^2 = 0.32$, $P < 0.0001$). The number of annual species recorded per plot was, as expected, strongly related to the rainfall of the preceding winter.

the ranges (Table 5.1). Of these, *Acacia woodmaniorum* had been previously collected from this range but not formally recognised and *Lepidosperma gibsonii* had no previous collections. Both species have restricted distributions and are being impacted by new mines. In addition to listed threatened species DEC maintains a register of species of conservation concern for which there is, as yet, insufficient data to formally list as threatened. New populations of species from this list were found on 23 of the 24 ranges and included 68 species (Table 5.1).

The first two years of the survey largely concentrated on ranges along the arid zone boundary and identified 20 new species (Table 5.2). These included species that had been previously collected but not recognised as distinct, and species which had not been previously collected. Many of these species were described in a concurrent DEC taxonomic project that has concentrated on formally naming threatened species and species of conservation concern being impacted by mining and other threatening processes (Wege *et al.* 2007).

The tabulation of distribution patterns of specialist ironstone species (those with occurrences restricted to, or centred on ironstone) by Gibson *et al.* (2007) suggested these species were concentrated along the arid zone boundary. Further work has confirmed these patterns (Gibson *et al.* 2012) though the reason for these patterns is not clear. It may be related to edaphic factors or geographical position in relation to previous climatic fluctuations during the Tertiary period.

The 44 specialist ironstone species come from a variety of families, are all perennial and are predominately shrubs. Possibly the most remarkable are the five leafless *Tetratheca* species that grow out of massive BIF on five adjacent ranges in the Mt Jackson–Helena and Aurora Range area. Phylogenetic analysis has shown that most of these species are not closely related and suggests that this life form has evolved several times from unrelated sections of the genus (Butcher *et al.* 2007). One of these species, *Tetratheca paynterae* subsp. *paynterae*, has shown significant genetic diversity over linear distances of as little as 80 m (Butcher *et al.* 2009) and has a total extent of less than 1 km along a series of ironstone ridgelines.

Vegetation

With few exceptions, individual analyses (detailed in Appendix 5.1) showed a very strong correlation between floristic composition (based on presence–absence data of perennial species) and topographic–edaphic variables and confirm that the topographical sequences of vegetation originally reported by Beard (1981) hold for all except the most subdued ironstone ranges across the study area. From the 24 ranges, 134 floristic communities have been described (Appendix 5.1) with between three and eight community types being identified on individual ranges. Increased density of sampling would be expected to identify further units, especially in the areas of highest perennial richness along the south-western boundary of the study area. Twenty community types were identified from three or fewer plots and several plots from eight ranges remained ungrouped in the classifications. Further sampling would be needed to achieve a fully comprehensive classification of the vegetation of the ranges. Limited comparisons of adjacent ranges in these papers suggest that β diversity (i.e. the rate of change in composition between the ranges) is high (see Markey and Dillon 2008a and Meissner *et al.* 2010a in Appendix 5.1).

Similar high levels of β diversity within and between the ranges have been reported for a series of ironstone ranges occurring to the south of the current study (Gibson *et al.* 2010). While the high β diversity across individual ranges is not unexpected (given the significant environmental gradients from the lower colluvial slopes to the skeletal soils on the massive BIF on the ridge tops), the high turnover in species composition between ranges was much higher than expected for communities in the arid zone and more reminiscent of patterns seen in the higher rainfall areas of the south-west.

Biogeographic patterns

The broad scale biogeographic patterning shown by ordination of the range centroids is remarkably

Table 5.1. Species of conservation concern found during the surveys of 24 ironstone ranges

Species marked with asterisk formally listed as threatened flora.

Family	Taxon	Threatened
Aizoaceae	*Gunniopsis divisa*	
	Gunniopsis propinqua	
	Gunniopsis rubra	
Asteraceae	*Calotis* sp. Perrinvale Station (R.J. Cranfield 7096)	
	Millotia dimorpha	
	Podotheca uniseta	
	Rhodanthe collina	
Boraginaceae	*Halgania gustafsenii* var. Murchison (R. Meissner and B. Bayliss 743)	
Celastraceae	*Psammomoya implexa*	
Cyperaceae	*Lepidosperma gibsonii*	*
	Lepidosperma sp. Blue Hills (A. Markey and S. Dillon 3468)	
	Lepidosperma sp. Koolanooka (K.R. Newbey 9336)	
Euphorbiaceae	*Beyeria lapidicola*	
	Euphorbia sarcostemmoides	
	Ricinocarpos brevis	*
Fabaceae	*Acacia burrowsiana*	
	Acacia graciliformis	
	Acacia karina	
	Acacia muriculata	
	Acacia sp. Jack Hills (R. Meissner and Y. Caruso 4)	
	Acacia sp. Muggon Station (S. Patrick and D. Edinger SP 3235)	
	Acacia speckii	
	Acacia subsessilis	
	Acacia woodmaniorum	*
	Mirbelia ferricola	
Hemerocallidaceae	*Caesia* sp. Koolanooka Hills (R. Meissner and Y. Caruso 78)	
Lamiaceae	*Hemigenia exilis*	
	Hemigenia tysonii	
	Pityrodia iphthima	
	Prostanthera ferricola	
	Prostanthera petrophila	
	Spartothamnella sp. Helena and Aurora Range (P.G. Armstrong 155–109)	
Malvaceae	*Sida picklesiana*	
Myrtaceae	*Aluta teres*	

(Continued)

Table 5.1. (Continued)

Family	Taxon	Threatened
	Baeckea sp. London Bridge (M.E. Trudgen 5393)	
	Baeckea sp. Melita Station (H. Pringle 2738)	
	Baeckea sp. Mount Barloweerie (J.Z. Weber 5079)	
	Baeckea sp. Parker Range (M. Hislop and F. Hort MH 2968)	
	Baeckea sp. Perenjori (J.W. Green 1516)	
	Calytrix erosipetala	
	Calytrix praecipua	
	Calytrix uncinata	
	Darwinia masonii	*
	Euryomyrtus patrickiae	
	Homalocalyx echinulatus	
	Melaleuca barlowii	
	Micromyrtus acuta	
	Micromyrtus placoides	
	Micromyrtus trudgenii	
Phyllanthaceae	*Phyllanthus baeckeoides*	
	Sauropus ramosissimus	
	Sauropus sp. Woolgorong (M. Officer s.n. 10/8/94)	
Poaceae	*Austrostipa blackii*	
Proteaceae	*Banksia arborea*	
	Grevillea erectiloba	
	Grevillea inconspicua	
	Persoonia pentasticha	
Rhamnaceae	*Polianthion collinum*	
	Stenanthemum mediale	
	Stenanthemum patens	
	Stenanthemum poicilum	
Rutaceae	*Drummondita fulva*	
	Drummondita rubroviridis	
	Philotheca citrina	
Sapindaceae	*Dodonaea amplisemina*	
	Dodonaea scurra	
Scrophulariaceae	*Eremophila congesta*	
Zygophyllaceae	*Tribulus adelacanthus*	

Table 5.2. Species identified as new during the survey of 24 ironstone ranges

Some species had been previously collected before these surveys but had not been recognised as distinct entities. Most of these species were collected from ranges along the south-west boundary of the study area.

Family	Taxon
Asteraceae	*Calotis* sp. Perrinvale Station (R.J. Cranfield 7096)
Campanulaceae	*Lobelia heterophylla* subsp. *centralis*
Chenopodiaceae	*Sclerolaena* sp. Koolanooka Hills (R. Meissner and Y. Caruso 437)
Cyperaceae	*Lepidosperma gibsonii*
	Lepidosperma sp. Blue Hills (A. Markey and S. Dillon 3468)
	Lepidosperma sp. Koolanooka (K.R. Newbey 9336)
Fabaceae	*Acacia graciliformis*
	Acacia karina
	Acacia muriculata
	Acacia sp. Jack Hills (R. Meissner and Y. Caruso 4)
	Acacia sp. Weld Range (A. Markey and S. Dillon 2994)
	Acacia woodmaniorum
	Indigofera fractiflexa subsp. Mount Augustus (S. Patrick and A. Crawford SP 4737)
Hemerocallidaceae	*Caesia* sp. Koolanooka Hills (R. Meissner and Y. Caruso 78)
Lamiaceae	*Pityrodia iphthima*
	Prostanthera ferricola
Malvaceae	*Sida picklesiana*
Portulacaceae	*Calandrinia* sp. Truncate capsules (A. Markey and S. Dillon 3474)
Rutaceae	*Drummondita fulva*
	Drummondita rubroviridis
	Dodonaea scurra

similar to the geographic distribution of the ranges, emphasising the rapid turnover between ranges and the strong compositional gradients across the entire study area (Fig. 5.4). This is confirmed by a plot of the Sørensen dissimilarities against the distance between ranges (Fig. 5.5). While there was a high degree of variability in the dissimilarity values between ranges, there was a general trend for this to increase with increasing separation between the ranges (Spearman rank order correlation = 0.42, $P < 0.0001$) as seen in the ordination.

The highest degree in variability of dissimilarity distance nonetheless occurred between nearby ranges (separation < 100 km: Fig. 5.5).

Both the ordination and the correlation plot showed uniform gradational change in species composition across the study area. Similarly no strong clustering was seen in the dendrogram (Fig. 5.6). Nonetheless at a high level, three broad groups can be discerned: a western group, a northern group and an eastern group. Each of these major groups can be subdivided into further geographic

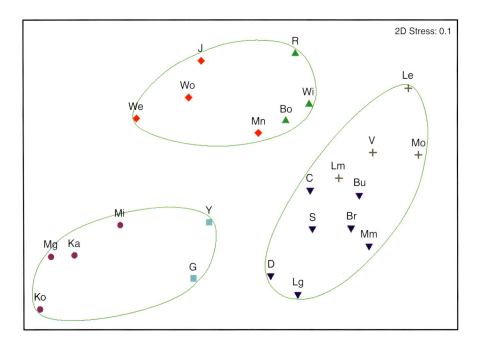

Figure 5.4: Non-metric multidimensional scaling ordination of the centroids for each range based on a Sørensen dissimilarity matrix. The ordination shows similar spatial distribution to the actual geographical spread of the ranges (Fig. 5.1) indicating the strong compositional gradient across the study area and the high β diversity between ranges. Circled ranges indicate geographic classification shown in Figs. 5.1 and 5.5.

subgroups. Only one range (Bo) is at variance with its geographic position clustering with a northern rather than eastern subgroup (Figs. 5.1 and 5.4).

Composition change over between ranges was very high as indicated by the ANOSIM analysis. The global test was highly significant (R = 0.773, $P \leq 0.0001$), all 264 pair-wise comparisons were significant at $P < 0.05$ after the conservative Bonferroni correction for multiple comparisons was applied. The R statistic for pair-wise comparisons

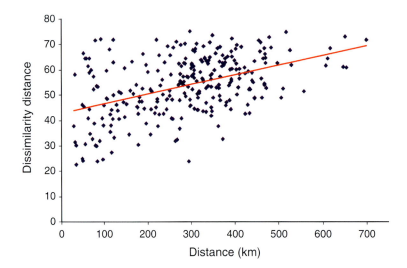

Figure 5.5: Dissimilarity between range centroids increased with increasing distance between ranges (Spearman rank order correlation = 0.42, $P < 0.0001$), however variability was highest with ranges less than 100 km apart varying from moderately (c. 20%) to highly dissimilar (c. 75%).

ranged from 0.109 for some adjacent ranges to 0.996 for the widely separate ranges. Low R values indicate considerable overlap in species composition while high R values indicate almost total change over. Results are consistent with the ordination and plots of dissimilarity versus distance that indicate turn over in species composition is closely related to geographical separation. What is surprising is that even adjacent ranges (< 30 km apart), although sharing a considerable proportion of their flora, are compositionally distinct. These findings are driven by patterns of perennial species (Gibson *et al.* 2012) and are consistent with patterns reported by Gibson *et al.* (2010) for other ironstone ranges south of our study area.

Edaphic patterns

Given the similar age and geology of the ironstone ranges it might be expected that soil chemistry would be broadly similar across the ranges. Gibson *et al.* (2010) suggested that edaphic factors may be correlated with the higher concentrations of specialist ironstone species near the arid zone boundary. If this hypothesis is true then the ranges with high numbers of specialist species should have similar soil chemistry and be dissimilar to those nearby ranges with few or no specialist species.

Detailed soil chemistry data are available for two adjacent ranges with a concentration of ironstone specialist (7–9 species on ranges: Ka and Ko) and four surrounding ranges with few or any ironstone specialists (0–3 species on ranges: G, Mg, Mi, Y; Fig. 5.1). The hypothesis was initially tested by comparing similarity in the soil chemistry of the skeletal ironstone soils from the crests of two ranges, then comparing the similarity of the soils of these two ranges to the other four ranges. The results of the ANOSIM showed little support for the hypothesis that the two ranges with concentrations of ironstone specialists have similar soil chemistry ($R = 0.19$, $P = 0.019$). Two of the comparisons between these ranges and the four

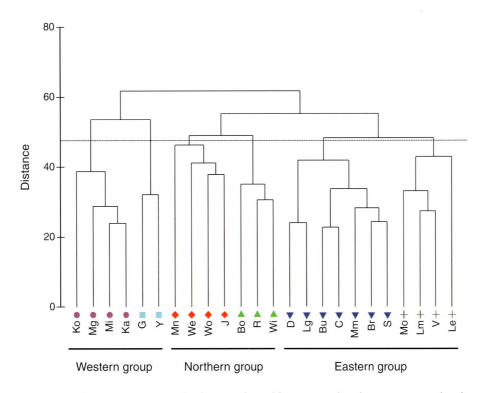

Figure 5.6: Cluster analysis of the range centroids showing broad biogeographical patterning at the three and six group levels. No strong clustering was seen indicating gradational change across the study area. Range locations shown in Fig. 5.1.

Table 5.3. Pair-wise comparison of the soil chemistry of six adjacent ranges

The analysis of similarity's R statistic is centred on zero when there is no difference between groups. Bonferroni correction for multiple comparisons applied. Range codes as in Fig. 5.1 and Appendix 5.1.

	Ranges			
	Mi	G	Mg	Y
G	0.22**			
Mg	0.19*	0.03 NS		
Y	0.28**	0.33**	0.56**	
Ka–Ko	0.17**	0.05 NS	−0.02 NS	0.41**

Significance of R statistic tested by permutation ($n = 9999$), * $P < 0.05$; ** $P < 0.01$; NS not significant.

surrounding ranges also showed highly significant differences while the other two did not (Table 5.3). The pair-wise comparisons of soils of the surrounding ranges were all significantly different except for one comparison (Table 5.3). It is somewhat surprising that the soils of adjacent ironstone ranges can be so different over short geographical distances.

A similar comparison of soils from all the plots on these six ranges was also carried out and gave essentially similar results. The comparison of just the skeletal soils of the crests of the ironstone ranges was run to eliminate any of the confounding factors such as soil depth, type of colluvial deposits and organic matter availability that may be present in the larger dataset. The results of these analyses indicated that the significant differences in soil chemistry were not related to confounding effects.

Discussion

The analysis of similarity showed that the composition of the vegetation on each range is unique and, within ranges, some very restricted vegetation units have been described along the arid zone boundary (Appendix 5.1). These findings have significant implications in achieving a comprehensive, adequate and representative reserve system where every range would need to be considered irreplaceable (*sensus* Pressey *et al.* 1994). While reservation of vegetation units on one range may reserve similar structural types found on other ranges, these are highly likely to be compositionally different. This spatial patterning of β diversity between the ranges is only partially explained by broad scale spatial and local environmental variables, with the long period of landscape and climatic stability also being suggested as important in their development (Gibson *et al.* 2010). Further analysis is continuing to better understand the relationship of β diversity with spatial, climatic, soil chemistry and local site variables across the ranges (Gibson *et al.* 2012).

Almost as surprising as the levels of β diversity encountered was the large number of new species identified during the survey. Of the 21 new species, 14 were recorded from just two ranges along the arid zone boundary where good access has been available for many years. These two ranges (Karara and Koolanooka Hills, Ka and Ko: Fig. 5.1) form a centre of specialist ironstone species not previously recognised (Gibson *et al.* 2012). These ranges and the Helena and Aurora–Mt Jackson Ranges, another area of high concentration of ironstone specialist species to the south of the study area, can provide a focus for conservation effort. It is in these areas that the evolutionary processes that lead to the unique ironstone flora can be conserved.

These findings are consistent with the conservation priorities previously identified in an interdepartmental strategic review of the conservation and resource values of the Yilgarn BIF (Government of Western Australia 2007). In that document, the Diemals–Die Hardy Range, Helena and Aurora Range, Blue Hills (which includes our Karara study area) and Mt Manning Ranges were identified as the highest priorities for conservation. Our more recent analysis indicates that the Koolanooka Hills should be added to this list (Gibson *et al.* 2012). That same review proposed that the most outstanding ranges be conserved in their entirety. This will not be possible in all cases since several of these areas have operating mines. On the more arid inland ranges, far fewer specialist ironstone species were encountered, and in these regions there will be significantly less conflicts between conservation values and resource development.

The surveys of the ironstone ranges over the last four years have highlighted the unique nature of the vegetation on each range and the occurrence of concentrations of species with distributions centred on this geology along the arid zone boundary. It has also provided information on the poorly known distribution of many species of conservation concern and has identified over 20 new species. The data illustrate the high biodiversity values of the ironstone ranges of the Yilgarn and will be integral in the assessment of future resource development proposals over these ranges.

Acknowledgements

We would like to thank S Dillon, G Owen, B Bayliss, J Wright, N Sheehy, J Allen and Y Caruso for a wide variety of technical assistance. The staff of the Western Australian Herbarium provided assistance with identifications and the curation of voucher material. The Western Australian Chemistry Centre undertook all soil analyses. This project was funded under the Western Australian Department of Environment and Conservation's Biodiversity Conservation Initiative.

References

Anand RR, Butt CRM (2010) A guide for mineral exploration through the regolith in the Yilgarn Craton, Western Australia. *Australian Journal of Earth Sciences* **57**, 1015–1114. doi:10.1080/08120099.2010.522823

Anderson MJ, Gorley RN, Clarke KR (2008) *PERMANOVA+ for PRIMER: Guide to software and statistical methods.* Primer–E, Plymouth.

Australian Bureau of Agricultural and Resource Economics (2010) *ABARE data – Australian mineral statistics. December quarter 2009. Historical data; Iron.* ABARE, Canberra. http://www.abare.gov.au/publications_html/data/data/data.html#acs

EPA (2002) Koolyanobbing Iron Ore Expansion. EPA Bulletin 1082, pp. 1–20. http://epa.wa.gov.au/EPADocLib/1539_B1082.pdf

Beard JS (1981) *Swan, vegetation survey of Western Australia.* University of Western Australia Press, Nedlands.

Butcher R, Byrne M, Crayn DM (2007) Evidence for convergent evolution among phylogenetically distant rare species of *Tetratheca* (Elaeocarpaceae, formerly Tremandraceae) from Western Australia. *Australian Systematic Botany* **20**, 126–138. doi:10.1071/SB06017

Butcher PA, McNee SA, Krauss SL (2009) Genetic impacts of habitat loss on the rare ironstone endemic *Tetratheca paynterae* subsp. *paynterae*. *Conservation Genetics* **10**, 1735–1746. doi:10.1007/s10592-008-9775-y

Clarke KR, Gorley RN (2006) *PRIMER v6: User manual/tutorial.* Primer–E, Plymouth.

Gibson N (2004) Flora and vegetation of the eastern goldfields ranges. Part 7, Middle and South Ironcap, Digger Rock and Hatter Hill. *Journal of the Royal Society of Western Australia* **87**, 49–62.

Gibson N, Lyons MN, Lepschi BJ (1997) Flora and vegetation of the eastern Goldfield Ranges. Part 1, Helena and Aurora Range. *CALMScience* **2**, 231–246.

Gibson N, Coates DJ, Thiele KR (2007) Taxonomic research and the conservation status of flora in the Yilgarn banded iron formation ranges. *Nuytsia* **17**, 1–12.

Gibson N, Yates CJ, Dillon R (2010) Plant communities of the ironstone ranges of south western Australia: hotspots for plant diversity and mineral deposits. *Biodiversity and Conservation* **19**, 3951–3962. doi:10.1007/s10531-010-9939-1

Gibson N, Meissner R, Markey AS, Thompson WA (2012) Patterns of plant diversity in ironstone ranges in the arid south western Australia. *Journal of Arid Environments* **77**, 25–31. doi:10.1016/j.jaridenv.2011.08.021

Government of Western Australia (2007) *Strategic Review of the Conservation and Resourse Values of the Banded Iron Formation of the Yilgarn Craton.* Department of Environment and Conservation & Department of Industry & Resources, Perth.

Pressey RL, Johnson IR, Wilson PD (1994) Shades of irreplaceability: toward a measure of the contribution of sites to a reservation goal. *Biodiversity and Conservation* **3**, 242–262. doi:10.1007/BF00055941

Wege JA, Shepherd KA, Butcher R (2007) Preface – A special edition of Nuytsia funded by the Western Australian Government's 'Saving our Species' biodiversity conservation initiative. *Nuytsia* 17, i.

Western Australian Herbarium (1998–). *FloraBase — The Western Australian Flora*. Department of Environment and Conservation, Perth. http://florabase.dec.wa.gov.au/

Appendix 5.1. Papers describing the flora and vegetation of selected Banded Iron Formation ranges of the Yilgarn Craton
First column indicates codes used in figures, note one paper covers two ranges. Papers are freely available for download from http://naturemap.dec.wa.gov.au/default.aspx.

Code	Reference
We	Markey AS and Dillon SJ (2008a) Flora and vegetation the banded iron formation of the Yilgarn Craton: Weld Range. *Conservation Science Western Australia* 7, 153–178.
Mi Ka	Markey AS and Dillon SJ (2008b) Flora and vegetation the banded iron formation of the Yilgarn Craton: the Central Tallering land system. *Conservation Science Western Australia* 7, 121–149.
Wi	Markey AS and Dillon SJ (2010a) Flora and vegetation of the banded iron formations of the Yilgarn Craton: Herbert Lukin Ridge (Wiluna). *Conservation Science Western Australia* 7, 391–412.
Bo	Markey AS and Dillon SJ (2010b) Flora and vegetation of the banded iron formations of the Yilgarn Craton: the Booylgoo Range. Conservation Science Western Australia. *Conservation Science Western Australia* 7, 503–529.
G	Markey AS and Dillon SJ (2010c) Flora and vegetation of the banded iron formations of the Yilgarn Craton: Gullewa. Conservation Science Western Australia. *Conservation Science Western Australia* 7, 531–556.
J	Markey AS and Dillon SJ (2011a) Flora and vegetation of the banded iron formations of the Yilgarn Craton: the Johnston Range. *Conservation Science Western Australia* 8, 137–161.
Lg	Markey AS and Dillon SJ (2011b) Flora and vegetation of the banded iron formations of the Yilgarn Craton: northern Yerilgee Hills. *Conservation Science Western Australia* 8, 163–186.
Y	Markey AS and Dillon SJ (2011c) Flora and vegetation of the banded iron formations of the Yilgarn Craton: Yalgoo. *Conservation Science Western Australia* 8, 113–136.
Ko	Meissner R and Caruso Y (2008a) Flora and vegetation of the banded iron formation of the Yilgarn Craton: Koolanooka and Perenjori Hills. *Conservation Science Western Australia* 7, 73–88.
J	Meissner R and Caruso Y (2008b) Flora and vegetation of banded iron formations of the Yilgarn Craton: Jack Hills. *Conservation Science Western Australia* 7, 89–103.
Mg	Meissner R and Caruso Y (2008c) Flora and vegetation of the banded iron formation of the Yilgarn Craton: Mt Gibson. *Conservation Science Western Australia* 7, 105–120.
C	Meissner R, Owen G and Bayliss B (2010a) Flora and vegetation of banded iron formations of the Yilgarn Craton: Cashmere Downs Range. *Conservation Science Western Australia* 7, 349–361.
R	Meissner R, Owen G and Bayliss B (2010b) Flora and vegetation of banded iron formations of the Yilgarn Craton: Robinson Ranges and Mount Gould. *Conservation Science Western Australia* 7, 363–376.
Bu	Meissner R, Owen G and Bayliss B (2010c) Flora and vegetation of banded iron formations of the Yilgarn Craton: Mount Forrest–Mount Richardson Range. *Conservation Science Western Australia* 7, 377–389.
Wo	Meissner R and Wright J (2010d) Flora and vegetation of banded iron formations of the Yilgarn Craton: Barloweerie and Twin Peaks greenstone belts. *Conservation Science Western Australia* 7, 557–570.
Bo	Meissner R and Owen G (2010e) Flora and vegetation of banded iron formations of the Yilgarn Craton: Brooking Hills area. *Conservation Science Western Australia* 7, 571–581.
Mm	Meissner, R. and Owen, G. (2010f) Flora and vegetation of banded iron formations of the Yilgarn Craton: Mt Ida greenstone belt and Mt Hope. *Conservation Science Western Australia* 7, 583–592.
V	Meissner R and Wright J (2010g) Flora and vegetation of banded iron formations of the Yilgarn Craton: Perseverance greenstone belt. *Conservation Science Western Australia* 7, 593–604.
S	Meissner R and Wright J (2010h) Flora and vegetation of banded iron formations of the Yilgarn Craton: south Illaara greenstone belt. *Conservation Science Western Australia* 7, 605–616.

(Continued)

Appendix 5.1. (Continued)

Code	Reference
Mn	Meissner R and Owen G (2010i) Flora and vegetation of banded iron formations of the Yilgarn Craton: western Narryer terrane. *Conservation Science Western Australia* 7, 617–628.
Lm	Thompson WA and Sheehy N (2011a) Flora and vegetation of banded iron formations of the Yilgarn Craton: the Lake Mason zone of the Gum Creek greenstone belt. *Conservation Science Western Australia* 8, 77–94.
Le	Thompson WA and Sheehy N (2011b) Flora and vegetation of banded iron formations of the Yilgarn Craton: the Lee Steere Range. *Conservation Science Western Australia* 8, 61–76.
Mo	Thompson WA and Sheehy N (2011c) Flora and vegetation of banded iron formations of the Yilgarn Craton: the Montague Range zone of the Gum Creek greenstone belt. *Conservation Science Western Australia* 8, 95–111.

6

Lessons drawn from iron ore mining in the Yilgarn region of Western Australia

Robert K Howard

Summary

Cliffs Asia Pacific Iron Ore operates the largest iron ore mining operation in the Yilgarn region of Western Australia. Mining commenced in 1994 at Koolyanobbing and expanded to include northern satellite mines at Windarling and Mt Jackson in 2004. Mining utilises multiple small, separate ore bodies to produce a blended product for export through Esperance port. Ore bodies are typically associated with low banded iron formations. Banded iron ranges are scattered across the region, forming islands of relatively distinct biodiversity, sometimes supporting flora with very restricted distributions. Where the location of ore bodies coincides with the distribution of such flora, this presents particular challenges. The hierarchy of 'avoid, minimise, mitigate' in the management of environmental impacts is universally applicable for any mining development in any setting, but its importance is heightened in this particular environment. Examples of the application of this approach and its success in achieving the environmental objectives to date are presented. The Cliffs experience provides useful lessons with relevance to prospective mining projects in both the Yilgarn and Midwest regions of Western Australia, where there are a large number of emerging projects based on geologically similar deposits with similar biodiversity issues. The planning aspects are discussed, both in terms of minimising impacts at the local level and in balancing resource extraction and biodiversity conservation at the broader regional level.

Introduction

Cliffs Asia Pacific Iron Ore Pty Ltd (Cliffs) operates the only significant iron ore mining operation in the Yilgarn region of Western Australia. Cliffs' operations are based at Koolyanobbing, 50 km north of the town of Southern Cross in Western Australia (Fig. 6.1).

Ore bodies are typically associated with low banded iron formations. Banded iron ranges are scattered across the region, forming islands of relatively distinct biodiversity, sometimes supporting flora with very restricted distributions. Where the location of ore bodies coincides with the distribution of such flora, this presents particular challenges.

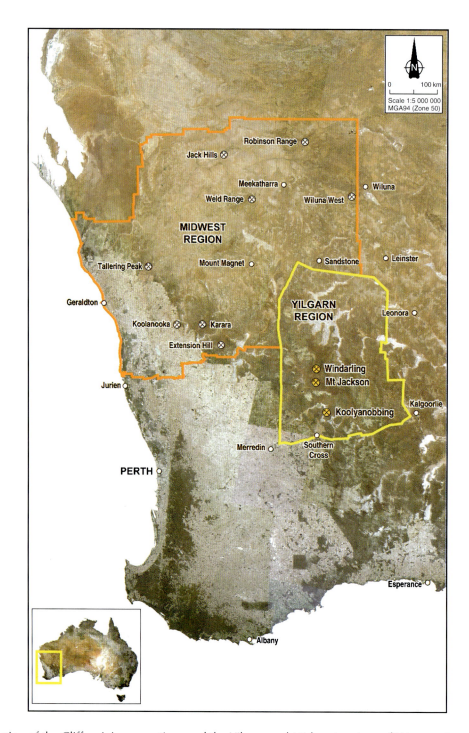

Figure 6.1: Location of the Cliffs mining operations and the Yilgarn and Midwest regions of Western Australia.

This paper discusses the issues associated with the opening up of new mining areas at Windarling and Mt Jackson where mining commenced in 2004, and presents information on the management of impacts on rare flora at these sites in the years since mining commenced. Examples of the application of the 'avoid, minimise, mitigate' management hierarchy approach, and its success in achieving the environmental objectives to date, are presented. The Cliffs experience provides useful lessons with relevance to prospective mining projects in both the Yilgarn and Midwest regions of Western Australia, where there are a large number of emerging projects based on geologically similar deposits with similar biodiversity issues. The paper also considers the strategic planning aspects, both in terms of avoiding and minimising impacts at the local level and in balancing resource extraction and biodiversity conservation at the broader regional level.

Mining history and current operations

Iron ore mining in the Koolyanobbing area dates back to 1950, when ore was mined to supply the Wundowie Charcoal Iron and Steel Mill near Perth. BHP later mined iron ore deposits at Koolyanobbing, principally to supply feed for its Kwinana blast furnace. The BHP mine at Koolyanobbing was active from 1960 to 1983, during which time it produced 25 million tonnes of iron ore.

The mine remained idle from 1983 until the tenements were acquired by Portman Iron Ore Limited in 1993. Portman commenced mining in 1994 at an initial rate of 1.5 million tonnes per annum (Mtpa). Production from the Koolyanobbing operations has progressively increased to a rate of 8.5 Mtpa. A significant step in the growth of the operations occurred in 2004, with the establishment of two new mining centres at Windarling and Mount Jackson, some 100 km north of Koolyanobbing. In 2009 the Portman operations were acquired by Cliffs Natural Resources. The Koolyanobbing mines are now operated by Cliffs Asia Pacific Iron Ore Pty Ltd.

Mining utilises conventional open pit mining methods, with the ore being hauled by road train from the northern mines to a crushing plant and rail head at Koolyanobbing. The ore is railed from Koolyanobbing to Esperance (via Kalgoorlie) from where it is shipped to steel mills in China and Japan. Multiple, separate ore bodies are mined concurrently. The ore product is of a high grade (typically greater than 61% Fe) and requires no processing other than crushing. However, ore from the various deposits does vary in the content of iron and minor elements such as phosphorus and sulphur. A sophisticated process of blending must be applied in order to achieve the product specifications required by customers. Current proven ore reserves are ~90 Mt.

The Cliffs operations are a major contributor to the state and regional economies. Operating expenditure is approximately A$480 million per year, and the state receives royalties from the Cliff's operations of around A$50 million annually. The Koolyanobbing operations support a direct workforce of 900 and an estimated 2300 flow-on jobs. Cliffs' exports represent 80% of the Kalgoorlie–Esperance rail tonnage and 70% of the revenue of the Esperance port.

Environmental setting

Cliffs' Koolyanobbing operations are located in a region of notable biodiversity and wilderness value. The project area is situated east of the Wheatbelt region of south-western Australia and east of the 'rabbit proof fence' that marks the transition from cleared farmland to the regions beyond where the native vegetation remains in good condition. This area, lying between the agricultural zone and the inland deserts, is regarded as the largest intact woodland landscape remaining in temperate Australia, and it is receiving increasing attention for its environmental significance (Watson *et al.* 2008).

The Yilgarn region is flat, geologically ancient and arid. Ore bodies are typically associated with low outcropping banded iron formations (BIF ranges) (Fig. 6.2). These features are scattered across the region, forming 'islands' of biodiversity where the vegetation often contains groupings and species that differ from one range to the next (Gibson *et al.* 2010). This situation is thought to arise from the spatial separation of these ranges resulting in the isolation and divergence of populations over a very long period of geological stability (Gibson *et al.* 2007). In some cases, species of flora are known from single populations at only one location. This presents particular challenges for mining projects as a result of the potential for ore bodies to overlap with the distributions of such populations.

This situation is repeated across the Yilgarn and the adjacent Midwest region where there are many developing iron ore mining proposals that are expected to form the basis of a major emerging mining province in Western Australia.

Lessons with broader application

Examples where mining proposals under development are encountering similar issues around BIF iron ore deposits in the Midwest include those at Mt Gibson, Koolanooka Hills, Jack Hills and the Weld Range. However, in contrast to the Cliffs Yilgarn operations, most of the Midwest projects are at the early stages of development. Because the Cliffs project has been operating for several years, and environmental impacts have been extensively monitored over this period, the information that is now available is highly relevant to the developing iron ore mining industry.

The absence of this type of information in the past has created difficulties for the environmental assessment of new mining proposals. For example, a 2007 government report that considered mining access and conservation issues in BIF landscapes of the Yilgarn and Midwest (DEC and DOIR 2007) expressed the view that *Tetratheca paynterae* ssp. *paynterae* that occurs at Windarling had suffered increased mortality due to the impacts of dust from

Figure 6.2: Mining is based on multiple small ore bodies that are typically located on or adjacent to banded iron formation ranges. The J3 mine pit shown here is located on the Mt Jackson range.

mining. The report went on to recommend that 'biodiversity impact issues experienced in operating mine development should be considered in the assessment of any future proposals to mine BIF ranges in this area'. The latter statement is perfectly valid, but in the context of the earlier statement about impacts from dust, it implied that this should weigh heavily against any decision to approve future mining proposals of deposits associated with BIF features.

The results of long-term monitoring work that is now available for plant populations adjacent to the Cliffs operations should aid in clarifying our understanding of the impacts of mining in these environments.

The Windarling and Mt Jackson mining developments

The 1993 environmental assessment and approvals for mining in the immediate vicinity of Koolyanobbing were relatively straightforward as a result of the mining footprint being largely restricted to areas that had been disturbed during the previous BHP operations. The opening up of new mining areas further north at Windarling and Mt Jackson (Fig. 6.1) some 10 years later was a different matter. These deposits were located in areas that had not been subject to previous mining, and baseline environmental surveys revealed the presence of flora species whose distributions appeared to be restricted solely to these ranges.

Three such species, which were subsequently given the status of Declared Rare Flora (DRF) under state legislation, were identified. *Tetratheca paynterae* is one of two species at Windarling that are classified as DRF. *Tetratheca paynterae* occupies a highly specialised habitat, growing only in the rock crevices of BIF outcrops (Fig. 6.3). The other species of DRF occurring at Windarling is *Ricinocarpos brevis*. *R. brevis* was initially nominated as DRF based on information that it only existed at Windarling, but

Figure 6.3: *Tetratheca paynterae* ssp. *paynterae* occurs on exposed banded iron outcrops at Windarling where it establishes a root system within rock crevices.

since that time has been found to occur in two other locations. It does not show the same narrow habitat specialisation displayed by *Tetratheca paynterae* and consequently is considered likely to exist over a much wider geographical range.

The mining proposal also included the development of mine pits at Mt Jackson (J2 and J3 deposits: Fig. 6.4), located 20 km south of Windarling, where a different species of Tetratheca (*Tetratheca harperi*) occurs that is endemic only to the Mt Jackson range. A notable difference was that *T. harperi* could be avoided in accessing the ore body at this location (Fig. 6.4).

At Windarling it was found that the location of the major ore deposit coincided closely with the distribution of a subspecies of *Tetratheca paynterae* (*T. paynterae* ssp. *paynterae*). Mining of the whole of the ore body would have resulted in the removal of around 90% of the 7500 individual plants recorded at Windarling.

During the environmental assessment and approvals process for the Windarling and Mt Jackson mines the Western Australian Environmental Protection Authority initially recommended against approving the Windarling component of the project on the basis of the impact on *T. paynterae* ssp. *paynterae*. Ultimately, the Minister for the Environment approved a modified version of the proposal that excluded mining from defined areas (Fig. 6.5), such that 70% of the population was retained. A further 20% of the population could potentially be removed subject to certain conditions being met, i.e. being able to demonstrate that the long-term viability of the population was not threatened.

For any mining development the impact management strategy follows a hierarchy of avoid → minimise → mitigate. Avoidance is always preferable if it can be achieved, but this was clearly not possible at Windarling if the valuable iron ore deposits there were to be accessed. In that case a government decision was required to define an acceptable balance between economic value and the impact on the Tetratheca population. The actual proportion of the original population that has been retained at Windarling is 73%. In achieving this

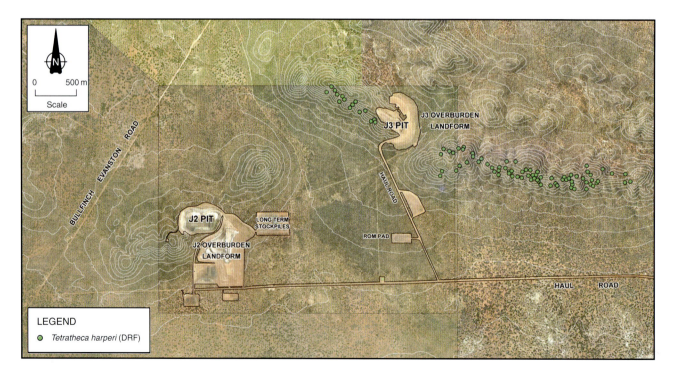

Figure 6.4: Aerial photograph of the Mt Jackson site showing mine development areas and the distribution of *Tetratheca harperi*. In this example the mine pits and associated infrastructure were able to avoid all *Tetratheca harperi* plants.

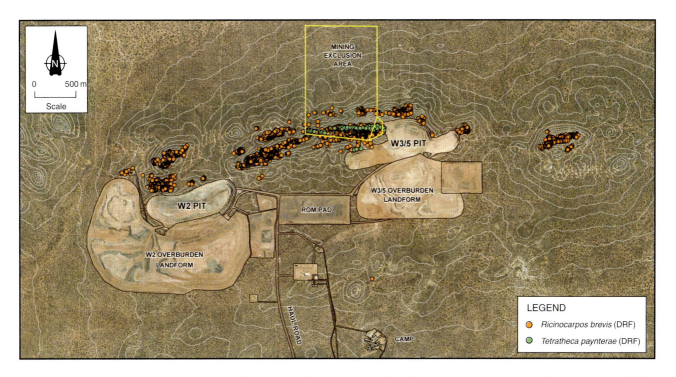

Figure 6.5: Aerial photograph of the Windarling site showing mine development areas, mining exclusion areas and the distribution of *Tetratheca paynterae* ssp. *paynterae* and *Ricinocarpos brevis*. The area shown by the smaller yellow polygon is an area that could potentially be mined in future.

result, access to ore worth around A$800 million, in today's terms, was foregone.

Despite this compromise, concerns about the impact of the loss of part of the population existed within the government environmental agencies, and the conditions of approval that were applied to the project required detailed and comprehensive research and monitoring programs.

Tetratheca paynterae at Windarling – Conclusions from six years of monitoring

Mining at Windarling commenced in early 2004 and involved the removal of the part of the *T. paynterae* population located at the eastern end of the Windarling range. The western boundary of the W3/5 pit that was located closest to the remaining *T. paynterae* population was established at this time (Fig. 6.6). A monitoring program was created to measure the condition of the remaining population with the aim of providing information about the health of the population over time and in relation to distance from the mining activity.

Condition monitoring has involved an annual assessment of a representative subsample comprised of 350 individual plants distributed across the population. Plant condition was assessed using five condition categories based on the percent of live stems present on each plant. Plants with greater than 80% live stems were placed in the highest condition category and plants with 20% or less live stems were assigned the lowest condition category.

The results of condition monitoring between 2003 and 2009 are presented in Fig. 6.7. The assessments indicate a similar condition between 2003 and 2004, followed by a significant deterioration in condition between 2004 and 2005. In 2005 the proportion of the population in the lowest condition category was more than twice that recorded in the previous two years. This was followed by a steady recovery in condition across 2005 to 2008. The condition in 2009 was similar to 2008.

The decline in condition of the Tetratheca population in 2005 was a cause for concern at the time due to its potential link with the commencement of

Figure 6.6: The Windarling site showing the W3/5 pit in the foreground. The unmined portion of the Windarling range extends west (right) from the pit.

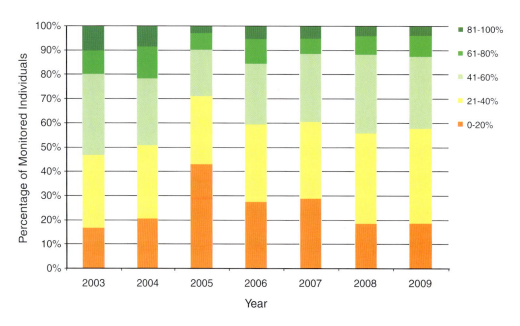

Figure 6.7: Results of condition monitoring of *Tetratheca paynterae* ssp. *paynterae* at Windarling, pre-mining (2003) to 2009. The five condition categories shown relate to the percentage of live stems on individual plants.

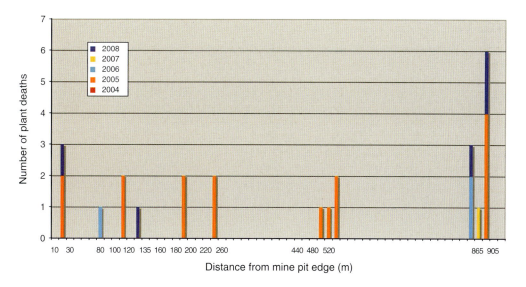

Figure 6.8: The distribution of *Tetratheca paynterae* deaths among the condition assessment population sample at Windarling from 2004 to 2008.

mining at Windarling. There was speculation that mine-generated dust could be the cause. Increased plant deaths were also observed in 2005 (Fig. 6.8). A continuing decline in plant condition or a spatial pattern of greater impact on plants located closer to the mine edge would support this conclusion.

However, data from the years following 2005 revealed that the pattern of decline did not continue, and suggested that factors other than mining were driving the observed patterns. The 2005 low point in plant condition was preceded by a period of very low rainfall. A subsequent return to higher rainfall conditions was accompanied by a strong recovery in plant condition. Yates *et al.* (2008) observed similar declines in plant condition at this time among three other Tetratheca species on BIF ranges elsewhere within the region. That rainfall is the dominant determinant of plant vigour is also supported by information on flowering and seed production, this being strongly related to rainfall in the preceding year (Yates *et al.* 2008).

The spatial pattern of plant condition and mortality at Windarling (Fig. 6.8) provides additional evidence that mining has not had a measurable effect on the health of the remaining Tetratheca population.

A parallel research program has examined the genetic structure, ecology and population dynamics of *T. paynterae*, as well as several other aspects including *in situ* and *ex situ* propagation techniques. This species is believed to be among the most intensively studied non-commercial native plant species in Australia. It has been successfully propagated from seed both in nursery conditions and in the field, and secure seed stores have been established. The improved knowledge of *T. paynterae* means that options are available to enhance population numbers should it be necessary to intervene in the future. It is emphasised, however, that the extensive research and mitigation work represents a major long-term commitment and comes at a cost – estimated at approximately A$0.5 million annually in the case of *T. paynterae* – underlining the point that avoidance is preferable wherever possible.

Planning for a new project – the Mt Jackson J1 deposit

The approach to mine development planning undertaken by Cliffs involves the determination of environmental values as early as possible to allow for both a full understanding of environmental impacts and to support decisions around the location of mine infrastructure. A good planning process should at least provide the opportunity to locate

Figure 6.9: Locations of Malleefowl nesting mounds and Priority flora in the vicinity of the Mt Jackson J1 deposit. The proposed mine pit, overburden landform and biodiversity areas are also identified.

infrastructure such as overburden landforms, stockpile areas and roads away from sensitive sites.

A current example of how biological information is applied to the mine design is the proposed development of the J1 deposit at the eastern end of the Mt Jackson range (Fig. 6.9).

Development of the Mt Jackson J1 deposit will require the clearing of 605 ha of native vegetation, primarily for the mine pit and the overburden landform. Mine design has taken into consideration the potential impact on vegetation and fauna values. The mine design considered the location of fixed mine infrastructure (mine pits) and non-fixed infrastructure (overburden landform and operational areas) and other mine constraints such as pit abandonment bunding. Key environmental factors were the presence of flora and fauna of conservation significance. In this case there were no DRF, but there were several species with the lower 'Priority' conservation status present in the area (Fig. 6.9). There was one significant species of fauna – the Malleefowl (*Leipoa ocellata*) – recorded that is protected under State and Commonwealth legislation.

By locating the overburden landform on the northern side of the mine pit it was possible to effectively avoid direct impacts on these species (Fig. 6.9). Based on a broader suite of environmental values, including vegetation associations, four 'biodiversity areas' were demarcated within the project area. The biodiversity areas contain 11 of the 19 identified vegetation associations within the project area and the key habitat of *L. ocellata*. These areas were nominated as areas to be avoided by mining-related activities.

The above examples at Windarling and Mt Jackson demonstrate that, despite the significant challenges that mining in BIF environments can present, the environmental issues can be successfully managed to meet avoidance and impact minimisation objectives that apply in these sensitive environments.

Regional conservation and strategic planning

A deficiency in the typical process that leads to mine development can occur as a result of major environmental issues emerging only at the 'eleventh hour'. Detailed surveys are time-consuming and costly, and it is unlikely that mining companies would invest in detailed biological surveys until there is a high degree of confidence that they have a viable ore deposit. In many regions of the state, including the Yilgarn, there is often very little baseline biological information available, so the presence of rare flora and fauna may not become known until plans for mining are well developed and large investments have been made in proving up mineral resources. This was certainly the case at Windarling, where the mining proposal was well advanced by the time that biological surveys were able to define the distribution and conservation significance of rare flora.

In an ideal world, exploration programs would be framed with a good knowledge of what are the more environmentally sensitive areas across a region, i.e. where environmental issues are likely to constrain or prevent mining. This would encourage the exploration effort to target less sensitive areas and effectively provide more certainty for both mining and conservation interests. In the absence of clear guidance on what are the areas of highest conservation significance and what areas are acceptable for mining, conflict and uncertainty are always likely. Some level of strategic land use planning is required in order to address this issue.

In 2007, the Western Australian Government undertook a strategic review of BIF ranges of the Midwest and Yilgarn regions (DEC and DOIR 2007). This was undertaken in recognition of the growing potential for conflict between conservation and industry objectives over BIF-associated areas of the Midwest and Yilgarn. The review indicated that, in addition to Cliff's existing operations in the Yilgarn, there were at least five significant projects under development in the Midwest representing a combined investment of around A$7 billion. The in-ground resource value for these projects was indicated to be between A$45 billion and A$75 billion with annual export revenues of A$3.6 billion to A$5.3 billion. The review asserts that 'a clear statement of Government's policy position in respect to access to the BIF ranges for development is required to give confidence, on the one hand, that the significant sums being expended in exploration and feasibility studies are worthwhile and the major financing arrangements currently being put in place can be brought to fruition through major project developments, and on the other hand, that the significant conservation values of BIF ranges will be adequately protected.'

While the aims of the review were sound, unfortunately the review fell down in its execution. It was undertaken by two government agencies, with no wider input or external consultation. The two parts of the report by the separate agencies expressed different views on what should be the extent of access to BIF areas for mining purposes, and the science and justification behind some of the positions taken in the report was unconvincing. For these reasons the review has not been particularly successful in meeting its objectives, and it has not translated into policy outcomes.

A further relevant example of land use decision-making, that in this case has had a direct impact on Cliffs' Koolyanobbing operations, can be found in the Western Australian Environmental Protection Authority (EPA) report titled 'Advice on areas of highest conservation value in the proposed extensions to Mount Manning Nature Reserve' (EPA 2007). This report arose from a condition placed on the environment approvals for the Windarling and Mt Jackson mines in 2003. On the basis of the EPA's assessment of biodiversity values, the report recommended a major expansion of conservation reserves in the Mt Manning area, including the establishment of 5665 km^2 as class A reserves from which mining would be excluded (Fig. 6.10). Significantly, the recommended class A reserves encompassed Cliffs' active mining tenements. If implemented, it would prevent any further development of the mining operations and have the

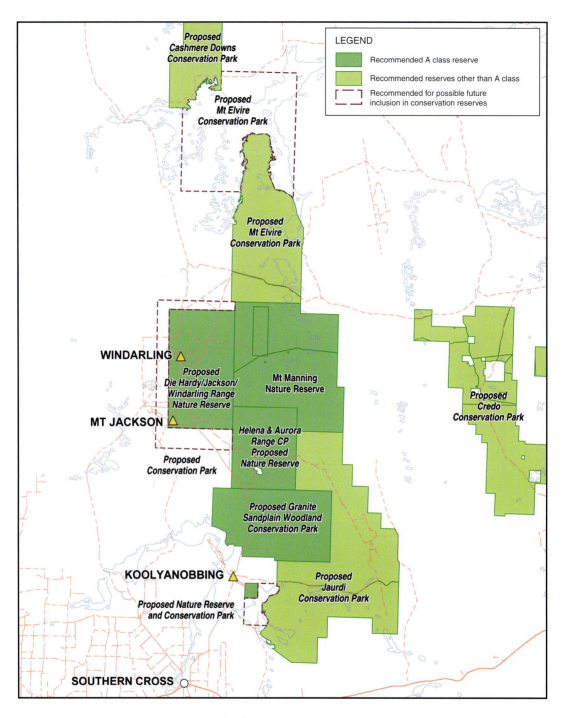

Figure 6.10: Conservation reserves as recommended by EPA Bulletin 1256.

effect of sterilising 43% of the identified ore resources. Although it could be argued that the report represented nothing more than recommendations for consideration by government, it had the effect of creating huge uncertainty around the value of Cliffs' assets and the future of the Koolyanobbing operations. This highlights the dangers in considering only the environmental dimension in framing land use recommendations. Social and economic impacts were not considered by the EPA, and the report was developed without consultation or input from stakeholders outside of the government environment agencies.

The above examples only serve to emphasise the importance of a balanced approach to planning that considers all factors – environmental, social and economic – in an integrated way. That the planning process must also be consultative, transparent and science-based is fundamental to its chances of success. There is a growing consensus within the mining industry and the non-government conservation organisations that this type of strategic planning across the regions of Western Australia is of critical importance.

Cliffs takes the view that, in gaining access to mineral resources, it accepts an obligation both in terms of managing its direct environmental impacts and in contributing to social development and conservation objectives more broadly (Cliffs Natural Resources 2010). In pursuit of this objective, Cliffs has recently formed a partnership with the WA Department of Environment and Conservation to provide increased on-ground management capability in the form of indigenous conservation officers, fire management, and flora and fauna research and management projects in the wider area around the companies' operations. This, and initiatives such as the Great Western Woodlands Collaboration (see Chapter 11), are examples of emerging models that provide the means for industry, conservation interests and other stakeholders to work towards common goals.

References

Cliffs Natural Resources Inc (2010) Committed to the Global Community – 2009 Sustainability Report. http://www.cliffsnaturalresources.com/EN/CorpResponsibility/ArchivedReports/ReportArchive2008/Documents/2009_ENG_Report.pdf

Department of Environment and Conservation and Department of Industry and Resources (DEC and DOIR) (2007) Strategic Review of the Banded Iron Formation Ranges of the Midwest and Goldfields. http://www.google.com.au/url?sa=t&rct=j&q=&esrc=s&frm=1&source=web&cd=1&ved=0CBwQFjAA&url=http%3A%2F%2Fwww.dec.wa.gov.au%2Fpublications%2Fdoc_download%2F1942-strategic-review-of-the-banded-iron-formation-ranges-of-the-midwest-and-goldfields-exec-summary.html&ei=PSDfU7m8Bs_k8AXZl4DADw&usg=AFQjCNFq4VjDa_o6Jxf1L7sy6e2C6nwxug

Environmental Protection Authority (2007) Advice on Areas of the Highest Conservation Value in the Proposed Extensions to Mount Manning Nature Reserve. EPA Bulletin 1256.

Gibson N, Coates DJ, Thiele KR (2007) Taxonomic research and the conservation status of flora in the Yilgarn banded iron formation ranges. *Nuytsia* **17**, 1–12.

Gibson N, Yates CJ, Dillon N (2010) Plant communities of the ironstone ranges of South Western Australia; hotspots for plant diversity and mineral deposits. *Biodiversity and Conservation* **19**, 3951–3962. doi:10.1007/s10531-010-9939-1

Watson A, Judd S, Watson J, Lam A, Mackenzie D (2008) The extraordinary nature of the Great Western Woodlands. The Wilderness Society.

Yates C, Pettit N, Gibson N, Dillon R, Palmer R (2008) The Population Ecology of *Tetratheca* (Eleaocarpaceae) on the Banded Iron Formation Ranges of the Yilgarn. Final Report to Portman Iron Ore Limited.

PROGRESS IN BAUXITE MINING

7

Mining and ecological restoration in the jarrah forest of Western Australia

John M Koch

Summary

The jarrah forest is a unique vegetation community restricted to the Darling Plateau in the south-west corner of Western Australia. It is flanked by the predominantly cleared coastal plain to the west and the almost totally cleared Wheatbelt to the east. Having a relatively undisturbed natural vegetation community close to Perth gives the area special significance. Alcoa is privileged to be allowed to mine in this sensitive environment and accordingly applies the highest standards of environmental management in its operations. Drinking water production, timber, conservation and recreation are the main land uses in the forest and restoration aims to reinstate these values after mining. Mining in this relatively intact natural ecosystem close to Perth requires high standards of environmental practice in all areas. This paper describes in detail, three significant aspects of environmental sensitivity and how they are managed: jarrah dieback disease, drinking water catchment and biodiversity impact.

Management of jarrah dieback disease is backed by a strong history of applied research, both within and external to Alcoa. Management includes knowing where the pathogen is present, restricting vehicle movement from infested to uninfested areas, cleaning vehicles before entering uninfested areas, preventing infested and uninfested soils mixing, preventing water draining from infested to uninfested areas, training all field staff and planners, monitoring the spread of the disease attributable to mining, and investigating the causes. Extremely small rates of dieback disease spread due to mining have resulted from these management practices.

Water catchment protection is underpinned by clear standards and protocols, government reporting requirements, continual monitoring, and high standards of earthworks and other engineering controls. Drinking water standards have never been compromised by Alcoa's operations in Perth's water catchment areas.

Successful management and restoration of biodiversity stems from the concept that Alcoa's mining is a transient land use and that mining is not complete until restoration has been carried out. Flora is successfully returned by maximising the diversity of species originating from the natural topsoil seed bank, applied seed and planting of difficult or 'recalcitrant' species. Fauna restoration requires the return of suitable habitat for fauna species as well as removal of any barriers to animal success, in particular feral animal predation. Alcoa funds feral fox control in the jarrah forest and is actively involved in rare species re-introductions in areas within and around its mining operations.

Key enablers to ensure high standards of environmental management are:

- Management support – Gaining financial, staff resources and moral support from senior management is probably the most important task a mine environmental scientist can achieve.
- Objectives, targets and auditing – Measurable targets agreed to by management are required for improvements to be made. These are standard components of any Environmental Management System (EMS).
- Research – Understand the biology/ecology of the system and use this knowledge for restoration.
- Applying the research – This is often more difficult than the research itself and also requires management support and staffing.

Introduction

Alcoa currently operates two mines in the jarrah forest in the south-west of Western Australia, Willowdale mine 140 km SSE of Perth and Huntly mine 80 km SSE of Perth. The total area mined and rehabilitated each year is 500–600 ha. A detailed description of the mining and restoration process is provided in Koch (2007) and previously in Gardner (2001).

More than 1.3 million people live within an hour's drive of the bauxite leases in the Darling Range. Recreational use of the forest region is extensive with several National Parks, State Parks, Nature Reserves and Forest Parks, many picnic sites, the long-distance walking and cycling tracks, the Bibbulmun Track and Mundi Bindi Trail – all important scenic and recreational attractions for the people of Western Australia. The jarrah forest has a long history of wood production and the mines are within Perth's drinking water catchments. Because of the proximity to the main concentration of population in Western Australia, Alcoa's mining operations are constantly under public scrutiny. As a consequence, Alcoa's environmental performance is strategically important to the company's future in the region. Alcoa currently commits ~8% of its mine operation costs to post-mining restoration processes and associated research. The average cost of restoration is approximately A$34 000 per hectare (Gardner and Bell 2007).

Alcoa's first mine restoration in 1966 returned soil to the mined out pits and planted exotic pine trees and eucalypts (Tacey 1979). The objective at that time was to restore the mined areas to a productive land use, primarily timber. Over the years the aim of the restoration has changed to become consistent with operating in a diverse forest ecosystem. In 1990, Alcoa published the current restoration objective: 'To restore a self-sustaining jarrah forest ecosystem, planned to enhance or maintain water, timber, recreation and conservation values.' (Gardner 2001, p. 4). Meeting and exceeding stakeholders expectations has led to continuous improvements such that Alcoa's restoration practices are generally considered successful and some of the best in the world (see Koch and Hobbs 2007, Fig. 7.1).

The jarrah forest is flanked by the predominantly cleared coastal plain to the west and the almost totally cleared Wheatbelt to the east. Having a relatively intact natural vegetation community close to Perth gives the area special significance. Alcoa is privileged to be allowed to mine in this sensitive environment and accordingly applies the highest standards of environmental management in its operations. The purpose of this paper is to describe Alcoa's mining operations, the

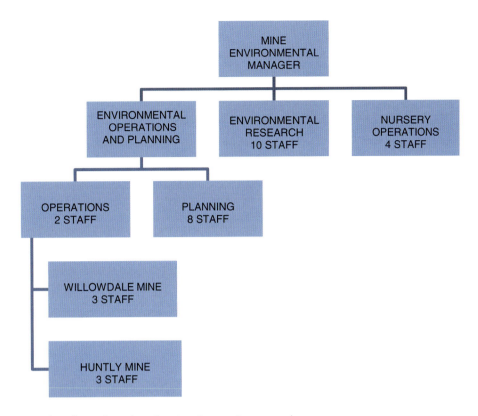

Figure 7.1: Structure and staff numbers for Alcoa's mine environmental group.

potential environmental impacts of the mining and give detailed accounts of how dieback disease, water and biodiversity are managed in this environment.

Environmental management requires resources

Responsible environmental management in a sensitive ecosystem requires a substantial input of resources by mining companies. Environmental management in the 21st century is a complex process including ongoing government reporting, compliance with various licences and permits, measurement of a wide range of environmental parameters and ongoing auditing of environmental performance. In addition there is a policing role, an internal consulting role, a training and educational role, and sometimes a research role for environmental staff. This large workload cannot be carried by one person and indeed will require at least several full-time staff. These staff also need to be adequately resourced and must have senior management support and empowerment to fight the inevitable battles that will develop across the divide of 'production and profit' versus 'environmental responsibility'. Alcoa's mining environmental staff currently number ~30 (Fig. 7.1). This number indicates recognition of the importance of the environmental function and also provides the environment with a strong voice within the company. All these components combine to produce high levels of environmental performance at Alcoa's Western Australian mine operations.

Alcoa's mining and restoration sequence

Current practices involve pre-mine surveys of heritage values, dieback mapping, flora and fauna surveys, vegetation mapping and bauxite exploration.

No drill lines are cleared for bauxite drilling. Once an area is proven for mining then timber is harvested, remaining vegetation is cleared, topsoil and overburden is stripped, caprock (part of the bauxite resource) is blasted or ripped and then the bauxite is mined.

The restoration process is: landscaping and pre-ripping, overburden and topsoil return, return of logs and rocks for fauna habitat, contour ripping, seeding, planting, fertilising and finally monitoring (Fig. 7.2). A detailed description of the mining and restoration operation is found in Koch (2007).

Potential environmental impacts of bauxite mining in the jarrah forest

The most important potential impacts from Alcoa's bauxite mining in the jarrah forest are:

- dieback disease spread
- water quality
- land clearing – loss of biodiversity
- loss of timber resource
- loss of recreation
- loss of ecosystem function
- effect on fauna
- fragmentation/wilderness values
- weed invasion
- neighbours (noise and dust).

All of these potential impacts are managed intensively at Alcoa's operations. Management of the first three (dieback, water and loss of biodiversity) is described in detail in this document.

Dieback disease management

Phytophthora cinnamomi is a microscopic soil-borne microorganism in the family Pythiaceae of the kingdom Chromista – in the past *P. cinnamomi* had been classified as a fungus in the kingdom Myceteae. The fungus is believed to have been introduced to Western Australia around the end of the 19th century, possibly with the importation of fruit trees by European settlers. Although the disease is colloquially known as jarrah dieback, in reality many species die suddenly when root or collar rot deprives them of water (Shearer and Tippett 1989; Colquhoun and Hardy 2000). The climate, soils and regolith characteristics of the jarrah forest suit survival and spread of the pathogen. The growth, reproduction and spread of *P. cinnamomi* are favoured by free water in the soil but especially by surface or sub-surface ponding (Colquhoun and Hardy 2000; Shearer and Tippett 1989). The inoculum levels in the regolith are highest on the surface of the bauxitic caprock at a depth of ~0.5 m and also in the topsoil (0–0.1 m) (Shearer and Tippett 1989). In the 1990s, better understanding of the pathogen and the disease it causes led to the implementation of greatly improved management procedures to minimise the spread and impact of the pathogen.

Mining and restoration processes not only move large volumes of soil, they also alter water movement patterns which could lead to the spread of the pathogen. In the jarrah forest, spores of *P. cinnamomi* are known to be present in streams and in the subsurface flow of water (Kinal *et al.* 1993), but movement of water on the surface is rare (Schofield *et al.* 1989). However, during mining the hydrological characteristics of a catchment can be altered by the creation of non-vegetated areas, some of which are compacted, e.g. haul roads and open mine pits. Water movement on these surfaces has the potential to spread spores of *P. cinnamomi*. Consequently, the risk of spreading spores in both soil and in water has to be addressed when developing intensive management procedures to minimise the spread of the pathogen.

Alcoa has been applying a range of disease management procedures for *P. cinnamomi* since the 1980s. Before mining in a predominantly uninfested region Alcoa developed a more intensive disease management program. A team of Alcoa employees developed a manual of disease management procedures following the principles of risk management (Standards Association of Australia 1999). The size of the threat of spreading the pathogen was estimated along with the likelihood of the threat occurring and the consequences if the threat occurred. The estimates used in this risk management approach were based on understanding

Figure 7.2: Photo sequence of a 1988 restored bauxite mined area from 6 months to 20 years old. The trees are jarrah (*Eucalyptus marginata*).

gained from scientific studies but also the more pragmatic and empirical evaluations from field supervisors as to what was practicable and economic. Procedures were developed for every stage of mining and restoration. The procedures made use of the following disease management principles:

1. Know where the pathogen is present.
2. Restrict vehicle movement from infested to uninfested areas.
3. Clean vehicles before entering uninfested areas.
4. Prevent infested and uninfested soils mixing.
5. Prevent water draining from infested to uninfested areas.
6. Train all field staff and planners.
7. Monitor the spread of the disease attributable to mining and investigate the causes.

Knowing where the pathogen is present by having reliable, up-to-date maps and field demarcation of diseased sites is the backbone of the management measures. Before commencing mining, the presence of *P. cinnamomi* is mapped within the entire envelope (~10 000 ha) and the disease boundaries are re-checked within a 12-month period before starting a major mining process. All the data on disease boundaries are stored on a geographical information system (Fig. 7.3). This information is updated frequently to ensure that mine planners and environmental scientists have the best maps for their planning tasks. The disease boundaries are marked in the forest by painting the stems of trees, and in cleared areas by establishing lines of painted pegs or lines of bunting.

All vehicles are cleaned before entering uninfested areas (Fig. 7.4). Operations occur in both infested and uninfested sites so procedures are

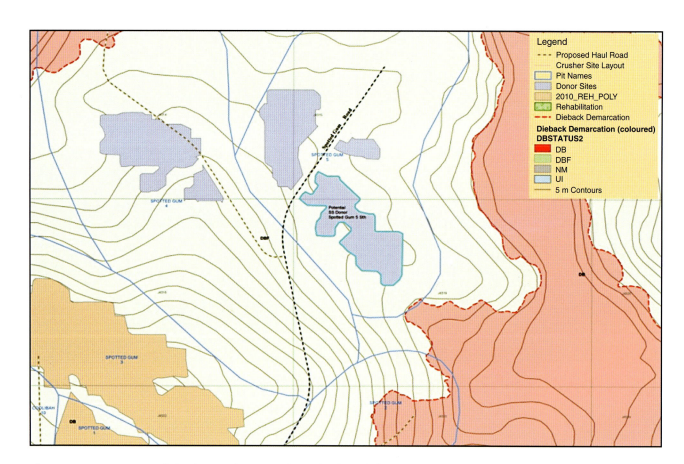

Figure 7.3: A pre-mining dieback map showing a future haul road, future mining (blue) and dieback presence (red).

Figure 7.4: A portable washdown trailer.

required to minimise the risk of spreading infested soil across boundaries. Machinery and vehicles are cleaned at all stages of mining where vehicles are required to cross disease boundaries. A permanent vehicle washing facility is located at the entrance to the mine where all vehicles are cleaned before entering the mine (Fig. 7.5). Large trailer-mounted high-pressure water pumps are located on disease boundaries to enable vehicles to be cleaned. Effective cleaning of vehicles in the field is difficult so the need for this task is reduced as much as possible by detailed planning of field operations.

The location of boundaries between infested and uninfested areas is known at all times during the mining and restoration processes. All uninfested soil is moved and stored separately from infested soil. All soil stockpiles are distinctively marked to differentiate infested (red) and uninfested (green) stockpiles (Fig. 7.6).

The risk of water draining from infested to uninfested areas is present at all stages of mining and restoration. Zoospores of *P. cinnamomi* are easily transported in water so surface water movement from infested to uninfested areas needs to be prevented. Surface water is never allowed to drain freely into uninfested forest. The wide roads constructed during mining shed a lot of water. A system of drainage channels and high bunds along the roads prevents surface water flowing directly into the forest. Instead, the water is directed into sumps which are designed to retain a 1:20 to 1:50 year rainfall event. In restored mine pits, the ripping pattern directs all surface water back into the mine pit away from the existing forest.

All operators are trained in the control procedures relevant to their duties. Audits are conducted every year to assess the level of compliance with the procedures. There is strong senior management commitment to the procedures so any breach to the rules is regarded as a reportable environmental incident that may result in a disciplinary response by the manager. Contractors are also trained in the procedures. Any breach by contractors could lead to the cancelling of a contract.

Alcoa monitors the spread of the disease attributable to mining and investigates the causes of all new infestations. They believed that it is essential to gauge the success of the disease management program and improve the procedures when there is evidence that the program is not effective. The cost

Figure 7.5: A permanent washdown station located at the entrance to the Huntly mine.

Figure 7.6: Topsoil from an area of cleared forest that was classified as 'dieback free'.

of implementing the intensive disease management procedure has been estimated to be in excess of US$1.5 million a year (Colquhoun and Hardy 2000). Consequently, a vital question is 'How effective are these disease management procedures?'

Monitoring the effectiveness of dieback management procedures

In a comprehensive monitor and re-monitor program from December 2003 to November 2004, 10 new infestations were identified at the two mines. This was equivalent to one new infection for every 192 ha cleared for mining (Table 7.1). The total area of the infections was 1.1 ha. This was equivalent to 0.0006 ha for every hectare cleared for mining. The size of the new infections ranged from 2 m^2 (one small plant) to 6620 m^2. The total potential spread was 3.56 ha, the equivalent to 0.002 ha of spread per hectare cleared for mining. Of the 169 km of clearing boundary monitored only 81 m was found to be infested.

The major causes of the new infections were judged to be:

- errors during the operation of clearing the vegetation
- placement of infested soil next to uninfested forest
- drainage from the restored mined area into the forest.

Many causes were not related to errors in the procedures but to errors in complying with the procedures, e.g. when water drained from the restored area into the forest due to ripping not being 'on contour', when earthmoving vehicles drove into the native forest from an infested area without removing soil from the vehicle. These findings were directly entered into training packages for operators to demonstrate what happens when the procedures are not properly implemented. In other cases Alcoa identified how to improve procedures or to give better guidelines to field supervisors, e.g. infested soil should be placed in the lowest part of the restored pit but if there is disease free forest adjacent to the pit then infested soil should be returned low in the pit but away from the edge.

Business systems throughout the world are based on the 'plan, do, check, revise' cycle as a way to continually improve the process. Without this systematic approach, Alcoa could have been spending millions of dollars every year on ineffective procedures.

There are five key factors that contributed to the success of the disease management program:

1. Commitment from all levels within the company from management to operators in the field.
2. Integrating environmental procedures as part of operating instructions.
3. Adopting a risk assessment process with input from a range of staff.
4. Supporting research and development into management of *P. cinnamomi*.
5. Implementing relevant monitoring and auditing programs.

Table 7.1. Summary of the spread of *P. cinnamomi* from mining operations at Huntly and Willowdale, monitored from November 2003 to December 2004

	Clearing (ha)	Infestation (ha)	Infestation/ clear (ha/ha)	Total potential spread[1] (ha)	Total potential/ clear (ha/ha)	Length of Interp (m)	Infestation length[2] (m)	% clearing length infested
Huntly	1261	1.06	0.0008	1.22	0.001	119 366	53	0.04
Willowdale	663	0.038	0.0001	2.34	0.0035	49 812	28	0.06
Total	1924	1.098	0.0009	3.56	0.0045	169 178	81	0.1

1. 'Total potential spread' refers to the area of infestation added to the area of high potential risk.
2. 'Infestation length' refers to the length of previously uninfested forest abutting clearing boundaries that is now infested as a result of mining.

Water protection

In conjunction with the Department of Water (DoW), Alcoa has developed working arrangements which detail all matters of practical relevance to water supplies in mining areas. In addition, Alcoa's Environmental Management Manual (EMM) provides Alcoa personnel with a description of the Environmental Management System used at the mines. A clear understanding of environmental policies, goals and management systems by personnel is crucial in achieving good environmental management. The management programs relevant to water quality and catchment protection are:

- catchment protection
    - runoff control
    - summer rainfall and rainstorm prediction
    - turbidity monitoring
    - access to bauxite in the Intermediate Rainfall Zone
    - hydrocarbon spill management
    - Willowdale – Samson Dam Causeway.
- water management
    - water supply
    - water conservation
    - groundwater.

The Huntly mining area (current crusher site) is drained to the west by the North Dandalup River, to the north and east by the Serpentine River and Big Brook, and to the south by the South Dandalup River. All these rivers and their tributaries flow through priority1 (P1) managed public drinking water source areas into storage reservoirs.

At Willowdale mine the catchment to the south is Samson Brook which flows into the Samson Reservoir. This is managed as a P1 protection area. Areas to the west of current mining drain to Lake Navarino and to the north and east drain to the Murray River. These are not drinking water sources. To the south and east several tributary streams flow into the Samson Brook and Murray River and through the Lane Poole Reserve.

Both mines receive ~1200 mm of rain each year, most of it falling between May and October. Rainfall declines rapidly in the eastern portion of the Darling Scarp.

The ground-water table is 10–30 m below the surface in upland areas, but in the valleys it is usually shallower. In the western, high rainfall parts of the Darling Ranges, the groundwater is often close to the surface in the valleys, and discharges to stream channels all year. Further east, as rainfall declines, the forest removes almost all of the rain absorbed deep into the soil each winter. As a result, the groundwater is often below the valley floor, so that streams are ephemeral and only flow in winter.

Reservoir and stream buffers

Alcoa has undertaken not to clear within the following zones:

- 100 m of the top water level of priority 1 reservoirs
- within the variable width buffers recommended by the DoW as summarised in Table 7.2.

The Water Corporation maintains, in principle, a 200 m buffer from the top water level (or to the ridge if closer), of P1 reservoirs. However, on a case by case basis the Water Corporation may approve mining on the Five Year Mine Plan with a minimum 100 m buffer where Alcoa has provided a plan for adequate risk management strategies. The final approval is determined at inspection when the clearing boundaries are marked and control strategies are agreed.

A Water Resource Sensitive Zone is self-imposed by Alcoa adjacent to the P1 reservoirs. Its dimensions are 500 m from the reservoir top water level and a 200 m buffer (from the centre of the stream zone) for 1 km upstream of the top water level. The Water Resource Sensitive Zone indicates areas within which Alcoa commits to apply risk assessment as part of the Clearing Advice, to ensure that mining or infrastructure has no impact on the

Table 7.2. Mining related buffer zones around streams and dams

Stream Order	Buffer
1st and 2nd Order	20 m from the centre of the stream channel
3rd Order +	30 m beyond stream zone riparian vegetation
1000 m upstream of Top Water Level of P1 reservoir. Note Samson, Serpentine, South Dandalup and North Dandalup dams are managed as P1 reservoirs	100 m – propose that Alcoa maintain at 50 m and complete a management plan for all areas within the 50–100 m buffer similar to those completed for all other Water Sensitivity Zones (100 m, 200 m and 500 m buffers)
Reservoir Top Water Level	100 m
Reservoir with adjacent slope of land > 15%	Buffer to extend to the ridgeline

water resource. The risk assessment methodology is documented in the Water Working Arrangements.

In 1999, the Regional Forest Agreement (RFA) delineated a series of reserves in order to fulfil the agreed commitment for a 'Comprehensive, Adequate and Representative' (CAR) reserve system. This included a series of both formal and informal reserves that are revised and updated periodically. The CAR process identifies many of the informal reserves along stream zones. Many of the stream zones that may be subject to disturbance for haul road crossings will require an additional approval process.

Stream monitoring

Groundwater and surface water monitoring is carried out in experimental catchments within the High and Intermediate Rainfall Zones. The monitoring is used to determine the long-term impact of mining and rehabilitation on water resources. The results of this research are reported in Alcoa's Annual and Triennial Environmental Reviews. Past monitoring of surface water courses within Alcoa's current mining operations demonstrated clearly that within the High Rainfall Zone, mining does not result in salinisation of water supplies. Turbidity is considered the surface water quality parameter with the greatest potential to be adversely affected by current mining operations.

Turbidity monitoring – operational

Stream turbidity monitoring points upstream of neighbours private water supply facilities or public drinking water reservoirs are referred to as Compliance Monitoring Points. The turbidity level at these points reflects the quality of water entering a reservoir, or becoming available for public use. Turbidity at these locations is monitored using Greenspan turbidity sensors, which are programmed to record turbidity readings every 30 minutes (Fig. 7.7).

Local monitoring points

Upstream from the compliance monitoring points are several local monitoring points. These are selected to provide short-term information on the performance of the drainage infrastructure of the mine, and are typically located in streams below haul road crossings or a series of large mine pits. Monitoring at these points may use turbidity sensors, stage rise samplers or grab samples. These points are monitored after 20 mm of rainfall is recorded within a 24-hour period.

Significant and moderate turbidity events that occur at compliance points as a result of Alcoa's operations are reported in the Annual and Triennial Environmental Reviews. If stream turbidity is caused by other factors unrelated to the mining operation the events will still be raised as incidents and investigated. Where appropriate the events

Figure 7.7: An automatic stream turbidity sampling station with solar panel and radiometric real time data transmitter.

will be reported to DEC and/or Water Corporation and Alcoa will assist where practicable to minimise the discharge of turbid water into streams.

Spill management

Alcoa are committed to minimising fuel, lubricant and hydraulic losses of containment (spills) associated with the use of heavy machinery. Mining has an implemented and maintained Emergency Response Plan (ERP) that includes Loss of Containment (LOC) response and management procedures and meets the recommendations of the DoW. A spill management trailer is located at each mine (Fig. 7.8).

The majority of losses of containment (LOCs) that occur are due to hydraulic hose failures and when equipment is being refuelled. Checks have been put in place to ensure hoses are inspected and replaced, if required, during routine maintenance work. Regular training and reinforcement of reduction of LOCs is carried out with all maintenance personnel. When LOCs do occur, they are cleaned up according to the following decision tree (Fig. 7.9). This decision tree is part of a training package deployed to all employees to ensure swift clean up of all LOCs.

Intermediate Rainfall Zone

A long-term research program has been developed to meet the commitment by Alcoa, as part of the revised 1978 Environmental Review and Management Program (ERMP), for the Wagerup Alumina Project, that 'mining will not take place in the eastern, lower rainfall portion of Alcoa's lease until research shows that operations can be conducted without significantly increasing the salinity of water resources' (Alcoa of Australia Ltd. 1978, p. 14). The 'eastern, lower rainfall portion of the lease' has been interpreted and accepted as that part of the mining lease east of the 1100 mm rainfall isohyet, and termed the Intermediate Rainfall Zone (IRZ). The Joint Intermediate Rainfall Zone Research Program (JIRZRP) details cooperative research between Alcoa, DoW and the Centre for Water Research based at the University of W.A.

The basis of the JIRZRP is Alcoa's requirement to access the IRZ for mining operations by about the year 2022. The JIRZRP is a long-term ongoing research program with activities scheduled until at least 2017 and has two main objectives:

1. To determine what impact bauxite mining in the IRZ will have on the water resources of the region; and
2. To document the forest, mine and rehabilitation management practices which should be used in the IRZ.

Figure 7.8: Spill control and containment trailer at Huntly mine.

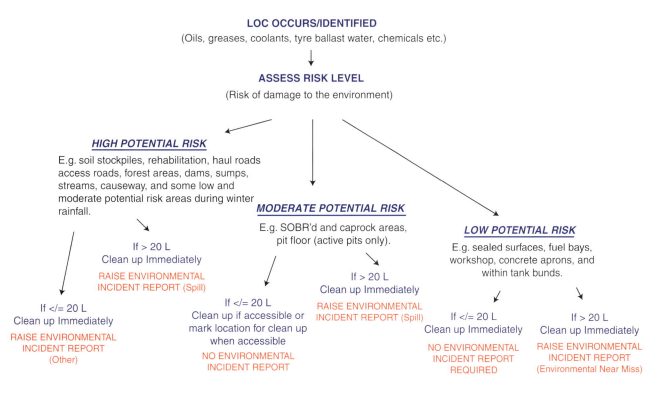

Figure 7.9: Decision tree for loss of containment (LOC) cleanup and reporting.

The Bauxite Hydrology Committee (BHC) oversees the JIRZRP and reviews research outcomes and is a technical advisory committee of the Mining and Management Program Liaison Group (MMPLG).

The JIRZRP schedule and research progress is reported in the Annual and Triennial Environmental Reviews submitted to the DoIR and circulated to members of the MMPLG. A recent document on 'Salinity risk of mining in the IRZ' has been presented to government. The conclusions of the document are that there is very little or no risk of sanitising water supplies due to the large depth to groundwater in these catchments (see Croton and Reed 2007). This is a direct result of low annual rainfall over the last several decades.

Haul roads and sumps

Runoff from haul roads and infrastructure is detained in sumps (Fig. 7.10). In sedimentation sumps, runoff is retained for sufficient time to allow the coarse sediment and some of the fine sediment to settle before the water is discharged to the environment. The minimum acceptable design criterion for short-term sumps is to retain water for a minimum of 5 minutes during rainstorm intensities up to a 20-year average recurrence interval. In areas of particular environmental sensitivity or areas where there is a risk to life or property, a 50-year or 100-year average recurrence interval is applied. These sumps should not discharge direct into streams but rather discharge through vegetated land (forest understorey) to allow further sediment removal.

In some locations infiltration sumps are used because discharge is not possible into a stream zone. These sumps are designed to allow the water to infiltrate and evaporate leaving the silt behind. Infiltration sumps are designed with emergency overflows or spillways for storm events that exceed the designed criterion for the sump.

Alcoa's Minesite Drainage Book and procedure called 'Design and Construct Haul Roads' are used by engineers as a reference for the construction of roads and sumps. In addition, the DoW has developed standards for roads near sensitive water resources which covers the recommendations for sealed and unsealed roads. These procedures and standards are applied to all road construction and maintenance operations.

Reporting of incidents

Alcoa has an environmental incident reporting and management system in place, which takes into account the need to report to government. An incident that has an impact on, or poses a significant risk to, water resources will be reported to the Water Corporation and DoW's Kwinana–Peel regional office regardless of the mine where it occurs. DEC will be notified if site contamination is likely to be an issue.

A list of environmental incidents that have been recorded in Alcoa's EHS IMS (Environment, Health and Safety Incident Management System) is submitted to the Water Corporation ranger on a monthly basis. This list includes hydrocarbon spills, elevated stream turbidity events (mining and non-mining related), operation of sewerage plants out of specification, and pit drainage failures. If additional information is required on the incidents this information is provided by environmental staff.

Environmental incidents that are reported at the time of occurrence include:

- wastewater non-compliances with potential to impact downstream users
- spills requiring major clean up works or ongoing remediation
- major spills inside containment areas where the spill could affect external environment
- turbidity events at compliance monitoring points ≥ 2 hours > 25 nephelometric turbidity units (ntu)
- drainage system failures such as sump wall failure or pit break out
- other turbidity events with potential to affect downstream users.

Figure 7.10: A three stage sump at Alcoa's Huntly mine. Sediment settles out of turbid water before flowing into the stream zone.

Restoration of biodiversity

Clearing and mining obviously removes all or most animal and plant life from the mined areas. Alcoa's belief is that mining is a transient land use and that the total biodiversity of the mined areas should be restored as far as practicable. Returning the conservation value of the forest is interpreted as requiring the return of fauna and flora diversity after mining. Restoration is aimed at replicating the biological composition of the unmined jarrah forest (the analogue community).

Flora

The jarrah forest is a dry sclerophyll open forest to woodland (Specht *et al.* 1974), unusual in Australia in that it is mostly dominated by a mixture of only two eucalypts over its entire extent, jarrah (*Eucalyptus marginata*) and marri (*Corymbia calophylla*) (Pryor 1959). In addition there is a small-tree component, with bull banksia (*Banksia grandis*), sheoak (*Allocasuarina fraseriana*), snottygobble (*Persoonia longifolia*) and woody pear (*Xylomelum occidentale*) the most common species. The undergrowth consists of sclerophyllous shrubs (to 3 m) predominantly from the families Anthericaceae, Dasypogonaceae, Leguminosae, Orchidaceae, Apiaceae, Epacridaceae, Asteraceae, Proteaceae, Restionaceae, Myrtaceae and Cyperaceae. Havel (1975) and Bell and Heddle (1989) classified the northern jarrah forest into site-vegetation types

Table 7.3. Brief description of some northern jarrah forest site-vegetation types including those that are impacted by Alcoa's mining activities

Adapted from Havel (1975) and Bell and Heddle (1989). Due to the gradient of vegetation communities there are often combinations of these site-vegetation types.

Site-vegetation type	Description	Key indicator species
A	Stream zone and swamp heath with small tree emergents	Melaleuca preissiana, Banksia littoralis, Hakea varia
C	Woodland along creeks and swamp margins	Eucalyptus patens, E. megacarpa
P*	Mid and lower slope forest of Eucalyptus marginata with occasional Corymbia calophylla Second storey of Allocasuarina fraseriana	E. marginata, A. fraseriana
S*	Mid slope to upland forest of E. marginata with C. calophylla and Banksia grandis	E. marginata, C. calophylla, B. grandis
T*	Ridge top and upper slope forest of E. marginata with C. calophylla	E. marginata, Pteridium esculentum, Leucopogon verticillatus
W*	Lower slope and valley floor forest	E. marginata. C. calophylla, E. patens

* Impacted by Alcoa's mining activities.

based on a set of canopy and understorey indicator species. They described 21 forest site-vegetation types. These not only covered jarrah and marri dominated forest communities which are typically site-vegetation types P, S and T (Table 7.3), but also the low open woodlands, open heaths and sedgelands of the stream zones and swamps (for example types A and C) (Bell and Heddle 1989). This 'Havel' site-vegetation type classification is the most widely used mapping system in the northern jarrah forest and is used by Alcoa in its pre-mining vegetation surveys.

Within a mine, ~40–50% of the landscape is cleared for mining, roads and infrastructure. Mine operations primarily impact upland areas of the forest in site-vegetation types S, P, T and combinations of these types (Fig. 7.11). Although stream zone and swamp site-vegetation types occupy approximately one-quarter of the area of a mine, very little of these site-vegetation types (A, C and W) is cleared.

Pre-mining vegetation monitoring

Detailed vegetation control plots are permanently established in forest site-vegetation types adjacent to mining areas based on the proportion of each site-vegetation type to be impacted by future mining (Fig. 7.11). In each new mining envelope, in areas that will not be mined, 12 to 15 plots (20 m × 20 m) are established. These plots are re-monitored every five years and provide controls which are compared to identical plots in the restored mines.

Species used in mine restoration as applied seed or planted greenstock are only natives found in upland jarrah forest areas in site-vegetation types T, S and P. Species are selected based on their presence in broad scale pre-mining vegetation surveys and the vegetation control plots. Restoration operations do not currently aim to reinstate the specific pre-mining site-vegetation types in each mine pit, although returning the topsoil seedbank does reinstate the same plant species in each pit to some extent. The major changes in the soil profile and landscape during mining and restoration (Kew et al. 2007; Koch 2007) are likely to override the more subtle soil and landscape differences which control the pre-mining site-vegetation types (Havel 1975).

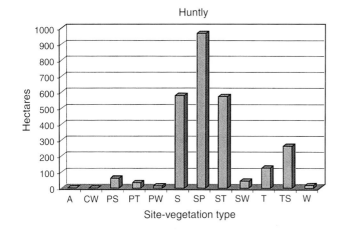

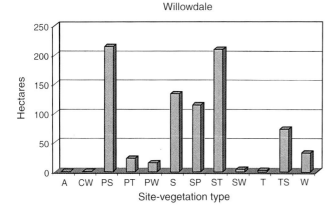

Figure 7.11: Hectares of forest of various vegetation site types predicted to be cleared for mining at Huntly and Willowdale mines over an approximate 5-year period.

Sources of plants

A plant propagation protocol was developed which classified jarrah forest species into the most cost efficient method for restoring them to mined areas (Table 7.4) (Koch *et al.* 1994). The protocol commenced with the 'easiest' method of restoration, that is species that return well from the soil seed bank, then those that invaded with time (mostly orchid species), those where seeds needed to be collected and broadcast, then various germination enhancing treatments and ultimately species that required more difficult propagation techniques such as tissue culture. At the time of publication (1994), of the 141 species, 90 (64%) were relatively easy to re-establish. A further 33 (23%) were likely to be amenable to vegetative propagation techniques, including tissue culture. The remaining 18 (13%) were species for which there was limited propagation knowledge.

In newly restored bauxite mines, most plant species originate from the seed bank in the returned topsoil and from applied seeds (Table 7.5). In 1994 there was no planting of nursery grown plants. There is some volunteering of windblown seeds but these are mainly introduced weeds or 'weed-like' species from the family Asteraceae (Koch and Ward 1994). Some tree species from the unmined forest edge drop seeds into the mine pits to a distance of ~20 m. Animal dispersal occurs in several fleshy fruited species (*Leucopogon nutans*, *Styphelia tenuifolia*, *Macrozamia reidlii* and *Persoonia longifolia*) (Vigilante 1996) but is not a major contributor of plants in newly restored areas. Orchids invade with time (Collins *et al.* 2005; Grant and Koch 2003).

Soil seed bank

The importance of the soil seed bank for mine restoration has been well documented (Bellairs and Bell 1993; Iverson and Wali 1982; Koch *et al.* 1996; Putwain and Gillham 1990; Read *et al.* 2000; Tacey and Glossop 1980; Vlahos and Bell 1986; Ward *et al.* 1997). In the jarrah forest of Western Australia the soil seed bank ranges from several hundred up to ~1500 native seeds m^{-2} (Koch *et al.* 1996; Smith *et al.* 2000; Tacey and Glossop 1980; Vlahos and Bell 1986; Ward *et al.* 1997). For bauxite mine restoration the soil seed bank can contribute more than 70% of the plant species richness (Koch and Ward 1994; Ward *et al.* 1996) and hence is an important resource for the restoration process. Studies have quantified the seasonal changes in the seed bank, (Ward *et al.* 1997), the depth distribution of the seeds (Koch *et al.* 1996), the ability of germinating plants to emerge from various levels of burial (Grant *et al.* 1996) and the effects of different soil handling procedures on the seed bank during bauxite mining

Table 7.4. Alcoa's plant propagation protocol

The numbers of species are from a total of 141 which are made up of the 100 most frequently occurring species in the unmined forest at Alcoa's three mines (from Koch et al. 1994). There are now two operating mines; Jarrahdale ceased mining in 1998 and restoration was completed by 2001.

Most cost efficient propagation method	Number of species (and %)	Example species
Establish from returned topsoil	28 (20)	Trymalium ledifolium Boronia fastigiata
Invade with time	10 (7)	Caladenia flava Pterostylis recurva
Germinate easily	24 (17)	Allocasuarina fraseriana Eucalyptus marginata
Germinate after heating	15 (11)	Acacia urophylla Acacia lateriticola
Germinate with smoke or other treatments	13 (9)	Grevillea wilsonii Stylidium schoenoides
Viable seeds do not germinate	16 (11)	Astroloma ciliatum Persoonia longifolia
Cuttings	7 (5)	Dampiera linearis Scaevola calyptra
Transplants or divisions	4 (3)	Tetraria capillaris* Loxocarya cinerea*
Tissue culture	6 (4)	Lepidosperma squamatum Lepidosperma leptostachyum
No method known yet	18 (13)	Lomandra hermaphrodita Drosera erythrorrhiza
All species	141 (100)	

* Species now propagated by tissue culture.

(Koch et al. 1996; Ward et al. 1996). These research findings have led to the following recommendations to maximise the contribution of the topsoil seed bank for restoration:

Table 7.5. Sources of plant species in 4320 m^2 of total sampling area in 9-month-old restored sites (from Koch and Ward 1994)

This does not include any planting of nursery stock.

Source of plants	Species in 4320 m^2	Plant density m^{-2}	Plant cover (%)
Topsoil seed bank	129	3.9	4.9
Applied seeds	39	2.9	18.6
Weeds	21	0.5	0.5

- Strip and return the topsoil in the dry season (summer to autumn in this climate type).
- Return the topsoil from new mining areas to restored areas directly with no storage period (avoid stockpiling the soil). This is called direct return of topsoil.
- Carry out the final soil tillage (deep ripping) in the dry season (summer–autumn).
- So that the seeds are not buried too deep to emerge, ensure the top layer of soil (~15 cm) is returned on the top in restored areas. This is called double stripping of soil.

Applied seed

More than four tonnes of native seed is used for Alcoa's mine restoration each year. About three tonnes of this is large *Macrozamia riedlei* (Zamia Palm) seeds which weigh ~15 g each. Since 1991, all of these seeds have been collected within ~20 km from the mine where they are to be used. Using local or 'provenance correct' seeds ensures the genetic diversity of the restored areas is maintained and matched to surrounding native vegetation. This conforms with the recommendations of McKay *et al.* (2005, p. 136) 'Collect locally if at all possible'. The recent availability of DNA analysis techniques has allowed testing of key plant species to determine the degree of genetic differentiation over their distribution range, and a more objective measurement of actual provenance size (Krauss and Koch 2004; Krauss *et al.* 2005). To date 13 species have been successfully tested. Six showed a broad provenance range over the whole northern jarrah forest and the other seven had smaller provenance sizes. Alcoa's seed collection strategies have been adjusted accordingly.

The seed mix composition varies from year to year depending on availability and includes the two dominant tree species and ~78 to 113 species of the ground flora. Many species from the Mimosaceae and Papilionaceae (legumes) require a heat treatment before they will germinate (Bell *et al.* 1993; Glossop 1980; Shea *et al.* 1979). This is a natural adaptation that ensures they will germinate only after a fire or other disturbances in the forest. Smoke is now applied to seed of appropriate species before being sown in the rehabilitated mine pits (see Dixon *et al.* 1995; Roche *et al.* 1997; Roche *et al.* 1998). Other jarrah forest species require more complex dormancy breaking treatments. For example *Persoonia longifolia* requires chipping, gibberellic acid and surface sowing to germinate (Mullins *et al.* 2002). The biology and propagation requirements of many other jarrah forest understorey species is unknown and is the subject of ongoing research by Alcoa and associated research groups. Research findings regarding germination requirements of native plant species are of interest to other researchers, students, plant nurseries, land care practitioners and other land reclamationists so wherever possible Alcoa makes this information available (Bell *et al.* 1993; Koch and Taylor 2000; Mullins and Koch 2001).

The timing of seed spreading is important in the establishment of plants from applied seed (Ward *et al.* 1996). In a study of 28 jarrah forest species, Bell and Bellairs (1992) found the highest germination rates occurred at temperatures of 10–15°C, temperatures most likely to occur during the winter rainfall period. Seeds could therefore be successfully sown during the dry summer – autumn season in newly restored areas and would remain dormant through summer and subsequently germinate in late autumn and winter when temperatures decreased and the rains came.

Planting of 'recalcitrant' species

In 1990, Alcoa set a target of achieving 80% species richness values in 1-year-old rehabilitated areas compared to identical control plots in the unmined forest. This was achieved in 1995 by using direct returned topsoil in over 60% of the post-mining area, carrying out topsoil handling and ripping in the dry season and by applying a seed mix of 78 to 113 species at the three mines. The balance of plant life forms was different to the unmined forest with an over-abundance of re-seeder and ephemeral species (see Bell 2001 for discussion of these life forms) and a shortage of resprouters (Fig. 7.12). The resprouters give the jarrah forest a high resilience to natural disturbances, particularly fire and grazing, and hence are a crucial component of the ecosystem. However because resprouting species generally have very low seed production, these species are difficult to restore from the soil seed bank or from collected seeds. Resprouters comprise most of those species that were identified as not amenable to conventional mine restoration techniques. A new target of 100% species richness was set which includes a requirement for 20% of the species to be difficult to propagate resprouter species. These are called 'recalcitrant' due to this difficulty and they are historically absent or under-represented in the restored mined areas. They are not to be confused with the true botanical

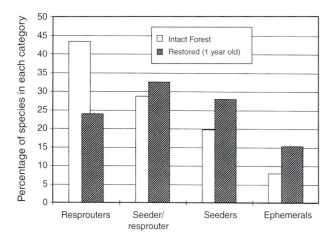

Figure 7.12: Percentage of different plant life forms in restored areas and intact forest at one year old. Restored areas have an over-representation of seeders and ephemerals and a shortage of resprouters.

meaning of 'recalcitrant' which is species (often tropical) having desiccation intolerant seeds.

Little was known about how to propagate or grow most of these species – they are not used in the horticultural industry. Tissue culture is the most difficult and expensive propagation method and is only used when there are biological barriers to other methods. Plants produced by tissue culture are often species that invest energy into underground biomass to ensure survival. They may be clonal over large areas within the forest (Meney *et al.* 1990; Pate and Dixon 1982; Perger *et al.* 1994) and have very low seed set, necessitating vegetative propagation methods.

A tissue culture laboratory was built at the Alcoa nursery to produce recalcitrant plants for mine restoration and to continue research into propagation methods. Production of clones began in 1992, and by 1999 significant numbers of plants were available for routine planting (Table 7.6). Approximately 12 plant species are now grown by tissue culture. Most of these are graminoids, from the sedges (Cyperaceae), rushes (Restionaceae), and certain species from the Anthericaceae (lilys) and Dasypogonaceae (lomandras). The development of tissue culture methods is a very slow process and it can take several years to learn how to mass produce some species.

Table 7.6. Numbers of recalcitrant plants and species planted into restored mines 1995 to 2010

Plantings from 1995 to 1998 were experimental.

Year	plants	spp.	ha^{-1}
1995	7226	16	14
1996	1912	14	4
1997	15 650	7	31
1998	4190	14	8
1999	67 600	15	170
2000	184 552	16	229
2001	213 678	19	230
2002	229 757	28	385
2003	184 000	23	354
2004	166 660	21	341
2005	137 208	20	319
2006	119 000	24	196
2007	219 800	25	352
2008	190 000	25	303
2009	135 000	30	276
2010	146 000	27	298

Approximately 100 000 plants are grown from tissue culture for mine site planting each year (Willyams 2005). A similar number are grown from cuttings and from seed where available. The cost of producing plants from tissue culture and cuttings and then planting them in the restoration sites is about A$5 per plant. Survival of a wide range of species grown from tissue culture and cuttings is between 55% and 90% after 18 months.

Many of these species, particularly the grass-like rushes and sedges and the grass-trees of the Xanthorrhoeaceae, are the favourite food source for kangaroos in the post-mining restored areas and can be killed by intensive grazing pressure (Koch *et al.* 2004; Parsons *et al.* 2006). This has two important implications: first, that these species have a significant function in the restored ecosystem as kangaroo food; and second, the plants need protection when they are small, otherwise they can be

Figure 7.13: Hair sedge (*Tetraria capillaris*) grown from tissue culture and planted in restored mined areas. The small mesh bag provides grazing protection.

killed by overgrazing. Small mesh bags (Fig. 7.13) are used to protect the grass-like species from overgrazing.

Vertebrate fauna

Fauna recolonisation of restored mine areas

Studies into the processes of vertebrate fauna colonisation of Alcoa's restored bauxite mines began around 1975. This recognised the key role of vertebrate fauna in jarrah forest ecosystem processes, and also the fact that some species were rare, so priority was given to determining their status in unmined forest, and promoting their return to restored areas following mining. Long-term studies have since taken place on mammals, birds and reptiles in both unmined forest and in restored areas of varying ages and techniques. Mammal recolonisation patterns vary, depending on species' food and shelter requirements, and their distribution and abundance in the surrounding forest. Birds rapidly recolonise and 95% of species have been recorded in restoration (Fig. 7.14) (Nichols and Grant 2007). Bird community structure changes with restoration type and age, and in current restoration it is similar to that of unmined forest by the age of 10 years. There are some avifauna differences in restored sites which are caused by the relative immaturity of the restored system.

Studies on reptiles have shown that 21 out of 24 known forest species have recolonised. The remaining three include one legless lizard and two snakes, all of whom feed on small vertebrates (e.g. skinks) and require shelter in the form of logs, stumps and coarse woody debris. Some other reptile species consistently occur in restoration in lower densities than in unmined forest and current studies are investigating the causes of this. Again the cause is the immaturity of the restored mines and the inherent lack of old ecosystem components such as very large trees and old rotting logs.

Regional fauna issues

Despite the fact that surveys on mammals in restored bauxite mines commenced in 1978, little has been published on this group. The main reason

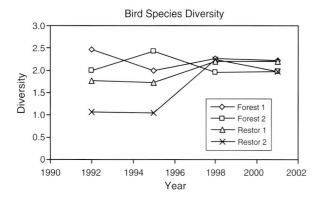

Figure 7.14: Bird species Shannon-Weiner diversity recorded in summer at Jarrahdale mine in 1990 restoration and in healthy unmined forest in long-term fauna monitoring sites. Surveys at each site were conducted along two 250-m transects on three occasions, recording all birds within 20 m of the transect (from Nichols and Grant 2007).

for this is because quantitative data have been difficult to obtain due to the low numbers of many species resulting from predation by the introduced fox, *Vulpes vulpes*. In order to clearly interpret patterns of mammal colonisation to restored bauxite mined areas, it has been necessary to understand that, concurrently with species reinvasion following mining, changes have been taking place in jarrah forest mammal populations.

Many of the medium-sized mammal species which occur in the jarrah forest suffered significant declines in the years leading up to the mid-1990s. Some, such as the Chuditch, *Dasyurus geoffroii*, originally occurred over large areas of Australia (Morris *et al.* 1998) but have declined to small remnant populations in the south-west. For others, such as the Common Brushtail Possum, *Trichosurus vulpecula*, and Mardo, *Antechinus flavipes*, the decline has not been as extensive. Research demonstrated that the major reason for the significant decline in abundance of many species in the jarrah forest was fox predation (Bailey 1996).

Research into the fox predation problem led to the development of a dried meat bait containing sodium fluoroacetate, a naturally occurring chemical found in jarrah forest plant species. Indigenous fauna species have developed a high tolerance to this toxin, but introduced species such as the fox are highly susceptible. Broad scale trials demonstrated that if large areas of the forest were aerially baited, within 2–3 years numbers of mammal species such as the Chuditch increased (Morris *et al.* 1998). On the basis of these results, the decision was made to implement fox baiting over 700 000 ha of the northern jarrah forest in order to halt the decline in numbers of those mammal species known to be susceptible to fox predation. Alcoa funds this program. Baiting in areas in which Alcoa's bauxite mining operations occur commenced in July 1994.

Following this baiting, Alcoa began a program of reporting all key fauna sightings by field staff. This showed that numbers of fox sightings recorded at Huntly mine decreased from 15 in 1994 to zero in 1999 and 2000. Percentage trap success results for key species such as Chuditch, Quenda and Possum in broad scale trapping conducted between 1980 and 1999 all showed evidence of an increase in the abundance of these species in the forest following fox baiting. Trapping success of both Chuditch and Possums in 2005 was much higher than in any of the previous years. For Quenda, trapping success was much higher in 1999 than in previous years, but by 2005 numbers of Quenda had declined. This result has been confirmed elsewhere and is thought to be due to a severe drought in early 2001.

On the basis of the results obtained, several steps have been taken to promote faunal conservation in areas close to mining operations and in mine restoration. These include:

- the continuation of fox baiting
- the construction of fauna 'habitats' and corridors. During pre-mining clearing, hollow logs, stumps and other material are usually burnt. Alcoa has now developed procedures for utilising them to construct habitats and corridors extending across narrow sections of pits. These provide mammals and reptiles with shelter for breeding purposes and protection

from predators. The procedures are now routinely implemented at all mines and Quenda, Mardo and Chuditch have been recorded using them (Mutzig 1998).

Overall, the results of 30 years of vertebrate fauna monitoring indicate that it is possible to establish restoration which caters for the habitat requirements of most species. However, for some species, further work is needed to understand how restoration practices might be improved to further promote their return. Research is currently being undertaken on the effects of long-term forest management practices on some vertebrate fauna species, particularly reptiles and mammals.

Threatened species management

Five vertebrate fauna that are found within Alcoa's mine lease are listed as 'Threatened' under *Western Australia's Wildlife Conservation Act 1950*. They are also listed as vulnerable under the federal *Environment Protection and Biodiversity Conservation (EPBC) Act 1999*:

- Western Quoll, *Dasyurus geoffroii*
- Quokka, *Setonix brachyurus*
- Noisy Scrub-bird, *Atrichornis clamosus*
- Baudin's Black Cockatoo, *Calyptorhynchus baudinii*
- Forest Red-tailed Black Cockatoo, *Calyptorhynchus banksii naso*.

Two species are listed as 'Specially Protected' under *Western Australia's Wildlife Conservation Act 1950*, but are not listed under the federal *Environment Protection and Biodiversity Conservation (EPBC) Act 1999*:

- Peregrine Falcon, *Falco peregrines*
- Carpet Python, *Morelia spilota imbricate*.

Alcoa has developed threatened fauna species management plans for all seven of these animals (see Stokes and Norman 2010). The combined recommendations and actions from these management plans include:

- Develop and implement specific monitoring programs for each animal species to obtain reliable measures of distribution and density within the mining lease and monitor mining impacts.
- Return substantial amounts of coarse woody debris (CWD), rocks and other den and nesting type materials to rehabilitated areas. Continue research into the best methods of returning these materials for various fauna groups.
- Remove fresh road kill from the mine access roads, so that carrion is not on the road attracting Chuditch. This will hopefully reduce the incidence of road kill of Chuditch on the mine-sites.
- Research the effect of prescribed burning and wildfires on den availability for animals in rehabilitation and unmined forest.
- Continue financial support for feral predator control in the northern jarrah forest through continued contribution to the Western Shield program.
- Survey and retain veteran marri trees in unmined areas to provide cockatoo nest sites.
- Investigate and trial translocations of noisy scrub-birds into rehabilitated mined areas.
- Fund feral pig control measures in mine areas and surrounding forest.
- Construct fauna underpasses where haul roads cross streams.
- Fund control of feral bees.
- Trial artificial nest boxes to replace veteran marri trees in mined areas.

Most of these recommendations and actions have been put into place. It is difficult to measure the effects of these measures on fauna populations but survey systems have been developed to gather such data.

Invertebrate fauna

Invertebrate fauna are not routinely monitored in pre- or post-mining forest. The exception is ants which are included with vertebrate fauna in long-term monitoring assessments in both unmined

Table 7.7. Invertebrate studies in jarrah forest and restored bauxite mines, showing the number of species/morphospecies found, the number present in mine pits and forest or both, and the proportion undescribed (from Majer et al. 2007)

Taxonomic group(s)	Reference	Number of species or morphospecies reported					
		In entire study	In unmined forest	In restored mines	Common to both unmined forest and restored mines	Total number undescribed or unidentifiable to species	% undescribed or unidentifiable to species
Soil and litter invertebrates	Postle et al. 1991	305	305	Not studied	–	283	93%
Collembola	Greenslade and Majer 1993	68	28	60	19	42	62%
Cryptostigmatid mites	Cuccovia and Kinnear 1999	49	37	36	25	49	100%
Termites	Nichols and Bunn 1980 Burns 1992 Langlands 2003	13 11 10	13 9 10	10 9 9	10 9 9	0 0 0	0% 0% 0%
Overstorey invertebrates	Recher et al. 1996	560	560	Not studied	–	523	93%
Herbivores	Casotti 1988	Not identified to species level					
Hemiptera	Moir 2005 Moir et al. 2005	382 115	240 68	299 75	176 28	304 87	80% 76%
Ants	Majer and Nichols 1998	63	52	50	39	10	16%
Predators	Nichols and Burrows 1985	Identified to species level, but not tabulated against plots in the publication					
Spiders	Brennan et al. 2004	287	–	–	–	204	71%
	Brennan 2002	168	54	56	58	–	–

forest and restored sites. Other invertebrate groups are studied in specific research projects, sometimes by PhD students and nearly always in conjunction with Universities. The role of terrestrial invertebrates in the restoration of Alcoa World Alumina Australia's bauxite mines has been the subject of more than 20 individual studies since 1974. Research projects range from arthropods in the soil and leaf-litter layer, to the understorey vegetation, and finally the tree canopy. Moreover, projects span a range of trophic groups, including decomposers (e.g. mites, springtails and termites), predators (e.g. ants, scorpions and spiders) and herbivores (e.g. true bugs and ants preying on seeds) (Majer et al. 2007) (Table 7.7). Elucidation of recolonisation trajectories uses both space-for-time substitutions and long-term regular sampling. Importantly, many studies are at species level rather than coarser taxonomic ranks such as Order. Data for other groups when measured by species richness (ants, spiders and hemipterans) and family composition (ants and spiders) show their re-assembly trajectories tracking towards unmined reference areas (Fig. 7.15). At a finer level (species or morphospecies), the

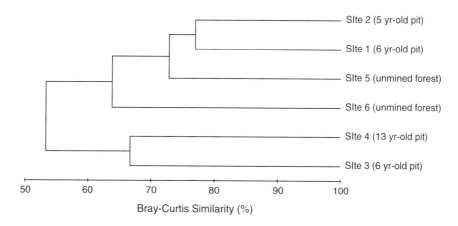

Figure 7.15: Cluster analysis of similarity in the composition of spider families between restored mine pits and surrounding unmined forest using data from Nichols and Burrows (1985).

specific plant/insect interdependencies can prevent re-establishment of an identical community in restored sites. Hemipteran species composition, for example, tracks the vegetation re-assembly trajectory but not towards unmined reference areas. In restored mine pits that have been burnt to reduce fuel loads (or through wildfire) the response of spiders to the double disturbance of mining plus burning was retrogression/alteration of the post-mining trajectory. Important groups not yet studied are the Coleoptera (beetles), Lepidoptera (butterflies and moths) and Orthoptera (grasshoppers and crickets).

Conclusions

Mining in the jarrah forest in the south-west of Western Australia requires high standards of environmental practice. Three potential high environmental impacts and the management of these impacts are described in detail: jarrah dieback disease, drinking water quality and loss of biodiversity.

Approximately 40% of the plant species of the jarrah forest are susceptible to jarrah dieback disease. Alcoa controls and manages this pathogen by comprehensive pre-mine dieback mapping, followed by effective quarantine, disease hygiene and vehicle washdown procedures. A resulting very low disease spread rate of 0.0006 ha per hectare of clearing was attributed to mining in 2003–2004.

Alcoa mines in Perth's drinking water catchments and manages this important resource by controlling sediment and turbidity, as well as potential hydrocarbon (fuel and oil) spills in catchments. Strict legislation, standards and procedures are in place to prevent pollution. Engineering and other earthworks are carried out to high standards. Alcoa's operations have never compromised the quality of Perth's drinking water supplies even after more than 40 years of mining and restoration.

Return of plant and animal diversity is considered essential if the conservation value of the jarrah forest is to be restored after mining. The flora of the forest is restored by maximising the contribution of plant species from the natural soil stored seed bank, by applying a wide range of local native seeds and by propagating and planting species not returning from the natural soil seed bank or applied seed. Most animal species return when the restored habitat becomes suitable for their requirements, some may require mature ecosystem components. The return of coarse woody debris is an intervention which can quickly return mature forest components and hence encourage a wider range of fauna species.

Adequate resourcing both financially and in staffing are required to responsibly mine in environmentally sensitive areas.

References

Alcoa of Australia Ltd. (1978) Wagerup Alumina Project. Environmental Review and Management Programme, Supplement. 145 pp. Alcoa of Australia Ltd and Dames and Moore, Perth, Australia, September 1978.

Bailey C (1996) Western Shield: bringing wildlife back from the brink of extinction. *Landscope* **11**, 41–48.

Bell DT (2001) Ecological response syndromes in the flora of southwestern Western Australia: Fire resprouters versus reseeders. *Botanical Review* **67**, 417–440. doi:10.1007/BF02857891

Bell DT, Bellairs SM (1992) Effects of temperature on the germination of selected Australian native species used in the rehabilitation of bauxite mining disturbance in Western Australia. *Seed Science and Technology* **20**, 47–55.

Bell DT, Heddle EM (1989) Floristic, morphologic and vegetational diversity. Pages 53–66 in The jarrah forest: A complex mediterranean ecosystem. Kluwer. Dordrecht, Netherlands.

Bell DT, Plummer JA, Taylor SK (1993) Seed germination ecology in Southwestern Western Australia. *Botanical Review* **59**, 24–73. doi:10.1007/BF02856612

Bellairs SM, Bell DT (1993) Seed stores for restoration of species-rich shrubland vegetation following mining in Western Australia. *Restoration Ecology* **1**, 231–240. doi:10.1111/j.1526-100X.1993.tb00032.x

Brennan KEC (2002) The successional response of spider communities following the multiple disturbances of mining and burning in Western Australian Jarrah forest. PhD Thesis, Department of Environmental Biology, Curtin University of Technology, Perth.

Brennan KEC, Moir ML, Majer JD (2004) Exhaustive sampling in a Southern Hemisphere global biodiversity hotspot: Inventorying species richness and assessing endemicity of the little known jarrah forest spiders. *Pacific Conservation Biology* **10**, 241–260.

Burns M (1992) Termite (Isoptera) abundance and diversity in rehabilitated bauxite mines. Biology Project report. Curtin University of Technology, Perth.

Casotti G (1988) The rate of herbivory on selected *Eucalyptus* in the Western Australian jarrah forest. Honours thesis, Murdoch University, Perth.

Collins M, Koch JM, Brundrett M, Sivasithamparam K (2005) Recovery of terrestrial orchids in the post-mining landscape. *Selbyana* **26**, 255–264.

Colquhoun IJ, Hardy GEStJ (2000) Managing the risks of Phytophthora root and collar rot during bauxite mining in the *Eucalyptus marginata* (Jarrah) forest of Western Australia. *Plant Disease* **84**, 116–127. doi:10.1094/PDIS.2000.84.2.116

Croton JT, Reed AJ (2007) Hydrology and bauxite mining on the Darling Range. *Restoration Ecology* **15**, S40–S47. doi:10.1111/j.1526-100X.2007.00291.x

Cuccovia A, Kinnear A (1999) Acarine (mite) communities colonizing rehabilitated bauxite mine pits in the jarrah forest of Western Australia. Pages 54–59 in W Ponder and D Lunney, editors. The other 99%: The conservation of biodiversity of invertebrates, Royal Zoological Society of New South Wales, Mosman.

Dixon KW, Roche S, Pate JS (1995) The promotive effect of smoke derived from burnt native vegetation on seed germination of Western Australian plants. *Oecologia* **101**, 185–192. doi:10.1007/BF00317282

Gardner J (2001) Rehabilitating mines to meet land use objectives: Bauxite mining in the jarrah forest of Western Australia. *Unasylva* **207**, 3–8.

Gardner JH, Bell DT (2007) Bauxite Mining Restoration by Alcoa World Alumina Australia in Western Australia: Social, Political, Historical and Environmental Context. Restoration Ecology. *Restoration Ecology* **15**, S3–S10. doi:10.1111/j.1526-100X.2007.00287.x

Glossop BL (1980) Germination responses of thirteen legume species to boiling. Alcoa Environmental Research Bulletin No. 5.

Grant CD, Koch JM (2003) Orchid species succession in rehabilitated bauxite mines in Western

Australia. *Australian Journal of Botany* **51**, 453–457. doi:10.1071/BT02127

Grant CD, Bell DT, Koch JM, Loneragan WA (1996) Implications of seedling emergence to site restoration following bauxite mining in Western Australia. *Restoration Ecology* **4**, 146–154. doi:10.1111/j.1526-100X.1996.tb00115.x

Greenslade P, Majer JD (1993) Recolonization by Collembola of rehabilitated bauxite mines in Western Australia. *Australian Journal of Ecology* **18**, 385–394. doi:10.1111/j.1442-9993.1993.tb00466.x

Havel JJ (1975) Site-vegetation mapping in the northern jarrah forest. I Definition of site-vegetation types. Forests Department WA. Bulletin Number 86.

Iverson LR, Wali MK (1982) Buried, viable seeds and their relation to revegetation after surface mining. *Journal of Range Management* **35**, 648–652. doi:10.2307/3898656

Kew GA, Mengler FC, Gilkes RJ (2007) Mine floor strength, water retention and its influence on ripping used to enhance forest restoration after bauxite mining in the Darling range of Western Australia. *Restoration Ecology* **15**, S54–S64. doi:10.1111/j.1526-100X.2007.00293.x

Kinal J, Shearer BL, Fairman RG (1993) Dispersal of *Phytophthora cinnamomi* through lateritic soil by laterally flowing subsurface water. *Plant Disease* **77**, 1085–1090. doi:10.1094/PD-77-1085

Koch JM (2007) Alcoa's Mining and Restoration Process in South Western Australia. *Restoration Ecology* **15**, S11–S16. doi:10.1111/j.1526-100X.2007.00288.x

Koch JM, Hobbs RJ (2007) Synthesis: Is Alcoa successfully restoring a jarrah forest ecosystem after bauxite mining in Western Australia? *Restoration Ecology* **15**, S137–S144. doi:10.1111/j.1526-100X.2007.00301.x

Koch JM, Taylor SK (2000) Seed germination records from Alcoa's Marrinup Nursery. Alcoa Environmental Research Note No. 18.

Koch JM, Ward SC (1994) Establishment of understorey vegetation for rehabilitation of bauxite-mined areas in the jarrah forest of Western Australia. *Journal of Environmental Management* **41**, 1–15. doi:10.1006/jema.1994.1029

Koch JM, Taylor SK, Gardner JH (1994) Research to maximise plant diversity in rehabilitated bauxite mines in the jarrah forest. pp. 41–50 In '*Proceedings of a National Workshop on Native Seed Biology for Revegetation*'. (Eds SB Bellairs and LC Bell) Perth, Western Australia 24–26 August 1994. Australian Centre for Minesite Rehabilitation Research: Brisbane, Australia and The Chamber of Mines and Energy of Western Australia Inc.

Koch JM, Ward SC, Grant CD, Ainsworth GL (1996) Effects of bauxite mine restoration operations on topsoil seed reserves in the jarrah forest of Western Australia. *Restoration Ecology* **4**, 368–376. doi:10.1111/j.1526-100X.1996.tb00189.x

Koch JM, Richardson J, Lamont BB (2004) Grazing by kangaroos limits the establishment of the grass trees Xanthorrhoea gracilis and X. preissii in restored bauxite mines in eucalypt forest of southwestern Australia. *Restoration Ecology* **12**, 297–305. doi:10.1111/j.1061-2971.2004.00335.x

Krauss SL, Koch JM (2004) Rapid genetic delineation of provenance for plant community restoration. *Journal of Applied Ecology* **41**, 1162–1173. doi:10.1111/j.0021-8901.2004.00961.x

Krauss SL, Koch JM, Vlahos S (2005) A novel approach for the rapid genetic delineation of provenance for minesite revegetation. *Ecological Management & Restoration* **6**, 153–155. doi:10.1111/j.1442-8903.2005.230-9.x

Langlands P (2003) Termite recolonisation of rehabilitated bauxite mines in the jarrah forest of Western Australia. Report to Alcoa World Alumina Australia Ltd.

Majer JD, Nichols OG (1998) Long-term recolonization patterns of ants in Western Australian rehabilitated bauxite mines with reference to their use as indicators of restoration success. *Journal of Applied Ecology* **35**, 161–182. doi:10.1046/j.1365-2664.1998.00286.x

Majer JD, Brennan KEC, Moir ML (2007) Invertebrates and the restoration of a forest ecosystem: thirty years of research following bauxite mining in Western Australia. *Restoration Ecology* **15**, S104–S115. doi:10.1111/j.1526-100X.2007.00298.x

McKay JK, Christian CE, Harrison S, Rice KJ (2005) "How local is local?" – a review of practical and conceptual issues in the genetics of restoration. *Restoration Ecology* **13**, 432–440. doi:10.1111/j.1526-100X.2005.00058.x

Meney KA, Dixon KW, Pate JS, Dixon IR (1990) Rehabilitation of mining affected flora. Minerals and Energy Research Institute of Western Australia, Report No. 66.

Moir ML (2005) Hemipteran assemblages of understorey habitats in rehabilitated bauxite mines and jarrah forest. PhD thesis, Curtin University of Technology, Perth, Western Australia.

Moir ML, Brennan KEC, Majer JD, Fletcher MJ, Koch JM (2005) Toward an optimal sampling protocol for Hemiptera on understorey plants. *Journal of Insect Conservation* **9**, 3–20. doi:10.1007/s10841-004-2351-y

Morris KD, Armstrong R, Orell P, Vance M (1998) Bouncing back. *Landscope* **14**, 28–35.

Mullins RG, Koch JM (2001) Field establishment of 40 jarrah forest species in response to gibberellic acid and smoke treatment. Alcoa Environmental Research Bulletin No. 30.

Mullins RG, Koch JM, Ward SC (2002) Practical method of germination for a key jarrah forest species: Snottygobble (Persoonia longifolia). *Ecological Management & Restoration* **3**, 97–103. doi:10.1046/j.1442-8903.2002.00101.x

Mutzig S (1998) Recolonization of rehabilitated bauxite mines by small mammals. Honours thesis, University of Western Australia, Perth, Western Australia.

Nichols OG, Bunn SE (1980) Termite utilization of rehabilitated bauxite mined areas. Alcoa of Australia Ltd Environmental Research Bulletin No. 9.

Nichols OG, Burrows R (1985) Recolonisation of revegetated bauxite mine sites by predatory invertebrates. *Forest Ecology and Management* **10**, 49–64. doi:10.1016/0378-1127(85)90013-1

Nichols OG, Grant CD (2007) Vertebrate fauna recolonisation of restored bauxite mines – key findings from almost 30 years of monitoring and research. *Restoration Ecology* **15**, S116–S126. doi:10.1111/j.1526-100X.2007.00299.x

Parsons MH, Koch JM, Lamont BB, Vlahos S, Fairbanks M (2006) Planting density effects and selective herbivory by kangaroos on species used in restoring forest communities. *Forest Ecology and Management* **229**, 39–49. doi:10.1016/j.foreco.2006.03.020

Pate JS, Dixon KW (1982) Tuberous, Cormous and Bulbous Plants. University of Western Australia Press.

Perger BA, Weaver P, Dixon KW (1994) Genetic diversity and restoration of a recalcitrant clonal sedge (*Tetraria capillaris,* Cyperaceae). *Conservation & Biodiversity* **3**, 279–294. doi:10.1007/BF00055943

Postle AC, Majer JD, Bell DT (1991) A survey of selected soil and litter invertebrate species from the northern Jarrah (*Eucalyptus marginata*) forest of Western Australia, with particular reference to soil-type, stratum, seasonality and the conservation of forest fauna. Pages 193–203 in D Lunney, editor. Conservation of Australia's forest fauna. Royal Zoological Society of NSW, Mosman.

Pryor LD (1959) Species distribution and association in Eucalyptus. *Monographiae Biologicae* **8**, 461–471.

Putwain PD, Gillham DA (1990) The significance of the dormant viable seed bank in the restoration of heathlands. *Biological Conservation* **52**, 1–16. doi:10.1016/0006-3207(90)90010-M

Read TR, Bellairs SM, Mulligan DR, Lamb D (2000) Smoke and heat effects on soil seed bank germination for re-establishment of a native forest community in New South Wales. *Austral Ecology* **25**, 48–57. doi:10.1046/j.1442-9993.2000.01031.x

Recher HF, Majer JD, Ganesh S (1996) Eucalypts, insects and birds: on the relationship between foliar nutrients and species richness. *Forest Ecology and Management* **85**, 177–195. doi:10.1016/S0378-1127(96)03758-9

Roche S, Koch JM, Dixon KW (1997) Smoke enhanced seed germination for mine rehabilitation in the southwest of Western Australia. *Restoration Ecology* **5**, 191–203. doi:10.1046/j.1526-100X.1997.09724.x

Roche S, Dixon KW, Pate JS (1998) For everything a season: smoke-induced seed germination and seedling recruitment in a Western Australian Banksia woodland. *Australian Journal of Ecology* **23**, 111–120. doi:10.1111/j.1442-9993.1998.tb00709.x

Schofield NJ, Stoneman GL, Loh LC (1989) Hydrology of the jarrah forest. Pages 179–201 in B Dell, JJ Havel, and N Malajczuk, editors. The Jarrah Forest. Kluwer Academic Publications, Dordrecht, The Netherlands.

Shea SR, McCormick J, Portlock CC (1979) The effect of fires on the regeneration of leguminous species in the northern jarrah forest of Western Australia. *Australian Journal of Ecology* **4**, 195–205.

Shearer BL, Tippett JT (1989) Jarrah Dieback: The dynamics and management of *Phytophthora cinnamomi* in the jarrah (*Eucalyptus marginata*) forests of south-western Australia. Research Bulletin No 3. Department of Conservation and Land Management, Perth, Australia.

Smith MA, Loneragan WA, Grant CD, Koch JM (2000) Effect of fire on the topsoil seed banks of rehabilitated bauxite mine sites in the jarrah forest of Western Australia. *Ecological Management & Restoration* **1**, 50–60. doi:10.1046/j.1442-8903.2000.00008.x

Specht RL, Roa EM, Boughton VH (1974) Conservation of major plant communities in Australia and Papua New Guinea. Australian Journal of Botany Supplementary Series No. 7:667 pp.

Standards Association of Australia (1999) Risk Management AS/NZS 4360:1999, Strathfield, NSW, Australia.

Stokes VL, Norman MA (2010) Threatened fauna species management plans for Alcoa's bauxite mining operations in the jarrah forest. Alcoa Environmental Research Bulletin No. 37.

Tacey WH (1979) Landscaping and revegetation practices used in rehabilitation after bauxite mining in Western Australia. *Reclamation Review* **2**, 123–132.

Tacey WH, Glossop BL (1980) Assessment of topsoil handling techniques for rehabilitation of sites mined for bauxite within the jarrah forest of Western Australia. *Journal of Applied Ecology* **17**, 195–201. doi:10.2307/2402974

Vigilante T (1996) A study of patch boundaries between forest and rehabilitated bauxite pits, at Jarrahdale, Western Australia. Honours Thesis. Department of Botany. The University of Western Australia.

Vlahos S, Bell DT (1986) Soil seed-bank components of the northern jarrah forest of Western Australia. *Australian Journal of Ecology* **11**, 171–179. doi:10.1111/j.1442-9993.1986.tb01388.x

Ward SC, Koch JM, Ainsworth GL (1996) The effect of timing of rehabilitation procedures on the establishment of a jarrah forest after bauxite mining. *Restoration Ecology* **4**, 19–24. doi:10.1111/j.1526-100X.1996.tb00103.x

Ward SC, Koch JM, Grant CD (1997) Ecological aspects of soil seed-banks in relation to bauxite mining. I. Unmined jarrah forest. *Australian Journal of Ecology* **22**, 169–176. doi:10.1111/j.1442-9993.1997.tb00656.x

Willyams D (2005) Tissue culture of geophytic rush and sedge species for revegetation of bauxite mine sites in the northern jarrah forest of Western Australia. pp. 226–241 in 'Contributing to a Sustainable Future' (Eds IJ Bennett, E Bunn, H Clarke, JA McComb). Proceedings of the Australian Branch of the International Association for Plant Tissue Culture & Biotechnology, Perth, Western Australia.

8

Recovery of terrestrial orchids in natural ecosystems after severe disturbance

Margaret Collins and Mark Brundrett

Summary

The ability of terrestrial orchids to recolonise ecosystems after total loss of vegetation is highly variable. Some orchids re-establish quickly, others are slow to return while some remain absent for at least several decades after disturbance. Orchid seed is minute, has been found to be readily dispersed and is not regarded as a limiting factor provided adequate rates of pollination occurs. A case study in the jarrah forest of Western Australia showed that compatible symbiotic fungi were present in a restored bauxite mine site, but inoculum levels were low and patchily distributed in soils. Comparative studies of orchids in disturbed ecosystems revealed key functional differences between early colonisers (disturbance opportunists) and disturbance intolerant species, especially concerning the degree of pollinator and fungus specificity. However more research is required to resolve the impact of microhabitats defined by soil and vegetation on orchid colonisation and how these factors change over time. In particular, there is a paucity of data on the ecological roles and habitat preferences of the fungi, and insects on which terrestrial orchids depend. Orchids that are early re-colonisers are indicators of disturbed habitats, while those that are disturbance intolerant and recolonise slowly are useful as indicators of habitats in excellent condition.

Introduction

Orchids belong to one of the largest and most diverse plant families estimated to contain in excess of 25 000 species (Chase *et al.* 2003). They represent almost 10% of the world's flowering plants and have adapted to a wide range of habitats and a wide geographical range, from the subarctic to subantarctic. The greatest diversity of genera and species occurs in tropical regions where stable warm humid conditions favour epiphytic life forms – ~73% of all orchid species are epiphytic (Atwood 1986). Terrestrial orchids predominate in the more climatically variable temperate regions and usually have perennial tubers or rhizomes that enable them to survive climatic extremes such as freezing winters or arid summers.

Most orchids are insect pollinated and attract pollinators by nectar, visual deception, fungus mimicry or sexual deception using insect pheromones (Adams and Lawson 1993; Brundrett 2007; Gaskett 2010; Phillips *et al.* 2009). These mechanisms involve varying degrees of pollinator specificity with the latter category of sexual deception often entailing pollination of an orchid by a single species of insect. Approximately 30% of orchids are deceptive and seed production in these species is dependent on the survival of particular insect species in numbers high enough to adequately pollinate orchids, which may be separated spatially due to fragmentation of natural vegetation.

Terrestrial orchid seed is minute and consists of a thin testa and a simple embryo, containing little if any endosperm (Arditti and Ernst 1984; Wirth and Withner 1959). Dependence on mycorrhizal fungi for germination and subsequent nutrition of seedlings has resulted in the absence of significant nutritional reserves in seeds. Symbiotic relationships with mycorrhizal fungi also continue in adult orchid plants (Rasmussen 1995). Successful germination and seedling establishment are critical stages in the life history of orchids. Orchids produce large amounts of dust-like seeds that are wind dispersed, but the recruitment of orchids is dependent on the distribution and availability of suitable microhabitats containing the mycorrhizal fungus, as explained below (Arditti and Ghani 2000).

Severely disturbed habitats can be used as model systems to examine succession from an initial state without orchids, in contrast to semi-disturbed systems (e.g. fire or weed invasion) where it is more difficult to separate survivors from colonists and species which benefit from disturbance from those which are adversely affected. Nevertheless, knowledge gained about disturbance impacts on orchids in new habitats such as mined lands is relevant to other less catastrophic types of disturbance, where ecological processes are severely disrupted.

In this review we will summarise knowledge of orchid recovery after disturbance and attempt to use these data to classify orchids into categories based on disturbance responses. This approach should also allow us to begin to investigate biological characteristics of orchids, their fungi and their pollinators that are linked to these ecological categories. This approach may help resolve some of the contradictions in the biology and ecology of orchids, which have an extremely large ecological range, from critically endangered species that only occupy a few square metres of habitat at one extreme, to those with weedy tendencies that rapidly colonise disturbed habitats and invade new continents on the other extreme. We have attempted to address key questions in this review:

1. How effectively and when do orchids recolonise after total loss of vegetation?
2. Is the absence of their symbiotic fungi a major limiting resource in disturbed soils?
3. Are some orchids more effective re-colonisers than others?
4. What can we learn from the study of orchids in disturbed ecosystems?
5. What are the ecological roles of orchids and their associated symbionts?

This review is primarily based on data from one of the best studied regions in Western Australia, rehabilitated bauxite mines in the jarrah forest, but it is anticipated that the discussion will be relevant to other disturbed habitats where orchids occur.

Orchids in disturbed habitats

Almost all studies reporting the reestablishment of orchids in severely disturbed sites do so as a part of observations of vegetation succession in newly created natural landscapes (Table 8.1). Examples of such studies include lava flows in Indonesia and Hawaii (Clarkson 1997; Gandawijaja and Arditti 1983) and landslides in Hawaii (Restrepo and Vitousek 2001). These studies have shown that orchids do return to these severely disturbed areas with time, but it often takes many years. In some cases these sites have become valuable conservation areas (Halliday and Johnson 1979). There have been major studies of orchid recolonisation after the cataclysmic volcanic eruption at Krakatau in 1883 (Partomihardjo 2003). As shown in Fig. 8.1, orchid diversity steadily increased over the first century in newly created habitats at Krakatau.

Table 8.1. Studies of orchid diversity in severely disturbed habitats

Disturbance	Site age	Orchid diversity	Reference
Landslide	4–130 years	3 spp.	Restrepo and Vitousek 2001
Lava flow	47–137 years	1–2 spp.	Clarkson 1997
Volcanic eruption	115 years	40 spp.	Partomihardjo 2003
Volcanic eruption	3–50 years	25 spp.	Gandawijaja and Arditti 1983
Volcanic eruption	3–100 years	21 spp.	Whittaker et al. 1989
Chalk and limestone quarry	40–70 years	0–3 spp.	Davis 1979
Limestone quarry	49 years	5 spp.	Lee and Greenwood 1976
Waste rock dump	40–50 years	*Thelymitra* sp.	Rufaut et al. 2006
Industrial waste dump	20–80 years	1–7 spp.	Ash et al. 1994
Mine tailings	10–40 years	3 spp. studied	Shefferson et al. 2008
Mining waste dump	20–58 years	2–4 spp.	Halliday and Johnson 1979
Open-cut bauxite mine	>10 years	16 spp.	Grant and Koch 2003
Open-cut bauxite mine	1–27 years	11 spp.	Collins et al. 2005
Former lignite mine	>25 years	*Epipactis palustris*	Esfeld et al. 2008
Cleared roadside	recently re-disturbed	*Ophrys apifera*	Gardiner and Vaughan 2009
Poplar plantations	13 years	2 spp.	Adamowski 2006

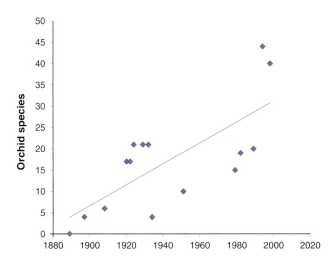

Figure 8.1: Diversity of orchids recorded in surveys of recovering ecosystems after the 1883 Krakatau eruption. Data compiled from different surveys by Partomihardjo (2003).

Orchid seed is minute and readily wind dispersed (Gandawijaja and Arditti 1983; Rasmussen 1995). However, most orchid seed is believed to fall within 10 m of the parent plant (Chung et al. 2005; Jacquemyn et al. 2007; Juersáková and Malinova 2007; Murren and Ellison 1998) and seed germination studies have found that germination and recruitment are most likely to occur close to adult plants (Batty et al. 2001; Diez 2007; Jacquemyn et al. 2007; Perkins et al. 1995). Orchid seed dispersal distances of tens or hundreds of kilometres have been reported (Table 8.1). For example, the Java mainland was found to be the predominant source of orchid seed for colonisation of Krakatau, 40 km away (Partomihardjo 2003) and many orchid species in New Zealand are believed to be derived from wind borne seed of closely related species from Australia ~1500 km away (Close et al. 1978).

Orchid recovery in manmade habitats

There are a few examples of vegetation surveys of abandoned mine sites, waste dumps and mine site rehabilitation areas in Germany, UK and Western Australia (Halliday and Johnson 1979; Grant and Koch 2003; Collins *et al.* 2005; Esfeld *et al.* 2008). Esfeld *et al.* (2008) found evidence of multiple independent colonisation events from molecular studies comparing genetic diversity in populations of the threatened orchid *Epipactis palustris* in former lignite mining areas and natural areas up to 125 km away. This study provides evidence of genetic exchange between these populations. The ease of dispersal of orchid seeds suggests that the seed will disperse to mine site rehabilitation areas provided that there are flowering and fruiting orchid populations in the surrounding undisturbed areas. Another example is *Dactylorhiza fuchsii* a long-distance colonist on industrial waste heaps (Ash *et al.* 1994).

Adamowski (2006) summarises data from over 40 reports of orchids that colonise disturbed habitats in Europe. This study revealed that ~40% of European orchid species are capable of colonising disturbed habitats such as roadsides, rail embankments, plantations and mined areas. The most commonly observed orchids in disturbed habitats were species of *Epipactis* and *Dactylorhiza*. The European species *Epipactis helleborine* has also become an exotic weed in North America (e.g. it was common in our garden in Canada, M. Brundrett Pers. Comm,). It was also common for orchids to form exceptionally large populations in disturbed areas and produce large amounts of seed (Adamowski 2006).

Case study: orchids in bauxite mined areas of the Western Australian jarrah forest

Comprehensive studies of orchids in rehabilitated bauxite mines in the Jarrah forest have provided considerable insights into the recovery of orchids in recreated forests (Collins *et al.* 2005, 2007; Grant and Koch 2003). Bauxite mining in the northern jarrah forest is an open-cut process that causes a severe localised impact on vegetation and a lesser secondary impact through fragmentation of surrounding unmined areas. The selective removal of particular vegetation types associated with bauxite deposits from the jarrah forest may also affect populations of plants and animals through both loss of habitat and fragmentation.

The main sources of plant propagules for revegetation of bauxite rehabilitation areas are respread topsoil, an applied seed mixture and natural seed dispersal (Grant and Koch 2007). Previous work on orchid soil seed banks in Western Australia has revealed a short-lived seed bank with viable seed not persisting through the wet season following dispersal (Batty *et al.* 2000). The soil seed bank in respread topsoil is therefore not expected to be significant in the recruitment of new individuals. Orchids belong to a group of plant species absent from seed mixes or topsoil that invades rehabilitation areas with time (Koch *et al.* 1994).

The Orchidaceae are one of seven poorly represented plant families in rehabilitation areas as only eight of 18 orchid species recorded in pre-mining surveys had re-established in young rehabilitation sites (Nichols *et al.* 1991). The youngest rehabilitation areas found to contain orchids were 27 months old, with the number of orchid species present in rehabilitation areas increasing with time since establishment (Nichols *et al.* 1991). Koch and Ward (1994) found that 48 plant species did not re-occur after mining. Of these, 14 species were tuberous geophytes including 10 orchid species. Grant and Loneragan (1999) found that fire in 11–13 year old rehabilitation areas resulted in increased orchid numbers. A positive effect of smoke treatment of soil on recruitment has been observed for the introduced orchid species *Disa bracteata* (Grant and Koch 1997).

Grant and Koch (2003) examined data from Alcoa World Alumina's long-term vegetation monitoring of unmined jarrah forest and bauxite mine rehabilitation areas in Western Australia, of between 1 and 31 years old, and found a total of 23 orchid species. Three species, *Cryptostylis ovata*, *Lyperanthus serratus* and *Prasophyllum elatum* occurred only in unmined forest while a *Diuris* sp. occurred only in rehabilitated areas. This study found that the overall density of orchids in unmined forest (13 800 plants ha^{-1}) was 10 times that of rehabilitation areas (1380 plants ha^{-1}) and the most abundant species were different for unmined forest and rehabilitation areas, with the exception of *Caladenia flava*, which is common in both areas.

Collins *et al.* (2005) conducted a comprehensive study of orchids in rehabilitated bauxite mines near Jarrahdale in Western Australia. They found that some terrestrial orchids colonised bauxite mined areas within five years of rehabilitation establishment, but orchid diversity was very low in most mined areas (see Fig. 8.2 and Tables 8.1, 8.2). Both the diversity and total number of orchids increased with rehabilitation site age (Fig. 8.2, Table 8.2), due to colonisation by a succession of orchids. However, the size, diversity and species richness of orchid populations were highly variable in both rehabilitation areas and unmined forest. Over the first 15 years, population densities of orchids were lower in rehabilitation areas than in unmined forest and were dominated by pioneer species (Table 8.2). Several species found in unmined forest were absent from the rehabilitation areas, as explained below.

The successful establishment of some orchid species in rehabilitation areas within five years indicates that seed dispersal is effective, at least for these taxa (Table 8.2). Orchids which are relatively common in the unmined Jarrah forest include the clonal species *Caladenia flava*, *Pterostylis* aff. *nana* and *Pyrorchis nigricans*, as well as the non-clonal orchid *T. crinita* (Table 8.3). Two of these orchids, *P. nigricans* and *T. crinita*, either failed to colonise or were extremely rare in rehabilitation sites compared to unmined forest (Table 8.3). There was a dramatic difference in recruitment of two of the most common orchids that occur in unmined forest and many other habitats in Western Australia. These are the blue sun orchid *T. crinita*, which was extremely rare in rehabilitated areas; in contrast with the common cowslip orchid *C. flava*, which

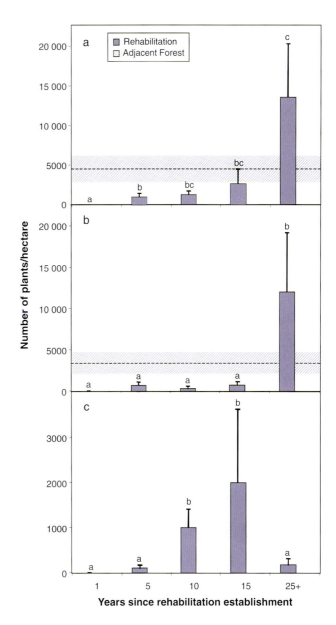

Figure 8.2: Mean density of orchids in a temporal sequence of transects in rehabilitated areas relative to adjacent unmined forest. (a) All orchids; (b) Clonal orchids; (c) Disturbance opportunists. Vertical bars represent SE. Columns identified by the same letter are not significantly different.

was relatively common in these areas (Tables 8.1, 8.2). Both orchids readily flower and produce seed, so the rarity of *T. crinita* in rehabilitated areas was surprising.

Pyrorchis nigricans is a common and widespread orchid that flowers in the spring following a hot

Table 8.2. A cumulative list of orchid species observed in forest and rehabilitation areas in the Jarrah forest of Western Australia during a comprehensive study by Collins (2008)

Orchid species	Forest	1 yr	5 yrs	10 yrs	15 yrs	>25 yrs
Pioneer species						
Caladenia flava	+	−	+	−	+	+
Disa bracteata*	+	−	+	+	+	−
Microtis media	−	−	+	+	+	+
Pterostylis sp. (aff. nana)	+	−	+	+	+	
Pterostylis recurva	+	−	+	+	−	+
Pterostylis sanguinea	+	−	+	−	+	−
Secondary colonisers						
Caladenia longiclavata	+	−	−	+	−	+
Diuris brumalis	+	−	−	+	+	+
Pterostylis barbata	+	−	−	+	+	−
Thelymitra macrophylla	+	−	−	+	−	+
Thelymitra benthamiana	−	−	−	+	−	+
Caladenia macrostylis	−	−	−	−	+	−
Cyanicula sericea	+	−	−	−	+	−
Pterostylis vittata	+	−	−	−	+	+
Thelymitra crinita	+	−	−	−	+	−
Climax species						
Eriochilus dilatatus	+	−	−	−	−	+
Caladenia reptans	+	−	−	−	−	−
Cryptostylis ovata	+	−	−	−	−	−
Cyrtostylis robusta	+	−	−	−	−	−
Prasophyllum parvifolia	+	−	−	−	−	−
Pyrochis nigricans	+	−	−	−	−	−
Number of species	**18**	**0**	**6**	**9**	**11**	**9**

* Alien species.

summer bushfire (Hoffman and Brown 1998). Wild fires were quickly extinguished and forest management burns, normally carried out by Department of Environment and Conservation (DEC), did not occur during the period that Alcoa World Alumina occupied the Jarrahdale mine site because of the need to protect personnel and infrastructure (J. Koch pers. comm.). Forest management burns to

Table 8.3. Mean density of orchids (plants ha⁻¹) in rehabilitated (rehab) and unmined jarrah forest areas in Western Australia (Collins 2008). Species are separated on growth habit

Scientific name	Common name	Forest	All rehab	>10 yr old rehab
Clonal species				
Caladenia flava	Cowslip orchid	613.3	1772.6	4771.4
Pyrochis nigricans	Red beaks	360.0	0.0	0.0
Pterostylis sp. (aff. nana)	Slender snail orchid	123.8	128.4	45.7
Pterostylis vittata	Banded greenhood	20.9	8.4	22.9
Diuris brumalis	Winter donkey orchid	19.0	181.1	445.7
Pterostylis sanguinea	Dark banded greenhood	0.0	12.6	19.1
Other species				
Thelymitra crinita	Blue lady orchid	817.1	2.1	0.0
Microtis media	Mignonette orchid	0.0	6.1	120.0
Eriochilus dilatatus	Common bunny orchid	53.3	0.0	0.0
Thelymitra macrophylla	Scented sun orchid	15.2	181.1	485.7
Pterostylis recurva	Jug orchid	66.7	61.1	28.6
Disa bracteata*	South African orchid	0.0	595.8	1057.1
Cyanicula sericea	Silky blue orchid	51.4	12.6	34.3
Pterostylis barbata	Dwarf bird orchid	13.3	4.2	5.7
Caladenia macrostylis	Leaping spider orchid	0.0	4.2	11.4
Caladenia longiclavata	Clubbed spider orchid	0.0	2.1	5.71
Prasophyllum parvifolia	Autumn leek orchid	1.8	0.0	0.0
Thelymitra benthamiana	Leopard orchid	0.0	65.3	171.4

* Alien weed.

reduce fuel load occur in spring, winter or autumn, and the timing of these burns does not stimulate flowering in *P. nigricans* (A. P. Brown pers. comm.). Consequently, a lack of local flower and seed production may be responsible for lack of recruitment of this species. Tardy colonisation of new sites is arguably of lesser long-term significance for this species than for non-clonal orchids such as *T. crinita*, since vegetative propagation results in large clones of *P. nigricans* (but it has to get there first).

In contrast to the recalcitrant orchid species, several orchids were much more abundant in disturbed areas than in adjacent forest plots (Fig. 8.3). The most common disturbance opportunist orchid in rehabilitation areas was *Disa bracteata*, an invasive weedy species originating in South Africa (Table 8.3). This orchid is short-lived, self-pollinating, produces large quantities of seed and has 'broad specificity' of its mycorrhizal associations, and these ecological characteristics are believed to have aided its rapid spread in Australia after its introduction near Albany ~60 years ago (Bonnardeaux *et al.* 2007; Erickson 1965). The appearance of this invasive South African orchid in natural habitats has

Figure 8.3: Examples of orchids from the West Austrian jarrah forest which fail to recover or are highly responsive to disturbance. *Pyrorchis nigricans* (a) and *Thelymitra crinita* (b) are common in forest, but were absent from rehabilitated areas; *Caladenia flava* (c) and *Pterostylis* sp. (d) are clonal orchids that relatively quickly colonised disturbed habitats; *Disa bracteata* (e, f) and *Microtis media* (g, h, i) are disturbance opportunists with self-pollination (arrow in f) and abundant seed set (i).

caused much consternation among local wildflower enthusiasts and orchidologists in the eastern Australian states because its weedy characteristics and sometimes dense populations are perceived to be a threat to indigenous species (e.g. Bates 1997). However, the r-strategist traits of D. bracteata combined with its absence from undisturbed forest and 25+ year old rehabilitation areas seems to indicate that it is unable to compete with established natural vegetation and is best considered to be an indicator of ecosystem disturbance rather than an serious weed.

Orchid mycorrhizal fungi in disturbed habitats

Mycorrhizas are symbiotic associations between specialised soil fungi and plants that are primarily responsible for nutrient transfer from plant to fungus (food) and fungus to plant (fertiliser) (see the website http://mycorrhizas.info/). On a global scale, the majority of plants in natural ecosystems have mycorrhizal associations (Brundrett 2009). Orchids differ from other plants by having a distinct type of fungal association that varies structurally and involves unique fungi. Many of the defining characteristics of orchids are linked to their fungal associations (Brundrett 2004). The need to encounter particular 'orchid fungi' is considered to have driven the evolution of tiny dust-like seeds for more efficient dispersal in patchy environments and this in turn would have lead to the need for efficient pollination mechanisms (Benzing and Atwood 1984).

Orchid mycorrhizas consist of coils of hyphae in cortical cells of the root, stem or protocorms (germinating seedlings) of orchids. These symbiotic associations are required to germinate the tiny seeds of orchids and are probably also essential for the nutrition of adult terrestrial orchids (Batty et al. 2002; Rasmussen 1995). These associations differ from the mycorrhizas of other green plants, as the fungus can provide both the energy and mineral nutrients required by orchids, and the fungi seem to receive very little or no benefit from the plant (Brundrett 2004). Most Australian terrestrial orchids are extremely dependent on mycorrhizal fungi to obtain nutrients from soil because they often have very coarse roots which result in a very limited capacity to gain soil minerals required for growth directly (Brundrett 2007).

Many Australian orchids require a particular fungus to germinate while others can form mycorrhizas with a wider diversity of them (Bonnardeaux et al. 2007). Unlike the mycorrhizal fungi of Eucalyptus trees, where mushrooms erupt conspicuously in winter, orchid fungi are usually invisible. In fact the easiest way to detect the presence of these fungi is by using orchid seed as bait, since it will only germinate in soil if a compatible fungus is present (Brundrett et al. 2003). Orchid seed germination is normally subterranean, but can be quantified in baiting trials as shown in Table 8.4. Knowledge of orchid fungus diversity and ecology is essential for us to understand orchid biology and ecology.

'Orchid seed baiting' methods that detect compatible fungi revealed a lower frequency of their occurrence in laterite soils than found in studies on sandy soils of the Swan Coastal plain or wheatbelt (Batty et al. 2001; Brundrett et al. 2003; Brundrett and Ager 2011). Detection of fungi that germinate a particular orchid taxon was not always correlated with occurrence of that orchid at a particular site, and it was interesting to note that a fungus compatible with P. nigricans was detected in a rehabilitation areas although the orchid was absent from all rehabilitation areas. Molecular identification of fungi in P. nigricans roots also revealed a relatively high diversity of fungi (Bonnardeaux et al. 2007). Therefore, the absence of the appropriate fungi is unlikely to be the cause of the failure of P. nigricans to colonise the post-mining landscape. In the jarrah forest, fewer types of orchid mycorrhizal fungi were found in rehabilitation areas and they were more sparsely distributed than in unmined forest in a similar manner to orchid plants (Table 8.4). Shefferson et al. (2008) found that similar mycorrhizal fungi were present in the roots of three orchid species in mine waste or ash hills as occurred to those in natural habitats in Estonia. Gardner and Malajczuk (1988) found that the

Table 8.4. Detection of orchid mycorrhizal fungi for six orchid species in three ages of rehabilitation (rehab) sites compared with adjacent unmined jarrah forest. Data are from Collins et al. (2007)

Orchid species	Rehab 1 year	Rehab 10 years	Rehab 26 years	Forest 1	Forest 2	Forest 3
Caladenia flava	0	2	4	4	2	1
Disa bracteata	0	0	0	1	0	0
Microtis media	0	0	0	1	1	0
Pterostylis recurva	3	0	0	0	2	0
Pyrorchis nigricans	0	1	0	1	2	0
Thelymitra crinita	0	0	1	1	3	2
% Baits detecting OMF*	15	15	25	30	42	16

* Baits detecting one or more fungus that germinates orchid seed ($n = 20$).

frequency of root colonisation by arbuscular and ectomycorrhizal fungi was initially lower in rehabilitated bauxite mined jarrah forest, but returned to unmined forest levels within seven years.

Molecular techniques have revealed diversity within the mycorrhizal rhizoctonias that was not evident using classical techniques. However, there is currently no clear concept of a species or genus within the rhizoctonia complex, and there is a need for more phylogenetic studies using multiple genes to clarify the taxonomy of these fungi. Australian orchids form mycorrhiza with fungi from different clades within the mycorrhizal rhizoctonias *C. flava* with Sebacinales and *T. crinita* with Tulasnellales (e.g. Bonnardeaux et al. 2007; Brundrett 2007). These fungi may respond differently to disturbance and may have diverse microhabitat and edaphic requirements. Further comparative studies of these two orchids in relation to compatible fungus diversity, vegetation associations and natural habitats may help elucidate why *C. flava* is a successful coloniser of rehabilitation sites, while *T. crinita* is not.

Caladenia flava associates with *Sebacina* spp. that may be involved in tripartite relationships with other plant taxa via their mycorrhizal roots (Collins 2008). A tripartite relationship has been shown to exist between the underground orchid *Rhizanthella gardnerii*, its mycorrhizal fungus and broom bush (*Melaleuca* spp.) (Bougoure et al. 2009). Such relationships could be important in helping orchids to colonise disturbed areas where ectomycorrhizal plants had already established and may also help explain the relative success of *C. flava*'s colonisation compared to *T. crinita*. However, there is insufficient evidence to confirm tripartite roles for this fungus at present and it is relatively difficult to assign ecological roles to orchid fungi due to their capacity to grow as endophytes in roots of non-host plants (Brundrett 2006).

Discussion

Rehabilitation processes provide plant propagules for revegetation from three main sources: the soil seed bank in respread topsoil, sown seed and re-planted recalcitrant species (Grant and Koch 2007; Willyams 2005). Adventitious plant species may also be distributed by birds, animals and wind. Orchid seeds are rarely included in seed mixes used for revegetation, often have a short life span in soil seed banks (Batty et al. 2000; Whigham et al. 2002; Whigham et al. 2006) and tubers are unlikely to survive topsoil disturbance and stockpiling. Colonisation of severely disturbed areas by orchids would therefore be dependent on the presence of reproductive adults in surrounding areas, pollinator availability and seed dispersal. Nevertheless orchids may have a better capacity to arrive on sites without human intervention than many other plants due to their wind dispersed seed.

Similar arguments concern the arrival and persistence of orchid fungi in disturbed habitats dispersal of fungi, as some can produce spores, but this has rarely been documented. Sclerotia may persist in respread topsoil, since related fungi persist in cultivated soils (where they can cause disease of crop plants). We are severely hampered by our lack of understanding of the biology, ecology and diversity of *Rhizoctonia* alliance fungi that associate with orchids (only related pathogenic fungi have been studied) and some of these fungi apparently have multifunctional roles as endophytes, pathogens or ectomycorrhizal fungi, in addition to being orchid fungi (Brundrett 2006).

There are many factors that must coincide before orchids can become established in disturbed habitats, as shown in Table 8.5. In addition to orchid seed availability and the presence and persistence

Table 8.5. Factors and processes limiting orchid establishment in disturbed habitats

Limiting process	Factors
1. Belowground habitat quality	Soil conditions
1.1. Soil organic matter	Leaf litter, branches, roots, etc.
1.2. Soil texture	Rehabilitation actions
1.3. Water	Soil water availability
2. Aboveground habitat quality	Vegetation
2.1. Light and shade	Vegetation and litter cover (competition and exposure)
2.2. Water	Competition for water
2.3. Climate	Rainfall and temperature
3. Pollination	Associated insects
3.1. Insect diversity	Insect habitats and food resources
3.2. Pollinator specificity	Associated insect diversity low *v.* high, or none
4. Seed dispersal	Spread within or into sites
4.1. Flowering	Environmental conditions or stimulus required
4.2. Seed sources	Local orchid abundance, distribution and proximity
4.3. Dispersal conditions	Understorey and canopy density
5. Seed germination and growth	Protocorms and seedlings
5.1. Soil and environment	Moisture, shade, etc.
5.2. Fungal inoculum levels	Fungus diversity and availability (see 6.)
5.3. Threats to seedlings	Grazing by vertebrates/invertebrates
6. Mycorrhizal fungi	Symbiotic associates of orchids
6.1. Inoculum levels	Patches of high fungal activity needed for seed germination
6.2. Fungal persistence	Survival as sclerotia or in soil OM in disturbed soils?
6.3. Fungal dispersal	Sexual spore states (rarely observed)?
6.4. Fungal specificity	Low for weedy orchids, high for some recalcitrant orchids

of compatible mycorrhizal fungi, these factors include the presence of suitable microhabitats, competition with other vegetation, as well as climatic and edaphic conditions. Our seed-baiting studies detected compatible fungi for recalcitrant orchid species and this confirmed that fungal inoculum is probably not the most important factor limiting recolonisation of bauxite mines for these species. However, the frequency of detection of fungi was low, so it seems likely that fungi were only present in certain patches of soil.

Litter and other soil organic matter are major food substrates for saprophytic fungi and have a direct effect on fungal growth, vigour and survival. Litter cover and/or depth increased in rehabilitation areas over time and was positively correlated with mycorrhizal fungus detection and orchid species richness and population (Collins et al. 2007). Therefore, fungal abundance and diversity also seemed to be increased with increasing age of rehabilitation areas. We agree with Koch et al. (1994) who postulated that orchids become re-established following litter and soil development in rehabilitation areas.

Orchids that are locally common in natural vegetation, but fail to re-colonise in the short-term require further investigation. In a few cases it is likely that seed availability (especially for uncommon or fire-stimulated flowerings species) may be a limiting factor for recruitment. However, if seed production occurred locally and fungi compatible with these orchids were detected in soils, it is likely that colonisation of orchids is inhibited by prevailing environmental conditions such as soil quality, competition and exposure, as listed in Table 8.5. Severe disturbance, which completely removes vegetation, results in loss of ecological processes and destroy the microhabitats where orchids occur. Consequently, recruitment of orchids in severely disturbed habitats requires the establishment of suitable microhabitats and is therefore expected to occur with time as vegetation cover is re-established and microbial populations develop in parallel with litter accumulation in soils.

Ecological categories of orchids

The Orchidaceae is an extremely species-rich plant family that is of interest ecologically because of their complex life cycles, highly specialised pollination strategies and wide variety of life forms (Chase et al. 2003; Dressler 1981). However, orchids are under-represented in the ecological literature when compared to other species rich plant families (Dearnaley 2007; Peakall 2007). The literature on terrestrial orchids generally focuses on either conservation issues such as population monitoring, protecting and/or managing natural habitats, transplantation trials, propagation techniques; or studies of the specificity of orchid associations with their mycorrhizal fungi (e.g. Brundrett 2007; Cribb et al. 2003; Janečková et al. 2006; Reinhammar et al. 2002). Most long-term demographic studies of terrestrial orchid populations have been limited to European and North American ecosystems (Light and MacConaill 2005; Shefferson et al. 2005; Tamm 1991; Willems 1982) and few of these have examined the recruitment of orchids in severely disturbed habitats. Studies of semi-disturbed habitats have shown that some orchids are adversely affected by loss of native vegetation and competition from weeds much more than others (Scade et al. 2006).

Like other large plant families there are ecological categories of orchids which include a gradient of species from those which are extremely rare with highly specific habitats, to opportunistic orchids that rapidly colonise disturbed habitats. Our studies revealed that orchids which occur in mined and unmined jarrah forest belong to three ecological categories: (i) pioneer species that rapidly invade disturbed sites; (ii) secondary colonisers common in both disturbed and undisturbed sites; and (iii) disturbance intolerant species absent from recently disturbed habitats (Table 8.6). Ecological characteristics of orchids in these three categories are summarised in Table 8.6. The major differences between these categories are that pioneer species tend to have a broad diversity of insect pollinators (or are self-pollinated) and often also

Table 8.6. Ecological characteristics of common orchids in rehabilitated bauxite mines and adjacent jarrah forest habitats in Western Australia

Scientific name	Disturbance[1]	Clonal	Pollination[2]	Fungi[2]
*Disa bracteata**	Pioneer	No	Low (self-pollination)	Low
Microtis media	Pioneer	No	Low (self-pollination)	Low
Pterostylis sp. (aff. *nana*)	Pioneer	Highly	Medium	High
Caladenia flava	Pioneer	Highly	Medium	High
Pterostylis sanguinea	Secondary	yes	Medium	High
Caladenia macrostylis	Secondary	No	High (wasp)	High
Cyanicula sericea	Secondary	No	Medium	High
Eriochilus dilatatus	Absent	No	Medium	Medium
Prasophyllum parvifolia	Absent	No	Low	Medium
Thelymitra benthamiana	Absent	No	Medium	Medium
Thelymitra crinita	Absent	No	Medium	Medium
Pyrochis nigricans	Absent	Highly	Unknown	Low

* Alien weed.
1. Pioneer = occurs in <10 year old rehab sites, secondary = occurs in >10 year old rehab sites, absent = not present in rehabilitated forest (Collins 2008).
2. Pollinating insect and mycorrhizal fungus specificity data summarised by Brundrett 2007 (low = wide diversity of organisms, medium = closely related insects or fungi, high = few or single organism).

associate with a relatively wide diversity of mycorrhizal fungi compared to disturbance intolerant species, but there are also expectations to these generalisations. It is also likely that disturbance tolerant orchids are relatively exposure tolerant/shade intolerant, drought resistant and more effective at growth in soils with relatively low organic matter, compared to disturbance intolerant orchids. The gradual loss of pioneer species from older rehabilitation sites suggest that these ecological specialisations that allow rapid colonisation of new habitats become detrimental over time due to increased shading and competition. It is less likely that the availability of inoculum of compatible fungi or pollinators are limiting factors for these species (e.g. *Disa bracteata* and *Microtis media* are self-pollinating and associate with a wide diversity of fungi (Bonnardeaux *et al.* 2007)).

Several orchids listed in Table 8.6 are potential indicators of undisturbed or previously rehabilitated areas in the jarrah forest of Western Australia. These include *Thelymita crinita* and *Pyrochis nigricans*, which are indicators of undisturbed jarrah forest, and *Disa bracteata* and *Microtis media*, which are very good indicators of recent ecosystem disturbance. It has also been observed that certain orchids require semi-disturbed habitats, such as meadows where woody plants are suppressed, in temperate regions of the world (Coates *et al.* 2006; Kull and Hutchings 2006; Reinhammar *et al.* 2002). These orchids also have some characteristics of pioneer species.

Research in rehabilitated bauxite mines in Western Australia has revealed that orchids and their associated fungi were sparsely distributed in rehabilitation areas with occasional, localised, high-density populations, especially of weedy orchids and clonal species. Thus, the unassisted colonisation of rehabilitation sites by recalcitrant orchid species may be a prolonged process. In the case of

Alcoa World Alumina, which aims to establish a self-sustaining jarrah forest ecosystem that maintains the functions of the previous landscape on areas that have been mined for bauxite (Koch 2007), improved post-mining land management could potentially also include the reintroduction of several recalcitrant orchid species. However, it is still too soon to tell if these species will eventually recolonise mined jarrah forest or not, as the process may take decades to conclude. Methods that have been successfully used for reintroducing orchids could include *in vitro* propagation and transplantation of recalcitrant orchids, or sowing rehabilitated areas with mycorrhizal inoculum mixed with seed (Batty *et al.* 2001; Brundrett and Ager 2011; Scade *et al.* 2006).

A key review article by Adamowski (2006) summarised data from many studies of European orchids in disturbed habitats. His review also concluded that there were particular characteristics of orchids linked to their capacity to colonise and thrive in disturbed areas. These included: (1) they tend to have relatively short life cycles; (2) they have broad ecological tolerances to environmental conditions; and (3) they are capable of rapidly expansion in population sizes. Another common phenomenon was for orchids to occur that normally appear in different habitats and for hybrid swarms to develop between several related species in disturbed habitats (Adamowski 2006). Several of these distance-opportunistic European orchids are also known to have relatively broad mycorrhizal fungus specificity and relatively-non-specific pollinators (Rasmussen 1995; Shefferson *et al.* 2008). Thus, there are ecological parallels between European and Australian orchids that allow them to be assigned to similar functional categories.

We have attempted to address six key questions in this review. In summary:

1. Some orchids are very efficient at recolonising ecosystems after total loss of vegetation, but many others are slow to return and some are absent in disturbed habits, at least for the first few decades.
2. The absence of compatible symbiotic fungi does not seem to be a major limiting resource in disturbed soils, but inoculum levels are often low and patchy in soils.
3. Some orchids are much more effective recolonisers than others and these opportunistic orchids can be used as indicators of disturbed habitats.
4. Comparative studies of orchids in disturbed ecosystems are beginning to reveal key functional differences between disturbance opportunists and disturbance intolerant species, especially concerning the degree of pollinator and fungus specificity. However there is more research required to investigate the impact of microhabitats defined by soil and vegetation on orchids and how these change over time.
5. The ecological roles of orchids and their associated symbionts require much further investigation. In particular, we lack data on the ecological roles and habitat preferences of the fungi, or the insects on which terrestrial orchids depend.

References

Adamowski W (2006) Expansion of native orchids in anthropogenous habitats. *Polish Botanical Studies* **22**, 35–44.

Adams PB, Lawson SD (1993) Pollination of Australian orchids: A critical assessment of the literature 1882–1992. *Australian Journal of Botany* **41**, 553–575. doi:10.1071/BT9930553

Arditti J, Ernst R (1984) Physiology of germinating orchid seed. In: *Orchid Biology; Reviews and Perspectives. III.* (Ed J Arditti). pp. 177–222. Cornell University Press: Ithaca, New York.

Arditti J, Ghani AKA (2000) Numerical and physical properties of orchid seeds and their biological implications. *New Phytologist* **145**, 367–421. doi:10.1046/j.1469-8137.2000.00587.x

Ash HJ, Gemmell RP, Bradshaw AD (1994) The introduction of native plant species on industrial waste heaps: a test of immigration and other factors affecting primary succession. *Journal of Applied Ecology* **31**, 74–84. doi:10.2307/2404600

Atwood JT (1986) The size of the Orchidaceae and the systematic distribution of epiphytic orchids. *Selbyana* **9**, 171–186.

Bates B (1997) The history of Monadenia. NOSSA Conservation Index. http://www.anos.org.au/groups/southaustralia/nossa-a.htm

Batty AL, Dixon KW, Sivasithamparam K (2000) Soil seed-bank dynamics of terrestrial orchids. *Lindleyana* **15**, 227–236.

Batty AL, Dixon KW, Brundrett MC, Sivasithamparam K (2001) Constraints to symbiotic germination of terrestrial orchid seeds in mediterranean woodland. *New Phytologist* **152**, 511–520. doi:10.1046/j.0028-646X.2001.00277.x

Batty AL, Dixon KW, Brundrett MC, Sivasithamparam K (2002) Orchid conservation and mycorrhizal associations. In *Microorganisms in Plant Conservation and Biodiversity* (Eds K Sivasithamparam, KW Dixon and RL Barrett). pp. 195–226. Kluwer Academic Publishers: Dordrecht, Netherlands.

Benzing DH, Atwood JT, Jr (1984) Orchidaceae: Ancestral habitats and current status in forest canopies. *Systematic Botany* **9**, 155–165. doi:10.2307/2418820

Bonnardeaux Y, Brundrett M, Batty A, Dixon K, Koch J, Sivasithamparam K (2007) Diversity of mycorrhizal fungi of Western Australian terrestrial orchids: Compatibility webs, brief encounters, lasting relationships and alien invasions. *Mycological Research* **111**, 51–61. doi:10.1016/j.mycres.2006.11.006

Bougoure J, Ludwig M, Brundrett M, Grierson PG (2009) Identity and specificity of the fungi forming mycorrhizas with the rare mycoheterotrophic orchid *Rhizanthella gardneri*. *Mycological Research* **113**, 1097–1106. doi:10.1016/j.mycres.2009.07.007

Brundrett MC (2004) Diversity and classification of mycorrhizal associations. *Biological Reviews of the Cambridge Philosophical Society* **79**, 473–495. doi:10.1017/S1464793103006316

Brundrett MC (2006) Understanding the roles of multifunctional mycorrhizal and endophytic fungi. In *Microbial Root Endophytes* (Eds B Schulz, C Boyle and TN Sieber). pp. 281–298. SPRINGER VERLAG: Berlin.

Brundrett MC (2007) Scientific approaches to terrestrial orchid conservation with particular reference to Western Australia. *Australian Journal of Botany* **55**, 293–307. doi:10.1071/BT06131

Brundrett MC (2009) Mycorrhizal associations and other means of nutrition of vascular plants: Understanding the global diversity of host plants by resolving conflicting information and developing reliable means of diagnosis. *Plant and Soil* **320**, 37–77. doi:10.1007/s11104-008-9877-9

Brundrett M, Ager E (2011) *Wheatbelt Orchid Rescue Project Final Report 7. Seed Collecting, Soil Baiting and Propagation of Orchids.* Wheatbelt Orchid Rescue Project, University of Western Australia.

Brundrett MC, Scade A, Batty A, Dixon KW, Sivasithamparam K (2003) Development of in situ and ex situ seed baiting techniques to detect mycorrhizal fungi from terrestrial orchid habitats. *Mycological Research* **107**, 1210–1220.

Chase MW, Cameron KM, Barrett RL, Freudenstein JV (2003) DNA data and Orchidaceae systematics: a new phylogenetic classification. In *Orchid Conservation* (Eds KW Dixon, SP Kell, RL Barrett and PJ Cribb). pp. 69–89. Natural History Publications: Borneo.

Chung MY, Nason JD, Chung MG (2005) Spatial genetic structure in populations of the terrestrial orchid *Orchis cyclochila* (Orchidaceae). *Plant Systematics and Evolution* **254**, 209–219. doi:10.1007/s00606-005-0341-5

Clarkson BD (1997) Vegetation succession (1967–89) on five recent Montane lava flows, Mauna Loa, Hawaii. *New Zealand Journal of Ecology* **22**, 1–9.

Close RC, Moar NT, Tomlinson AI, Lowe AD (1978) Aerial dispersal of biological material from Australia to New Zealand. *International Journal of Biometeorology* **22**, 1–19.

Coates F, Lunt ID, Tremblay RL (2006) Effects of disturbance on population dynamics of the threatened orchid *Prasophyllum correctum* DL Jones and implications for grassland management in south-eastern Australia. *Biological*

Conservation **129**, 59–69. doi:10.1016/j.biocon.2005.06.037

Collins MT (2008) *Factors Affecting the Recovery of Orchids in a Post-mining Landscape*. PhD Thesis. The University of Western Australia.

Collins M, Koch J, Brundrett M, Sivasithamparam K (2005) Recovery of terrestrial orchids in the post-mining landscape. *Selbyana* **25**, 255–264.

Collins M, Koch J, Brundrett M, Sivasithamparam K (2007) Colonisation of bauxite mine rehabilitation areas of south-west Western Australia by orchid mycorrhizal fungi. *Australian Journal of Botany* **55**, 653–664. doi:10.1071/BT06170

Cribb PJ, Kell SP, Dixon KW, Barrett RL (2003) Orchid conservation: a global perspective. In *Orchid Conservation*. pp. 1–24. Natural History Publications, Kota Kinabalu, Sabah.

Davis BNK (1979) Chalk and limestone quarries as wildlife habitats. *Minerals and the Environment* **1**, 48–56. doi:10.1007/BF02010717

Dearnaley JDW (2007) Further advances in orchid mycorrhizal research. *Mycorrhiza* **17**, 475–486. doi:10.1007/s00572-007-0138-1

Diez JM (2007) Heirachical patterns of symbiotic orchid germination linked to adult proximity and environmental gradients. *Journal of Ecology* **95**, 159–170. doi:10.1111/j.1365-2745.2006.01194.x

Dressler RL (1981) Evolution. In *The Orchids, Natural History and Classification*. pp. 121–141. Harvard University Press: London, England.

Erickson R (1965) *Orchids of the West*. Paterson Brokensha, Perth.

Esfeld K, Hensen I, Wesche K, Jakob SS, Tischew S, Blattner FR (2008) Molecular data indicate multiple independent colonizations of former lignite mining areas in Eastern Germany by *Epipactis palustris* (Orchidaceae). *Biodiversity and Conservation* **17**, 2441–2453. doi:10.1007/s10531-008-9391-7

Gandawijaja D, Arditti J (1983) The orchids of Krakatau: evidence for a mode of transport. *Annals of Botany* **52**, 127–130.

Gardiner T, Vaughan A (2009) Scrub clearance and soil disturbance increases bee orchid *Ophrys apifera* frequency in calcareous grassland at Norton Heath roadside verge, Essex, England. *Conservation Evidence* **6**, 39–41.

Gardner JH, Malajczuk N (1988) Recolonisation of rehabilitated bauxite mine sites in Western Australia by mycorrhizal fungi. *Forest Ecology and Management* **24**, 27–42. doi:10.1016/0378-1127(88)90022-9

Gaskett AC (2010) Orchid Pollination by sexual deception: pollinator perspectives. *Biological Reviews of the Cambridge Philosophical Society* **86**, 33–75.

Grant CD, Koch JM (1997) Ecological aspects of soil seed-banks in relation to bauxite mining: II. Twelve year old rehabilitated mines. *Australian Journal of Ecology* **22**, 177–184. doi:10.1111/j.1442-9993.1997.tb00657.x

Grant CD, Koch J (2003) Orchid species succession in rehabilitated bauxite mines in Western Australia. *Australian Journal of Botany* **51**, 453–457. doi:10.1071/BT02127

Grant CD, Koch JM (2007) Decommissioning Western Australia's first bauxite mine: co-evolving vegetation restoration techniques and targets. *Ecological Management & Restoration* **8**, 92–105. doi:10.1111/j.1442-8903.2007.00346.x

Grant CD, Loneragan WA (1999) The effects of burning on the understorey composition of 11–13 year-old rehabilitated bauxite mines in Western Australia. *Plant Ecology* **145**, 291–305. doi:10.1023/A:1009821128075

Halliday RJ, Johnson MS (1979) The contribution of derelict mineral and industrial sites to the conservation of rare plants in the United Kingdom. *Environmental Geochemistry and Health* **1**, 1–7.

Hoffman N, Brown A (1998) *Orchids of South-west Australia*. University of Western Australia Press, Perth.

Janečková P, Wotavová K, Schödelbauerová I, Jersáková J, Kindlmann P (2006) Relative effects of management and environmental conditions on performance and survival of populations of a terrestrial orchid, *Dactylorhiza majalis*. *Biological Conservation* **129**, 40–49. doi:10.1016/j.biocon.2005.09.045

Jacquemyn H, Brys R, Vandepitte K, Honnay O, Roldan-Ruiz I, Weigand T (2007) A spatially explicit analysis of seedling recruitment in the terrestrial orchid *Orchis purpurea*. *New Phytologist* **176**, 448–459. doi:10.1111/j.1469-8137.2007.02179.x

Juersáková J, Malinova T (2007) Spatial aspects of seed dispersal and seedling recruitment in orchids. *New Phytologist* **176**, 237–241.

Koch JM (2007) Alcoa's mining and restoration process in Western Australia. *Restoration Ecology* **15**, S11–S16. doi:10.1111/j.1526-100X.2007.00288.x

Koch JM, Ward SC (1994) Establishment of understorey vegetation for rehabilitation of bauxite-mined areas in the jarrah forest of Western Australia. *Journal of Environmental Management* **41**, 1–15. doi:10.1006/jema.1994.1029

Koch JM, Taylor SK, Gardner J (1994) Research to maximise plant diversity in rehabilitated bauxite mines in the jarrah forest. In *Proceedings of the National workshop on native seed biology for revegetation* (Eds SM Bellairs and LC Bell). pp. 41–50. The Chamber of Mines and Energy: Perth.

Kull T, Hutchings MJ (2006) A comparative analysis of decline in the distribution ranges of orchid species in Estonia and the United Kingdom. *Biological Conservation* **129**, 31–39. doi:10.1016/j.biocon.2005.09.046

Lee JA, Greenwood B (1976) The colonisation by plants of calcareous wastes from the salt and alkali industry in Cheshire, England. *Biological Conservation* **10**, 131–149. doi:10.1016/0006-3207(76)90057-4

Light MHS, MacConaill M (2005) Long-term studies: A case for orchid species survival. *Selbyana* **26**, 174–188.

Murren CJ, Ellison AM (1998) Seed dispersal characteristics of *Brassavola nodosa* (Orchidaceae). *American Journal of Botany* **85**, 675–680. doi:10.2307/2446537

Nichols OG, Koch JM, Taylor S, Gardner J (1991) Conserving biodiversity. In *Australian Mining Industry Council environmental workshop, Perth*. pp. 116–136. Australian Mining Industry Council: Perth, Australia.

Partomihardjo T (2003) Colonisation of orchids on the Krakatau Islands. *Telopea* **10**, 299–310.

Peakall R (2007) Speciation in the Orchidaceae: confronting the challenges. *Molecular Ecology* **16**, 2834–2837. doi:10.1111/j.1365-294X.2007.03311.x

Perkins AJ, Masuhara G, McGee PA (1995) Specificity of the associations between *Microtis parvifolia* (Orchidaceae) and its mycorrhizal fungi. *Australian Journal of Botany* **43**, 85–91. doi:10.1071/BT9950085

Phillips RD, Faast R, Bower CC, Brown GR, Peakall R (2009) Implications of pollination by food and sexual deception for pollinator specificity, fruit set, population genetics and conservation of *Caladenia* (Orchidaceae). *Australian Journal of Botany* **57**, 287–306. doi:10.1071/BT08154

Rasmussen HN (1995) *Terrestrial orchids, from seed to mycotrophic plant* Cambridge, UK: Cambridge University Press.

Reinhammar L-G, Olsson EGA, Sørmeland E (2002) Conservation biology of an endangered grassland plant species, *Pseudorchis albida*, with some references to the closely related alpine *P. straminea* (Orchidaceae). *Botanical Journal of the Linnean Society* **139**, 47–66. doi:10.1046/j.1095-8339.2002.00041.x

Restrepo C and Vitousek P (2001) Landslides, alien species, and the diversity of a Hawaiian montane mesic ecosystem. *Biotropica* **33**, 409–420.

Rufaut CG, Hammit S, Craw D, Clearwater SG (2006) Plant and invertebrate assemblages on waste rock at Wangaloa coal mine Otago, New Zealand. *New Zealand Journal of Ecology* **30**, 311–319.

Scade A, Brundrett MC, Batty AL, Dixon KW, Sivasithamparam K (2006) Survival of transplanted terrestrial orchid seedlings in urban bushland habitats with high or low weed cover. *Australian Journal of Botany* **54**, 383–389. doi:10.1071/BT04025

Shefferson RP, Kull T, Tali K (2005) Adult whole-plant dormancy induced by stress in long-lived orchids. *Ecology* **86**, 3099–3104. doi:10.1890/05-0586

Shefferson RP, Kull T, Tali K (2008) Mycorrhizal interactions of orchids colonizing Estonian mine tailings hills. *American Journal of Botany* **95**, 156–164. doi:10.3732/ajb.95.2.156

Tamm CO (1991) Behaviour of some orchid populations in a changing environment. Observations on permanent plots, 1943–1990. In *Population ecology of terrestrial orchids* (Eds TCE Wells and JH Willems). pp. 1–13. SPB Academic Publishing: The Hague, The Netherlands.

Whigham DF, O'Neill J, McCormick M, Smith C, Rasmussen H, Caldwell B, Daniell T (2002) Interaction between decomposing wood, mycorrhizas, and terrestrial orchid seeds and protocorms. In *Trends and fluctuations and underlying mechanisms in terrestrial orchid populations* (Eds P Kindlmann, JH Willems and DF Whigham). pp. 117–131. Backhuys Publishers: Leiden, Holland.

Whigham DF, O'Neill JP, Rasmussen HN, Caldwell BA, McCormick MK (2006) Seed longevity in terrestrial orchids – Potential for persistence *in situ* seed banks. *Biological Conservation* **129**, 24–30. doi:10.1016/j.biocon.2005.10.029

Whittaker RJ, Bush MB, Richards K (1989) Plant recolonization and vegetation succession of the Krakatau islands, Indonesia. *Ecological Monographs* **59**, 59–123. doi:10.2307/2937282

Willems JH (1982) Establishment and development of a population of *Orchis simia* Lamk. in the Netherlands. *New Phytologist* **91**, 757–765. doi:10.1111/j.1469-8137.1982.tb03355.x

Willyams D (2005) Tissue culture of geophytic rush and sedge species for revegetation of bauxite mine sites in Northern Jarrah forest of Western Australia. In *Contributing to a sustainable future: Proceedings of the Australian Branch of the IAPTC&B*. (Eds I Bennett, E Bunn, H Clarke and J McComb). pp. 226–241. Australian Branch of the IAPTC&B: Perth, Australia.

Wirth M, Withner CL (1959) Embryology and development in the Orchidaceae. In *The Orchids*. (Ed CL Withner). pp. 155–188. The Ronald Press Company: New York.

9

The mining-restoration system and ecosystem development following bauxite mining in a biodiverse environment of the seasonally dry tropics of Australia

Alister V Spain, Mark Tibbett, Dieter A Hinz,
John A Ludwig and David J Tongway

Summary

The system of mining and restoration developed and used at the Rio Tinto Alcan Gove bauxite mine between 1973 and 2005 is described. By 2009, this system had been used to restore successfully more than 3000 ha of mined lands to ecosystems dominated by local native open-forest species. The major features of post-restoration development in these restored ecosystems are described, largely based on chronosequences of study sites extending to ~30 years. Over this period the plant community developed from a grassland to a shrubland to a closed forest and finally to an open forest. By 26 years, it possessed a similar three-tiered structure to the local native open forests (including the same dominant eucalypt species), together with diverse understory and herbaceous strata colonised by more than 170 species beyond the ~39 species sown. The few reported studies of the animal community have indicated rapid colonisation of the restored sites and ongoing community development within three key environmental indicator groups: ants, termites and birds.

Due to fire exclusion, substantial litter layers have developed at the longer-restored sites and because of their magnitudes they may constitute a risk to the developing keystone canopy eucalypt species when eventually burned. Rapid soil development occurred contemporaneously, with substantial changes in physical properties and large increases in near-surface soil organic matter concentrations, from c. 10 g kg^{-1} C at planting to values greater than those present in the adjacent native forests. A wide diversity of nutritionally important ecto and arbuscular mycorrhizas were reported and appear to change with ongoing plant community development. Continuing development of these novel ecosystems seems assured but their future trajectory is likely to depend partly on management decisions relating to the institution of a firing regime concilient with a landscape subject to regular firing.

Introduction

Surface mining processes necessarily entail destruction of the existing ecosystems and soils associated with the resource being extracted. Bauxite mining is no exception to this and the scale of land disturbance is large in comparison to that of most other mining types. The legacies of past mining have included unstable eroding post-mining landscapes that may damage adjacent ecosystems and waterways and are detrimental to future users of the mined areas, nearby inhabitants and other land users, down-stream and down-wind. In the past, mined areas were often left with the negative long-term environmental legacies of these actions. Now most societies require such lands to be restored to standards acceptable to the stakeholders. This can be a particularly challenging exercise in the fragile, biodiverse and poorly studied ecosystems that occur in many regions of the tropics.

The restoration process at what is now the Rio Tinto Alcan Gove bauxite mine began in 1973, two years after the mining operation commenced. Its generalised aim is to return the post-mined landscape to a local historical system (Hobbs et al. 2009): a diverse and self-sustaining native forest community dominated by large trees, notably *Eucalyptus tetrodonta* and other eucalypts (D. Hinz unpublished). The restored mature forests should therefore be structurally similar to local unmined native forests and comprise three strata: an herbaceous stratum dominated by native grasses; a sparse mid-stratum of shrubs and small trees; and an upper-stratum dominated by large trees. Also required is the establishment of species that have cultural importance or are otherwise useful to local stakeholders (Hinz 2007). It has been company policy to restore the mined areas progressively and by 2009, more than 3000 ha of mined land had been restored at the Gove mine (Rio Tinto Alcan 2009).

As with restored environments in disturbed natural landscapes (Society for Ecological Restoration 2004), it is recognised that ecosystems restored post-mining are unlikely to exactly replicate those existing before mining. Therefore, the general aim of the restoration process has been interpreted as requiring that the ongoing trajectory of early ecosystem development (including soil development or pedogenesis) should be towards that of the surrounding unmined landscape. Because it is intended that the restored areas will have no (or minimal) inputs of labour or materials following lease relinquishment, there is a further need to ensure that the restored areas are substantially resistant to damage by external (cyclones, fire, drought, tornadoes) or internal agents (disease organisms, pest animal species) and that these systems have sufficient resilience to recover when damage does occur. A particular consideration for current rehabilitation practice is the need to have a broad genetic base in light of expected climate change (Banning et al. 2011). Additionally, objective and numerically based systems for broadly monitoring all major components of ecosystem development on restored lands are also required (see, for example, Tongway and Ludwig 2011).

This paper presents an account of the mining and restoration procedures as they were developed and practised at Gove between the years 1973 and 2005. It also presents the results of quantitative, spatially integrated studies of plant and soil development in restored areas, including site colonisation by animals and other organisms, notably mycorrhizas and termites. A tentative integration of the results of these studies defines a sequential series of variable and overlapping developmental stages, which are based on the changing but interdependent characteristics of the plant community, the soils and the general biota. Lastly, some possible future trajectories for the restored ecosystems are suggested, based on the different management decisions that may be taken.

The information presented in this paper is largely based on assessment, sampling and subsequent analyses conducted within 16- and 26-year chronosequences of sites during 1991 and 1992 (Reddell et al. 1992; Reddell et al. 1993) and in 2001 and 2002 (Spain et al. 2009) and by studies linking conventional scientific investigations with parallel studies of 'ecosystem function analysis' (EFA), an objective monitoring system for assessing the health of restored ecosystems based on formally verified indicators (Tongway and Ludwig 2011). It is further based on on-site observations made by D. Hinz over 32 years of development and management of the restoration works (Hinz 1992; Hinz 2007). To provide comparative information on the unmined environments, parallel studies were also conducted in adjacent mature native forest sites.

During field work carried out in 1991–92 and 2001–02 (Reddell *et al.* 1992; Reddell *et al.* 1993; Spain *et al.* 2009) and where scale permitted, studies were integrated between disciplines through the use of a common core of sites and locations. Much of what is reported in this paper relates to the redevelopment of this unique forest system and the extent to which the developing ecosystems are on a trajectory towards the surrounding native forest ecosystems and where components of the ecosystem succeed or fail to match parameters measured in the unmined native forest. Further comparisons are made with Woinarski *et al.* (2009), Brady and Noske (2010) and Cook (2012).

The Gove environment

The Gove bauxite mine is located at 12° 16′S, 136° 51′E on the Gove Peninsula in the far north-eastern corner of Australia's Northern Territory. Ore mined from the extensive lease area undergoes local preliminary crushing and is transported approximately 12 km by conveyor to a near-shore processing facility in which alumina is produced for export.

The Gove mine is located within the 'Tropical and sub-tropical grasslands, savannas and shrublands' ecoregion of Australia (Department of Sustainability, Environment, Water, Population and Communities 2011) and, within this, in the Arnhem Coast Bioregion (Department of Sustainability, Environment, Water, Population and Communities 2008). The Arnhem Coast Bioregion is entirely Aboriginal land; no pastoral usage occurs and local industries are limited to bauxite and manganese mining and to tourism and related industries. Extensive fires are a regular feature of this region:

it has been estimated that a median value of 31% of the land area was burned each year between 1997 and 2005 (Department of Sustainability, Environment, Water, Population and Communities 2008). A general guide to the natural history of northern Australia is presented by Woinarski *et al.* (2007). Biogeographical aspects of plant and animal distributions in the region are presented by Bowman *et al.* (2010).

Climate

The climate of the Gove Peninsula is classified as 'tropical savanna' in the modified Köppen classification used by the Australian Government Bureau of Meteorology (Bureau of Meteorology 2011). The area has a seasonally variable climate with marked contrasts in mean monthly rainfall and monthly mean maximum and minimum temperature conditions (Fig. 9.1). Based on complete records between the years 1944 and 2011, mean annual precipitation at the adjacent Gove Airport was 1485 mm with extreme monthly values of 6 mm (August) to 288 mm (February) (Bureau of Meteorology 2011). Large rainfall events are episodic and are related to the presence of cyclones and their associated depressions; cyclones occur during the months from November to April (Bureau of Meteorology 2011). Mean maximum temperatures are less elevated than further inland due to maritime effects. Total annual pan evaporation is estimated at 2153 mm (Bureau of Meteorology 2011).

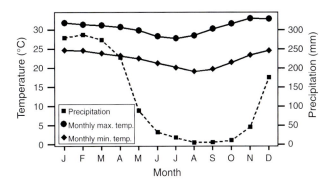

Figure 9.1: Mean monthly rainfall and mean monthly maximum and minimum temperatures, Gove Airport (Bureau of Meteorology 2011).

The marked seasonality of rainfall controls many biological features of the Gove environment, including the growth of the dominant eucalypts. Prior *et al.* (2004) found that most of the growth in open-forest (and other) species occurs during the wet season, December to May. Similarly, most (70% or more) of the CO_2 efflux from the soils occurs during the wet season (Chen *et al.* 2002; Livesley *et al.* 2011).

Natural vegetation

The pre-European vegetation of much of northern Australia was open forest or savanna dominated by *Eucalyptus tetrodonta* and *Eucalyptus miniata* (Fox *et al.* 2001) and remains so today. *Eucalyptus tetrodonta* is a widespread species occurring in the tropical part of Australia from west of Darwin to the east coast of Cape York Peninsula (Atlas of Living Australia 2012a). *Eucalyptus miniata* has a similar distribution but is less widespread on Cape York Peninsula (Atlas of Living Australia 2012b). Open eucalypt forest is the most widespread vegetation type on deep, well drained sites on the Gove Peninsula, which lies within the Arnhem Coast bioregion. Although these forests appear uniform in composition and structure, they are floristically diverse (Yunupingu *et al.* 1995), varying greatly in both their shrubby mid- to lower-understory and grass-dominated herbaceous layers (Wilson *et al.* 1990). Aboveground tree biomass is relatively low in these forests. Estimated values reported by O'Grady *et al.* (2000) ranged from about 43 to 72 t ha^{-1}, with a median mass of 56 t ha^{-1}. As shown below, basal area appears to vary inversely with soil profile depth in forests associated with the Gove bauxite mine.

Unmined soils

Under the Australian soil classification (Isbell 2002), the dominant soils of the Gove mining area are classified as bauxitic, dystrophic red kandosols. These soils were formerly known as bauxitic red earths (Stace *et al.* 1968) but would probably be classified as orthoplinthic ferralsols (McKenzie *et al.* 2004) under the World Reference Base (IUSS Working Group WRB 2007) and as oxisols under Soil Taxonomy (Soil Survey Staff 1999). The profile comprises a variably developed A horizon overlying a B horizon which increases in clay content with depth; this, in turn, overlies the indurated bauxitic layer (Fig. 9.2). Limited soil variation occurs on the mine lease area. One variant has less gravel and a deeper profile while small areas have a very shallow A horizon over the indurated layer, which may outcrop at the surface in some locations. Similar soils occur in the Weipa area of Cape York Peninsula and in the Mitchell Plateau of north-western Australia (McKenzie *et al.* 2004).

The term 'kandosol' implies soils that lack strong texture contrast, have massive or only weakly structured B horizons and are not calcareous (Isbell 2002). They are widespread throughout the continent and occupy approximately 17% of the Australian land area. Red kandosols (McKenzie *et al.* 2004) are typically well drained, permeable soils with a clay mineralogy dominated by kaolinite and with small amounts of iron oxides (confirmed for the Gove site by Reddell *et al.* 1993). The term dystrophic refers to soils with a base status of less than 5 cmol kg^{-1} of clay and a bauxitic horizon is one that has more than a visually estimated 20% of bauxitic nodules or concretions (Isbell 2002).

Determination of post-mining land use

The goal of the system of post-mining restoration eventually adopted resulted from the decision handed down in 1971 on the local Aboriginal peoples' petition against the Commonwealth Government and Nabalco Pty Ltd. This petition sought freedom from occupation of their lands, damages and the prevention of further mining activities (Anonymous 2011). The lease agreement between the mine and the traditional owners requires proof that the mined land will be returned to its original condition when it is handed back to them after restoration (Hinz 1979). There are several groups of traditional owners of clan lands within the mining lease who are directly affected by the mining and related operations (Federal Law Report 1971; Kauffman 1998). The Gove mine site and the townships

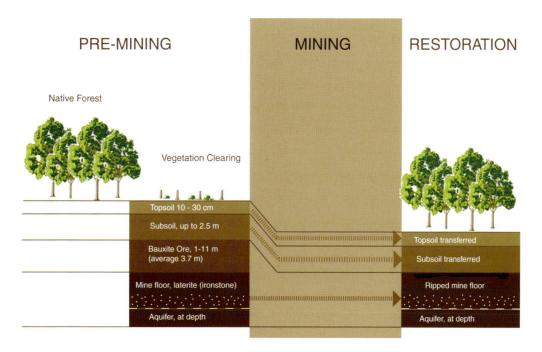

Figure 9.2: Schematic of the processes of vegetation clearing, mining, soil replacement, new profile ripping and restoration of the native forest community at the Gove mine.

of Nhulunbuy and Yirrkala are all on traditional Rirratjingu lands (Yunupingu *et al.* 1995).

Pastoral agriculture was initially proposed as an appropriate post-mining land use and this lead to the establishment of several experimental-scale plots containing pasture grasses and woody and herbaceous legumes (Richards and Calder 1978). Although pastures were not sown at paddock scale, these experiments included several fertiliser trials, which provided early guidelines for fertiliser application. A legacy of this experimentation has been the spread of *Leucaena leucocephala*, *Melinus minutiflora* and *Melinis repens* into the restored areas (D. Hinz unpublished). In addition, limited areas of several timber species were planted in preliminary experiments designed to evaluate the potential for commercial wood production in this environment (D. Hinz unpublished). The best-performing species were *Khaya grandifoliola* and *Khaya senegalensis*; other species did not survive firing (D. Hinz unpublished).

Pastoralism and timber production were rejected by the traditional owners as appropriate end uses for the mined land and the first restoration vegetation plantings were made in 1973 using directly seeded, locally collected native plant species. This followed requests to this effect from a council spokesman from the Yirrkala Dhanbul Community Association at a meeting conducted in 1972 with members of the local and Northern Territory governments and with senior management staff from Nabalco Pty Ltd (Hinz 2007, unpublished).

The requirements considered above have been subsequently interpreted as necessitating a development trajectory trending towards the structure, composition and function of the local mature three-layered open forests. Other necessary qualities are assumed to be ecosystem resistance to damaging influences and resilience when it does occur. Further requirements include the presence of certain food plants and others with medicinal, cultural and other values.

The mining-restoration sequence

Mining commenced in 1971 (O'Keefe 1992) and the mining-restoration process described here was developed and used at the Gove mine between the years 1973 and 2005. It is likely that at least some details of the restoration process described here have changed since that time.

Effective restoration is dependent on detailed planning and a close integration with mining activities (O'Keefe 1992) and entails the rapid and separate movement of large amounts of topsoil and subsoil (Fig. 9.2). The forest is felled and the timber is windrowed and burned. The land is then left fallow for two or three years, until mining begins. Initial studies indicated that, where cleared areas were allowed to remain undisturbed for a period before topsoil was stripped, mined and restored, flora and fauna would recover more quickly (Hinz 1981). During this period, vegetation regrowth occurs from propagules contained in the soil providing some protection of the soil surface. Partial decomposition also occurs of much of the soil organic matter and of litter and the smaller near-surface coarse woody debris remaining at and near the soil surface. It is possible that the successional status of the ecosystem is reset to an earlier phase capable of better facilitating vigorous growth of sown and other species after topsoil replacement.

Soil handling and profile reconstruction

Immediately before mining, the soil is removed from the site by a process known as 'double-stripping'. Under this system, the topsoil (A and upper B horizons, ~0.1 to 0.3 m) is removed from areas cleared for two to three years and is placed directly on a previously mined area (O'Keefe 1992) already prepared to receive topsoil before seeding (Banning *et al.* 2011; Bell 2001; Mulligan *et al.* 2006; Tibbett 2010). The subsoil (remaining soil material above the bauxitic layer) is removed and separately stockpiled (for up to eight months) until mining has been completed. It is subsequently placed on the mine floor in preparation for receiving the topsoil. To minimise compaction and structural damage, soil handling normally takes place during the dry season.

The topsoil is of particular importance in that it provides suitable chemical and physical conditions for plant growth and comprises a valuable source of nutrients and native plant seed (Tacey and Glossop 1980), notably of the understorey species (Nichols *et al.* 1985). It is also a source of micro-organisms and organic matter (Jasper 2007; Koch 2007; Tibbett 2008). Stockpiling topsoil for extended periods is known to reduce seed viability and to diminish soil physical, chemical and biological fertility (Johnson *et al.* 1991; Schäffer *et al.* 2007); topsoil is therefore moved directly to a new site that has previously been mined and the subsoil replaced (Banning *et al.* 2011; Hinz 1992; Koch 2007; Vlahos *et al.* 1999) (Fig. 9.2). This process has become known as 'direct-return' soil handling and aims to maximise ecological benefit in terms of retaining fertility and life in the soil.

Subsequent to soil replacement on the lowered post-mining landscape (Fig. 9.2), the restored soil profile is deeply ripped ('sub-soiled') at a lateral interval of ~1.7 m (O'Keefe 1992) in order to decompact soil materials compacted by heavy machinery traffic and to break up the upper part of the mine floor (Croton and Ainsworth 2007; Kew *et al.* 2007; Szota *et al.* 2007). In this process, a straight tine is dragged through the reconstructed soil profile (both top and sub-soil) and to a depth into the upper part of the mine floor that depends on its degree of induration, typically 0.2 to 0.3 m. Passage of this tine through the soil thereby creates physical conditions more amenable to root penetration and growth and to free drainage (Banning *et al.* 2011; Kew *et al.* 2007; Lardner *et al.* 2011; Schäffer *et al.* 2008). Subsequently, a chain is dragged behind the tine to topple any unstable larger blocks of laterite that may have been dragged to the surface.

Ecosystem establishment

Prior to mine establishment, it had been expected that the propagules contained in the topsoil would provide an appropriate suite of species and that return of the topsoil would therefore be sufficient to revegetate the mined areas satisfactorily. Whereas this was perhaps adequate to establish acacias and many grasses and forbs (Scott *et al.*

2010), it was clearly insufficient to restore the keystone eucalypts and other species (Hinz 2007; see also Norman *et al.* 2006). From this, a decision was made to sow local provenance seeds of the keystone tree and shrub species harvested from unmined forests located within the mine lease area. All batches of eucalypt seeds were tested independently for viability.

Although seeds were sown by hand during the first three years of restoration works, native plant seeds (Table 9.1) have subsequently been sown towards the end of the dry season using an agricultural 'air-blow' seeder modified for spreading larger seeds. Seeds were sown in two separate procedures using a total of 1.2 kg ha^{-1} of seed. In the first seeding, native plant seeds are mixed in a 1:20 ratio with seeds of a sterile cultivar of *Sorghum almum*, broadcast and the mixture is tined into the soil. The second application comprised native plant seeds in the same ratio as above with *S. almum* and 0.5 kg ha^{-1} of the seeds of *Chloris gayana* cv. *Pioneer* mixed with granulated single superphosphate. This was applied at the rate of 200 to 300 kg ha^{-1} (*c.* 25 kg ha^{-1} P) using an open-bucket, power-take-off-driven fertiliser spreader. Seeds were sown in a ratio of 6:2:7 of, respectively, eucalypts, acacias and other species.

Up to 39 species were planted although, as indicated in Table 9.1, some species were not sown in all years because of seed shortages due to the inter-annual variation in local seed set that occurs in these species (Setterfield and Williams 1996). The seeds of some large-fruited species and those temporarily in short supply (including *Buchanania obovata*, *Pandanus spiralis*, *Persoonia falcata*, *Syzigium suborbiculare* and *Livistona humilis*) were sown by hand (Hinz 2007). Supplementary hand sowing was also carried out in already restored sites where seeds of certain species had not been available at the time when the remaining seeds were sown. *Callitris intratropica* was initially included with the seed mix but its use was later discontinued. *Melaleuca viridiflora* seed was sown in some years depending on seed availability; this species was also planted as container-grown seedlings in a few waterlogging-prone locations in the years 2002–2005.

Table 9.1. Species sown as part of the restoration process at the Gove mine site

Acacias	Other species
Acacia aulacocarpa	*Alphitonia excelsa*
Acacia holosericea	*Buchanania obovata***
Acacia leptocarpa	*Brachychiton diversifolius*
Acacia multisiliqua	*Brachychiton paradoxus*
Acacia oncinocarpa	*Callitris intratropica*†
Acacia simsii†	*Calytrix exstipulata*†
Acacia sublanata†	*Cycas angulata*†
Acacia torulosa	*Grevillea heliosperma*
Acacia yirrkallensis†	*Grevillea mimosoides*†
Eucalypts	*Grevillea pteridifolia*
Corymbia clavigera	*Grevillea pungens*†
Corymbia foelscheana	*Grewia retusifolia*
Corymbia grandifolia	*Hakea arborescens*†
Corymbia papuana†	*Livistona humilis***
Corymbia polycarpa	*Melaleuca viridiflora****†
Corymbia ptychocarpa†	*Pandanus spiralis****†
Corymbia setosa†	*Persoonia falcata****†
*Eucalyptus alba**	*Syzigium suborbiculare****†
Eucalyptus miniata	
Eucalyptus tetrodonta	**Grasses**
	Chloris gayana cv *Pioneer*
	Sorghum almum

* used selectively.
** hand sown.
† dependent on seed availability.

The grasses *Chloris gayana* cv *Pioneer* and *S. almum* stabilise the soil surface and provide an early protection against surface erosion. Further, they provide the initial inputs of organic matter to the upper soil layer through litter decomposition and fine root turnover. By reducing site irradiance they reduce the otherwise high surface soil temperatures and provide microhabitats for the developing meso- and microfauna (Hinz 2000).

Fertilisation is thought to be necessary to initiate ecosystem development by replacing the nutrient losses associated with plant removal and subsequent burning and with the oxidative losses of organic matter that occur during the fallow period. Fertilisation may also help to redress the nutrient dilution involved in physical mixing of the soil during handling (Banning *et al.* 2011; Hinz 1992; Koch 2007; Short *et al.* 2000; Standish *et al.* 2008; Standish *et al.* 2010; Vlahos *et al.* 1999) although there is little empirical evidence to support such assertions.

Establishment success and early growth was carefully observed. Normally, no further direct intervention would take place from this point onwards, except for local re-seeding where establishment failure may have occasionally occurred, as with seasonal water logging.

A particular feature of post-establishment management of these ecosystems has been the exclusion of fire from the restored sites. As indicated below, this has had considerable ramifications for many aspects of the post-mining ecosystem development but particularly in relation to the putative risk of damage to the plant communities from the firing of the substantial litter layers that develop (Cook 2012; Reddell *et al.* 1992).

Post-mining ecosystem development

Ecosystem development was assessed using a chronosequence approach or space-for-time substitution (Huggett 1998; Johnson and Miyanishi 2008; Walker *et al.* 2010), where sites of different ages are studied to construct the temporal sequence of biotic community and soil development. The assumptions that underlie this space-for-time approach include homogeneity of the soil materials and close similarity of the range of cultural procedures used during clearing of the vegetation, the mining process and restoration of the post-mining profiles. Assumptions extend to include the uniformity of the seed mixture sown and to the amount and type of fertilisers applied. Although these assumptions have been largely met at Gove, where not met, interpretation has necessarily been adjusted.

From both short- and long-term viewpoints, plant community success is dependent on soil development to provide continuing stability and to promote an ongoing physical, chemical and biologically based fertility. Where long-term ecosystem sustainability is required under circumstances where no further restoration intervention is likely to be practicable, it is critical to verify that the biological agencies mediating decomposition and nutrient recycling processes are established and that these are functioning adequately.

Based on these requirements, integrated studies of plant community development (including root distributions) and soil formation were conducted, the latter with an emphasis on soil organic matter accumulation and distribution in the profile. Included in the biologically based studies, considerable attention was paid to litter mass dynamics and composition. No direct measurements of litter or soil organic matter decomposition rates have yet been conducted although Cook (2012) estimated a decomposition constant based on litter fall and litter masses. Termite colonisation was studied in relation to the food and nesting resources present at different ecosystem stages. Because of their importance as key ecosystem components, studies of initial mycorrhizal colonisation rates were also made.

Further comparisons were made with the limited published works available on restoration of the Gove site, notably those relating to the post-establishment colonisation and use of the restored landscapes by larger organisms, particularly by the birds and mammals responsible for the transport of plant propagules (Brady and Noske 2010; Woinarski *et al.* 2007; Woinarski *et al.* 2009). The invasive ant species *Anoplolepis gracilipes* ('yellow crazy ant') has been reported from the mine lease area (Rio Tinto Alcan 2009) although no information has yet been published on its effects on the restored environments.

Plant community development

Plant community development was studied at a range of restored sites during 1991 and 1992 (Reddell *et al.* 1992, 1993) and in 2001 and 2002 (Spain *et al.* 2009). During both study periods, community composition and structure were assessed across 16- and 26-year chronosequences of restored sites and at unmined reference sites of differing profile depths. Further studies have been published by Brady and Noske (2010) and Cook (2012). Additional unpublished studies of plant community development at these sites are indicated in Woinarski *et al.* (2009).

Data from restored sites were examined to determine if trends in the establishment and development of plant species were towards those expected on the basis of reference sites. These trends helped to identify those compositional and structural attributes of the vegetation most useful as indicators of revegetation success in mine site restoration.

The following questions were addressed:

- Does the composition of the mid- and upper-stratum vegetation develop over time towards that found in the reference sites?
- Are trends in tree growth measured, for example, by basal area, stem density and height, towards those found on reference sites?
- Do species' composition and abundance in the herbaceous stratum change as expected towards values found in the reference sites?

Composition of upper- and mid-stratum vegetation

The development of eucalypt dominance is a critical indicator of whether mine site restoration is progressing successfully on the Gove mine site. The basal area of the eucalypts developed approximately as a sigmoidal progression so that after 26 years it had reached a value of about 12 m² ha⁻¹ (Fig. 9.3); that is, within the range expected from reference site data (Reddell *et al.* 1993; Spain *et al.* 2009). Although the acacias developed as co-dominants with eucalypts for the first 10 years, their basal area

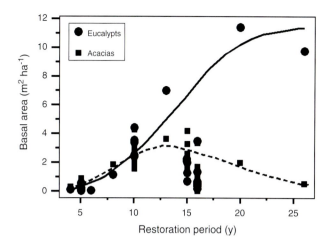

Figure 9.3: Trends in basal areas for eucalypts and acacias over a 26-year chronosequence of restored sites at the Gove mine site (plot means from Reddell *et al.* 1993 and Spain *et al.* 2009).

then declined to less than 1 m² ha⁻¹, which was also to be expected from reference sites. After 26 years, restored sites at Gove were therefore eucalypt-dominant in terms of basal area.

The two keystone eucalypt species *E. tetrodonta* and *E. miniata* are important components of all restored ecosystems (Brady and Noske 2010; Reddell *et al.* 1993). In terms of species composition, after 26 years the restored sites had a greater diversity and a more developed mid-stratum than the reference native forests (Brady and Noske 2010).

Trends in tree and shrub growth

Mean total tree and shrub basal areas increased with time at the restored sites (Fig. 9.4a) with a high correlation between basal areas and restoration periods for the 16 plots measured (r = 0.763, $P < 0.001$). Although the basal area at the longest restored site (26 years) exceeded those of the unmined native forest sites, most basal areas fell within those expected from native forest sites (Fig. 9.4b). The basal areas reported by Brady and Noske (2010) were slightly greater in the early years than those reported by Reddell *et al.* (1993) and Spain *et al.* (2009).

Over the 26-year chronosequence of restored sites, tree stem densities achieved high values

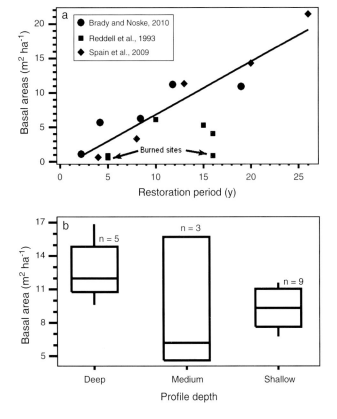

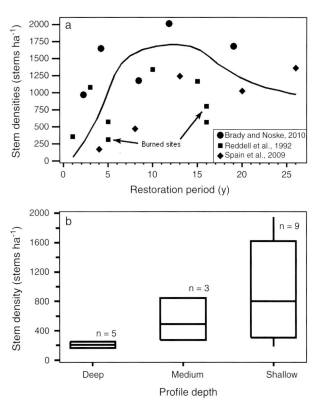

Figure 9.4: (a) Trends in total tree and shrub basal areas at 16 sites over a 26-year chronosequence of restored sites; and (b) at 17 native forest sites (medians, interquartile ranges, outlying observations) of three profile depths at the Gove mine site (data from Brady and Noske 2010; Reddell et al. 1993; Spain et al. 2009).

Figure 9.5: (a) Suggested trends in stem density with restoration period (Brady and Noske 2010; Reddell et al. 1993; Spain et al. 2009); contrasted with (b) stem densities (medians, interquartile ranges, outliers) in local native forests on soils of differing profile depths (Reddell et al. 1993; Spain et al. 2009).

(Fig. 9.5a) soon after canopy closure (9–11 years) and declined thereafter. Stem densities fell within the range of values found in local native forest sites (Fig. 9.5b), which ranged from 150 to nearly 2000 stems ha^{-1}, depending on soil depth. Over time, tree stem densities are likely to continue to decline due to deaths associated with firing and perhaps by self-thinning (Sea 2012). Tree canopy heights increased steadily following site restoration (Fig. 9.6a), with the eucalypts initially growing more slowly than the acacias. However, by 10 years, eucalypts overtopped the acacias, and were ~10 m tall by 20 years. Stem heights of mature trees in local native forest eucalypts appear to be related to profile depth and can exceed 25 m on deep soils (Fig. 9.6b). Nonetheless, heights of the eucalypts in the oldest restored forests were still at the lower end of the range of average tree heights found in local forests.

Burning may effect a substantial and differential reduction in basal areas between plant groups. The basal area of woody plants at an inadvertently burned site restored for five years was only slightly lower than that of an unburned site of the same restoration period; in contrast, a substantial difference in basal areas was apparent (Kruskal–Wallis test: $H = 6.82$, $P = 0.009$) between burned and unburned sites restored for 16 years, as assessed three years following inadvertent firing (see Fig. 9.4a). The inadvertently burned site had (at < 1 m^2 ha^{-1}) ~22% of the basal area of the equivalent unburned site (Reddell et al. 1993). For the eucalypts and acacias, basal areas at the burned site

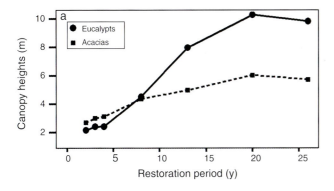

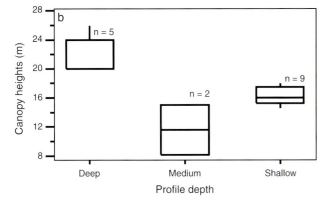

Figure 9.6: (a) Trends in the development of canopy heights for acacias and eucalypts with restoration period (Spain et al. 2009); and (b) in local forests of three soil profile depths (Reddell et al. 1993).

were 45% and 5%, respectively, of those at the unburned sites; for the remaining species, the basal area of the burned site was 80% of that in the unburned site. Despite these differences, total stem densities were not significantly different (Kruskal–Wallis test, H = 1.60, $P = 0.206$, $n = 5$ for both groups), due to the dense regrowth of *Grevillea pteridifolia* in the burned plot (Reddell et al. 1993).

Composition of the herbaceous stratum

As expected from native forest reference sites, perennial grasses dominated the herbaceous stratum throughout the restoration chronosequence. However, during the early years exotic perennial and annual grasses such as *Sorghum almum* and *Chloris gayana* tended to dominate the restored sites because they were sown as part of the seed mixture to assist in site stabilisation and initial soil development processes (Hinz 2007). These sown exotics gradually diminished in abundance over time, but were not totally replaced by tall tropical native grasses such as *Alloteropsis semialata*, *Heteropogon triticeus* and *Sorghum plumosum*.

Perennial grass abundance (measured as basal area) on restored sites of different ages was highly variable (Wedd 2002), but appeared to peak at three to four years before declining to lower values (Fig. 9.7). This trend was expected from reference site data and is consistent with the litter studies presented below; this peak-decline pattern would also be expected to occur as trees grow and mature. However, as open forest communities develop over time, and fires occur in these savannas, the compositions and relative abundances of the perennial tropical grasses will remain an important characteristic of these restored sites.

Post-establishment recruitment

Another useful indicator of ecosystem development is whether plant species, not included in the returned topsoil or in the seed mix broadcast sown at restored sites, have successfully established onto the restored sites. Such plant species are introduced through dispersal of their propagules by wind, water and by birds, bats and other mammals (Brady and Noske 2010; Hinz 1992, 1999, 2007 and unpublished information; Richards and Calder 1978; Woinarski et al. 2009). Some plant species have also

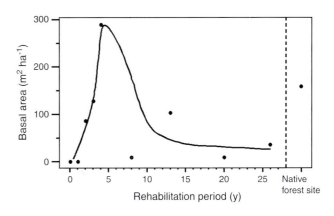

Figure 9.7: Suggested trend for basal areas of perennial grass tussocks across a 26-year chronosequence of restored sites at Gove and at a native forest site with a medium profile depth.

been introduced by humans, both deliberately (Richards and Calder 1978) and accidentally on vehicles or in association with other transported materials, as considered below.

In addition to the 39 species applied in the broadcast seed mixtures and the single species occasionally planted as container-grown seedlings (Hinz 1992, 1999, 2007 and unpublished), an undoubtedly incomplete total of 170 plant species (see Appendix 9.1 Table) has been reported from mostly unburned areas restored for periods up to 30 years (Brady and Noske 2010; Hinz 1992, 1999, 2007 and unpublished; Reddell *et al.* 1993; Richards and Calder 1978). This list contains 133 genera classified into 49 families. The plant families most commonly represented among species colonising the restored sites were the Poaceae (35 species), Fabaceae (32 species) and the Euphorbiaceae (13 species); the remaining families all contained less than 10% of the total species records.

It is likely that seeds contained in the replaced topsoil and those transported by birds or mammals (Scott *et al.* 2010) contributed to the initial establishment of a range of species, notably the acacias. This includes the single species of this genus, *Acacia gonocarpa*, present that was not included in the seed mix. Neither of the two keystone eucalypts, *E. tetrodonta* and *E. miniata*, are likely to establish from seeds transported in the topsoil: neither species retains seeds in the canopy and seeds do not survive in the soil beyond the first wet season (Setterfield and Williams 1996). Further, few of the eucalypts initially established appear to have originated vegetatively from underground organs transported with the topsoil on to the restored sites (see also Richards and Calder 1978). The sparseness of their establishment by both these means confirms the inappropriateness of relying on either method of propagation under this system of restoration.

Establishment of the next and subsequent generations of the keystone eucalypts is a critical indicator of sustainability. While all the eucalypt species in the restored sites had flowered or set fruit in sites restored for 10 or more years (Reddell *et al.* 1993), it is likely that regeneration of subsequent generations of the eucalypts will occur largely through vegetative means (see, for example, Setterfield and Williams 1996).

Apart from the sown species, other species introduced to the restored areas during agricultural experimentation (Richards and Calder 1978) have persisted after its abandonment (see also Mack and Lonsdale 2001). While some sown species such as *Axonopus fissifolius* may not have survived, others such as *Cenchrus ciliaris*, *Panicum maximum* and *Urochloa mossambicensis* have persisted in the restored sites; further species such as *Calopogonium mucunoides* and *Vigna sinensis* have persisted in disturbed industrial sites where an enhanced water supply exists. As elsewhere, the exotic woody leguminous species *Leucaena leucocephala* has become a notable weed on the mine site, requiring ongoing control. Finally, other species (e.g. *Themeda quadrivalvis*) are likely to have been introduced on vehicles and perhaps other equipment.

Cultural importance

Many of the native species planted in, or colonising, the restored areas possess value for aboriginal peoples, including the traditional Yolngu owners (Yunupingu *et al.* 1995); this may be in terms of providing materials for practical use and for medicinal and cultural reasons. For example, sheets of the bark of the major canopy species *E. tetrodonta* are used for construction purposes (as covering for the roofs and walls of shelters), for the hull materials of canoes, to make containers for carrying solids and liquids, and for paintings (D. Hinz, unpublished). Didgeridoos are made from the termite-hollowed branch and stem wood of both *E. tetrodonta* and *E. miniata* (Werner *et al.* 2008). Infusions of the leaves of *E. tetrodonta* are used as an antiseptic and a source of vapours to stop post-parturition bleeding (D. Hinz unpublished). Both species are sources of honey and the seeds are eaten.

Several of the species naturally colonising the restored environments provide edible fruits and other foods (Hinz 1999). Two food plants include the nutritious Bush Potato, *Eriosema chinense*, which can be harvested all year round, even during the dry season when only 30 cm high dried stems are visible. Two further yams (*Discorea bulbifera* and *Discorea transversa*) are also found in the restored areas (Hinz 1999).

Animal colonisation

As with all but a few mines (Cristescu *et al.* 2012), little published information is available on usage of the restored areas by most animal groups beyond simple records of the presence of several species and higher taxa. Published studies of invertebrates colonising these areas have been limited to preliminary studies of ants and termites. Unpublished records exist of sporadically occurring populations of unidentified pigmented earthworms present seasonally at low densities (up to *c.* 100 individuals m^{-2} in the 0–0.05 m soil layer) in several sites restored for five and more years (Reddell *et al.* 1993). Among the vertebrates, birds are the only group to have received systematic attention (Brady and Noske 2010; Woinarski *et al.* 2009) although unpublished records of other vertebrates are indicated by Woinarski *et al.* (2009). A feature of the few studies conducted to date is the occurrence of regular changes in community structure and the increasing biodiversity that occur with restoration period as more niches become available in these immature and rapidly developing ecosystems.

Ants

The important roles fulfilled by ants in the restored environments of the Gove mine may be attributed to their predatory and related interactions with other organisms, their influences on the survival and dispersal of seeds and their specific interactions with plants (Hölldobler and Wilson 1990). They are also likely to play a part in soil development (Paton *et al.* 2003), although it is currently unknown to what extent this occurs directly or through their effects on other animal groups (such as termites) that more directly impact on the soil environment.

Ant community composition and species abundance have been widely used as general indicators of ecosystem condition because of their correlations with other ecosystem components. Such measures have frequently been used to infer restoration success and biodiversity in mine sites and natural environments in Australia and elsewhere (Andersen and Majer 2004; Andersen *et al.* 1998; Andersen *et al.* 2004; Andersen *et al.* 2012; Majer *et al.* 2007).

The earliest work reported from restored environments in the Gove mine was that of Majer (1984a) who studied colonisation by ants of environments restored for periods of up to 7.5 years at this and two other tropical Australian strip mining operations. This work indicated that ants rapidly colonised the restored areas during the first 1.5 years, achieving in this period 43 to 68% of the species richness found at the longest restored sites. The rate of colonisation subsequently slowed although species richness attained values similar to those found in adjacent unmined forests in sites restored for 7.5 years.

Majer (1984a) found a successional trend in the composition of the ant community that was related to the changing character of the ecosystems present: the initial communities of soil-nesting generalists and some seed feeders giving way to shrub- and tree-associated species at the longer-restored sites. However, even at the longest restored sites studied, species composition differed from that of the adjacent unmined forest. Majer (1984a) considered that many of the species normally associated with litter and dead logs had not yet colonised the site due, particularly, to the absence of the latter resource. As considered below, the litter layers formed in the longer-restored sites are quantitatively and qualitatively dissimilar to those found in the local native forests; in the absence of fire, it seems likely that such contrasting resources imply ongoing differences between the ant communities of the restored and unmined forests. From Andersen *et al.* (2012), it also appears likely that, under a continued fire exclusion regime, ongoing ant diversity will remain lower in the restored forests than in adjacent regularly burned areas; this is expected to be due to the probably lower representation of arid-adapted taxa in this densely vegetated environment and the favouring of generalised more shade-tolerant species, including the fire-sensitive weaver ant, *Oecophylla smaragdina*.

Although other organisms also influence the fates and distributions of seeds, ants are considered the predominant post-dispersal seed predators in the region (Andersen and Lonsdale 1990). Many ant species disperse seeds, particularly of plants that possess arils or other structures that make them attractive as food sources to ants (and to birds). The majority of these movements take place over relatively short distances, with a mean seed dispersal distance over several species of 3.04 m in the experiments of Parr *et al.* (2007); distances generally increased after firing. This is rather greater than the mean distance of seed dispersal estimated at 1.25 m for the southern hemisphere (Gómez and Espadaler 1998). Nonetheless, the frequency distribution given by these last authors was highly right skewed with distances extending to 77 m.

Majer (1984b) conducted experimental studies of seed removal by ants in the restored environments at the Gove mine and in the two other mines considered above. The seeds of *Grevillea* species possess aril-like structures and were removed more frequently than those of the arillate *Acacia* species; in contrast, lesser amounts of the non-arillate seeds of two *Eucalyptus* spp. were taken. The seeds of *E. tetrodonta* (and also *E. miniata*, S. Setterfield and A. Andersen pers. comm. October 2011) may be subject to extensive predation in natural environments (Andersen and Lonsdale 1990; Setterfield and Williams 1996) and Majer (1984b) also reported removal of *E. tetrodonta* seeds in a five-year restored environment at the Gove mine. Grass seeds are widely available in the first few years following restoration and were extensively utilised by a particular suite of ant species (Majer 1984a). Majer (1984b) provided anecdotal evidence of a reduced density of the seedlings of *Grevillea* species in newly seeded areas close to previously restored sites and attributed this to seed removal; a lesser edge effect may also occur adjacent to native forest areas.

An additional impact has arisen at the Gove mine (and surrounding areas) in north-eastern Arnhem Land due to the presence of the invasive ant species *Anoplolepis gracilipes*, the 'yellow crazy ant', at restored sites. This ant species has been spread widely throughout the world by man from its African or Asian origins (Holway *et al.* 2002). It is considered to be among the 100 top invasive species in the world (Lowe *et al.* 2000) and has been reported to cause substantial damage to many components of native and agricultural ecosystems wherever it has invaded. It is a generalised feeder and forages actively on vegetation and on the ground. It forms extremely large colonies (supercolonies), often displacing important (keystone) native species (including other ants, reptiles and invertebrates), attacking wildlife and other vertebrates and disrupting natural ecosystem processes such as ant-plant mutualisms (Lach and Hoffmann 2011).

Anoplolepis gracilipes has been known from a range of environments (including unmined native forests) in north-east Arnhem Land for some years (Hoffmann and Saul 2010). It was initially reported from restored sites at the Gove mine by Majer (1984a) and has since been recognised as a major pest species in the restored areas of the mine (Hoffmann 2007; Rio Tinto Alcan 2009). Hoffmann (2010) reported successful attempts to eradicate infestations from several locations external to the mine. Other invasive ants are known to occur in the vicinity of the Gove mine and these may comprise sources for invasion or reinvasion following control efforts (Hoffmann 2011).

Termites

As major decomposers of a range of dead plant materials and effective soil modifiers, termites play important roles in nutrient cycling and soil development and maintenance throughout the world's tropics and subtropics (Lavelle and Spain 2001). In particular, they are considered to process much of the world's dead wood (Cornwell *et al.* 2009) although grass and litter feeding termites also occur widely in most tropical terrestrial Australian environments. Termites are also known as 'ecosystem engineers' due to their capacity to modify their own environments and to create habitat for other organisms within decomposing woody materials, the soil and through their construction of near-surface mounds and other structures (Bignell 2006). A few termite species also attack living plants, reducing their vigour or killing them, depending on the severity of their attacks.

Previous studies published on termite activities at the Gove mine include those of Hinz (1997, 2000), Reddell *et al.* (1992, 1993) and Spain *et al.* (2010). The most recent of these studies extended over a 26-year chronosequence of sites and included comparisons with native forest reference sites.

Strategies represented

The termites reported from restored sites at the Gove mine represent a range of ecological strategies. Table 9.2 lists the species reported so far from these sites (Spain *et al.* 2010); it is certainly incomplete.

The ecological groupings not so far observed in the restored areas include the grass-harvesting species of the genus *Drepanotermes*, those species of the '*Termes*' group (Watson and Gay 1991) that feed on highly humified soil organic matter and the widespread Australian species *Coptotermes acinaciformis*. This last species is commonly associated with eucalypt tree species outside the restored areas of the Gove mine where it creates central pipes within the trunks of these trees, leading to reduced growth rates and impaired longevity (Werner and Prior 2007; Werner *et al.* 2008).

Temporal sequence of colonisation

Termites progressively colonise the restored environments from the time of soil placement, initially being transported within woody materials and perhaps in soil clods brought on to the restored areas. Subsequent colonisation is likely to occur as a consequence of annual mating flights although the territories of some species may be sufficiently large (Watson and Gay 1991) to extend from the edges of undisturbed areas into the restored sites.

Table 9.2. Termite species reported from restored sites at the Gove mine (Spain *et al.* 2010), with data on feeding habits from Lee and Wood (1971), Watson and Gay (1991) and other sources

Species	Food habits
Amitermes laurensis, *Amitermes vitiosus*	Grass litter, animal faeces
Ephelotermes melachoma	Decomposing wood, often at soil interface
Heterotermes vagus, *H. venustus*	Decomposing wood
Mastotermes darwiniensis	Highly polyphagous, including faeces, sound and decomposing wood, living plant tissues
Nasutitermes graveolus	Sound and decomposing wood, plant detritus
Nasutitermes longipennis	Polyphagous, including dung, grass, wood
Schedorhinotermes breinli, *Schedorhinotermes actuosus*	Wood
Tumulitermes sp.	Poorly known, grass and herbaceous species
Xylochomitermes melvillensis	Decomposing wood, often at soil interface

Species richness was lower than recorded from unmined Northern Territory environments (Braithwaite *et al.* 1988).

The wider range of termites likely to be found in the adjacent unmined environments and the short distances of the restored sites from local native forests suggest that termite species from the local area could readily disperse onto the restored sites through their seasonal mating flights. While dispersal may be successful, it is likely that the persistence of species from some trophic groupings colonising the restored sites will be limited by the absence of particular feeding or nesting resources at specific developmental stages or by the changing environmental conditions that occur within these immature ecosystems (Spain *et al.* 2010).

Within the first year following restoration, general feeders on decomposing wood (including the highly polyphagous species *M. darwiniensis*) are active within woody materials transported onto the site with the topsoil. Litter harvesting *Tumulitermes* sp. occurs in the soil within three years following restoration. Populations of the mounds of the litter harvesting species *A. vitiosus* appear sporadically at different sites in the third year following restoration and may occur locally at densities of up to *c*. 200 mounds ha^{-1}. This species persists until close to the time of canopy closure but dies out contemporaneously with the decline of grasses and the supervention of a forest environment.

Mounds of *Microcerotermes* spp. were present at sites restored for four or more years and those of *E. melachoma* and *X. melvillensis* from *c*. five years although these last two species may colonise sites earlier where they appear to be feeding on decomposing transported wood. The characteristic arboreal nests of *N. graveolus* were initially found at sites restored for 20 years although it is also likely that this species colonises such sites at an earlier stage in ecosystem development. Colonies of this species initially establish in the soil but later move their nests into locations in the crowns of the trees; nonetheless, they always retain contact with the soil (Hill 1942).

As shown by the temporal sequence of intensity of attacks on paper baits exposed at sites of different restoration periods, 'effective' populations occurred at sites restored for periods of *c*. three or more years but showed little subsequent increase over the remaining years of the chronosequence. Substantial differences occurred between replicate plots of the same restoration period.

Termites as potential pests of restoration

Few termite species have caused significant damage to the plants of the restored environments, at least during the early stages of ecosystem development considered here. The damage noted to date has been due to *Mastotermes darwiniensis* and has comprised sporadic, minor attacks on a few plants; damage is localised and of no ecological significance. Nonetheless, this species has the capacity to cause severe damage, especially under conditions of plant stress (Cowie *et al.* 1991) or when a large source of potential food material becomes available (Lenz 1994). Severe damage to experimental tree plantings on restored tailings ('red mud') ponds has been observed at the Gove mine, probably due to the stress caused by the saline sodic materials used as a growing medium and perhaps exacerbated by the substantial volumes of wood chips employed as mulch at the site.

As considered above, attack by *Coptotermes acinaciformis* on eucalypts in unmined native forests in the Northern Territory leads to the formation of central pipes in the stems. This type of attack was not apparent in the restored areas but may develop following firing where the fungal colonisation of burn scars may make the trees more susceptible to termite attack (Perry *et al.* 1985), or when the first generation of eucalypts begins to senesce.

Effects on soils

In the restored sites, the importance of termites in soil development follows from their physical and enzymatic breakdown of dead plant materials and the incorporation of their breakdown products into the upper soil layers. Termites also redistribute and modify soil materials, form voids in the upper profile, build near-surface mounds and other structures at and above the soil surface.

Termites of many types attack dead plant materials including plant parts damaged by other agencies, dead wood and other litter lying on the soil surface or buried within the profile. Many species transport food and constructional materials to nests located within the soil through a near-surface network of subterranean galleries, storage chambers and nests. Inorganic soil components are admixed with faeces and other secretions that comprise the unique materials that occur in their mound materials, subterranean nests and in the organic matter-rich linings of galleries and other voids. Over long periods, termites are responsible for depositing substantial amounts of soil material on the surface in the form of the feeding covers built over food materials, as mounds or as soil packed within trees; this process eventually leads to finer textured surface soils and buried stony materials (see, for example, Holt and Lepage 2000; Tardy and Roquin 1992).

At the Gove mine, gallery systems are often associated with the substantial quantities of decomposing woody materials transported with the topsoil and distributed largely throughout the upper parts of the reconstructed soil profiles. Termites often nest within such buried materials. At sites restored for less than one year, near-surface gallery systems were found to have developed under several the larger laterite fragments, branches and logs found at the soil surface. As indicated in more detail below, voids constructed by termites progressively comprise substantial and increasing volumes within the near-surface soils. Termites, and thus their galleries, were found at maximum depths of 0.4–0.5 m in a site restored for 16 years (Spain *et al.* 2010). Where soil depths exceed this, the maximum depth of termite activity is also likely to be deeper: Miller *et al.* (1996) found that several of the wood feeding termites considered here (including *C. acinaciformis*, *M. darwiniensis* and species of *Schedorhinotermes* and *Heterotermes*) were able to penetrate to at least 0.7–0.8 m to attack wooden baits buried in natural soils.

As indicated above, termite mounds may be locally populous in the restored areas and this implies substantial physical and chemical modification of their associated soils. Such changes are most marked beneath and close to the mounds and occur in addition to the subterranean galleries and other voids built by non-mound building species.

Vertebrates

Native vertebrates are an important component of natural ecosystems and provide a wide range of ecosystem goods and services, both generally and in the Gove environment (Hinz 1992, 1999; Woinarski *et al.* 2009). Secondary forests of all types (including those growing on such modified substrates as the incipient soils created after mining restoration) have also been recognised as having conservation importance (Woinarski *et al.* 2009), with their value in this sense increasing commensurately with their state of ecosystem development.

Published information on vertebrate populations and community structure at the Gove mine is largely limited to studies of bird colonisation of the restored sites (Brady and Noske 2010). However, Woinarski *et al.* (2009) make reference to unpublished studies of vertebrates conducted within restored environments at this mine.

Birds are significant components of ecosystems and provide important seed spreading, pollination, predation and other services (Whelan *et al.* 2008; Woinarski *et al.* 2007). Birds, fruit bats and other mammals provide important pollination services to the *Melaleuca* species (Beardsell *et al.* 1993; Ford *et al.* 1979) and to the eucalypts, including the keystone canopy species, *E. tetrodonta* and *E. miniata*. Previous records of bird species using the resources of the restored sites are those of Hinz (1999, 2007) and Brady (2008); further casual records (D. Hinz unpublished) exist of such species as the Australian bustard (*Ardeotis australis*), brolgas (*Grus rubicunda*), the orange-footed scrub fowl (*Megapodius reinwardt*) and various raptors. Birds may also be significant predators of the seeds of *E. tetrodonta*, *E. miniata* (Setterfield and Williams 1996) and other species (Brady and Noske 2010).

On the basis of a two-year study of bird abundance and community composition along a 24-year chronosequence of sites, Brady and Noske (2010) found that both species richness and abundance increased evenly and approached those of the native forest sites after restoration periods of 15 to 24 years. However, during this period, bird community composition changed in ways that were related to the development of the structure and function of the plant community.

For the first three years, the bird community was dominated by open country species although this altered as grasses declined and the coverage of shrubs and small trees increased, with a transition to those occurring after canopy closure. Over longer restoration periods, the post-closure communities became increasingly similar to those of local unmined native forests but retained significant differences from them, even at the longest restored site. Such differences were attributed to the absence of fire, which has allowed a greater development of the shrubby mid-stratum which possesses its own characteristic complement of bird species. A further difference is related to the lack of nesting resources available for tree hole nesting bird species at this relatively early stage of ecosystem development.

In tropical Australia, a range of mycophagous and other mammals have been reported to be effective dispersers of the propagules of ectomycorrhizal fungi in their faeces (Reddell *et al.* 1997), although they may be less effective in the specific environments represented by the Gove mine (Reddell *et al.* 1999). Animals may also disperse propagules of the arbuscular mycorrhizal fungi associated with the grasses and many other plants of the herbaceous strata of the restored sites; however, propagules of these organisms are also likely to have initially been transported onto the restored sites with the topsoil (Jasper *et al.* 1987; Jasper *et al.* 1989).

Casual observations of vertebrates in the restored areas of the Gove mine (D. Hinz) have included a range of native and feral animals. These include reptiles such as lizards including goannas (*Varanus* sp.), frill neck lizards (*Chlamydosaurus kingii*), snakes (including the death adder, *Acanthophis* sp.), and amphibians (including frogs and the introduced toad, *Bufo marinus*). Among the mammals, the echidna (*Tachyglossus aculeatus*) (Hinz 1999), agile wallabies (*Macropus agilis*), native 'rats' (*Melomys burtoni*) and the black flying fox (*Pteropus alecto*) have been observed in the restored areas. Feral animal species reported from the restored areas include water buffalo (*Bubalus bubalis*), cats (*Felis catus*), dingoes (*Canis lupus dingo*), hybrids between domestic dogs (*Canis lupus familiaris*) and dingoes, horses (*Equus caballus*) and cattle (*Bos primigenius*).

In a general summary of vertebrate communities in variably disturbed habitats in the Northern Territory, Woinarski *et al.* (2009) indicated that regrowing forests were intermediate in species composition and richness between the cleared areas and intact native forests. In these properties, the communities of disturbed areas increasingly resembled those of local intact forests as canopy height (and thus time of development) increased. By the time that canopies reached 9–12 m, there were few differences between the more developed secondary communities and the variable and disturbance prone native forests. However, as found elsewhere (Bowen *et al.* 2007), a major difference between the restored and unmined forests was the few tree hollow-associated species present in the former. It is concluded that some years further will be needed before sufficient hollows are present in the major eucalypts of the restored sites to support such species.

Roots and mycorrhizas

Roots are major components of the developing soils and, in most plants, act in conjunction with several fungal groups to form mycorrhizas. Such associations are, in the ecological sense, necessary to promote normal growth and development of the vegetation.

Roots

Roots perform the essential functions of acquiring nutrients and water and of anchoring plants within their growing media. Roots also perform several other ecosystem functions – grass roots in particular play a role in stabilising the soil surface. Where trees are present in the landscape, their roots provide a considerable reinforcement of the upper profile and are of particular importance in surface and slope stabilisation (Ali 2010; Fattet *et al.* 2011; Gray and Leiser 1982; Leavitt *et al.* 2000). Roots are an important factor in soil development processes and the progression of root penetration with time into virgin substrates alters such critical properties of incipient soils such as hydraulic conductance (DeJong *et al.* 2015). Roots may also 'lift' water from deeper in the profile and leak moisture into the surface soils (and perhaps the litters) through a process known as hydraulic lift

(Brooksbank *et al*. 2011; Pang *et al*. 2013; Prieto *et al*. 2012). They also contribute to the development of soil organic matter status through root necrosis (Pinno and Belanger 2008) and stimulate microbial activity through rhizosphere carbon flow (Jones *et al*. 2004). Hence the progression of root growth at Gove is of vital importance in understanding how the restored sites are performing.

Little is known of the root ecology of *E. tetrodonta* forests (Janos *et al*. 2008) and only a few studies have considered their root dynamics (Chen *et al*. 2002; Chen *et al*. 2004; Eamus *et al*. 2002; February *et al*. 2012; Janos *et al*. 2008). Data from these studies are variable and inconsistent although there is persuasive evidence that seasonal fluxes in fine root dynamics, with a wet season peak and a dry season trough, are characteristic of these forests (Chen *et al*. 2002; Janos *et al*. 2008). Both the finer (February *et al*. 2012) and coarser lateral roots seem more stable and populous in surficial layers.

The development of roots is fundamental to the success of restoration schemes and in this tropical ecosystem both climate and plant diversity combine to promote rapid development of roots. Through a chronosequence of sites restored from 1–15 years, fine root lengths (per unit volume of soil) increased rapidly for the first three years, after which they became stable (Reddell *et al*. 1993). In the restored soils root densities tended to decrease with depth (partitioned into three 0.10 m intervals to 0.30 m). In contrast, in the native forests, root density was greatest between 0.10 and 0.20 m (Fig. 9.8).

Fine roots in the restored sites were always more superficial and maximum root lengths always occurred in the 0–0.10 m layer, although after 15 years differences were smaller and the maximum root length was in the 0.20–0.30 m depth range. Root lengths were of similar magnitude in the native forest and restored sites, although the structuring with depth was somewhat different. This probably relates to the nature of the developing forest where grasses, forbs, shrubs and trees all occupy the surface layer in the early years of development. Changing plant species composition in a

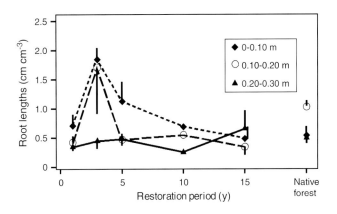

Figure 9.8: Volumetric fine (< 2 mm diameter) root length at three depths in five restored sites and in an unmined native forest site ($n = 5$, except for the 10-year restored site where $n = 1$; vertical bars = se).

redeveloping ecosystem was clearly reflected in the development of root density. This is supported by an analysis of roots of grasses (Reddell *et al*. 1993), which could be differentiated from those of dicotyledonous plants (Fig. 9.9). Grasses dominated during the first five years, followed by a steep decline between five and 10 years to densities approaching those of the native forest sites. This change is related to the suppression of grasses with the closures of the canopy and the litter layer that occur between about 9 and 12 years after restoration, and the emerging dominance of the *Eucalyptus* spp. No major changes in the relative

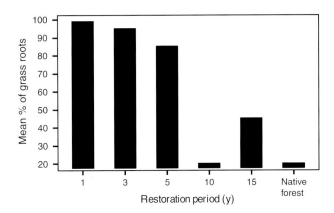

Figure 9.9: The percentages of grass roots in samples of fine roots (< 2 mm diameter) over depths of 0–0.30 m in five restored sites and in one unmined native forest site, April 1992 (Reddell *et al*. 1993).

proportions of the roots of grasses and dicotyledonous species were recorded at longer restored sites nor did the proportions differ among the depths sampled at any restoration period.

At sites restored for three years or more and in the native forests, two major classes of coarser roots (> 2 mm diameter) occur: sinker and lateral roots. Sinker roots penetrate deeply below the profile and their role is primarily to acquire water from deep profile reserves. These roots include the main tap roots although they may also originate from lateral roots. In these lateritic environments such roots may be deflected by the indurated mine floor or lateritic rock (Fig. 9.10), although many penetrate to access water below these layers (Reddell *et al.* 1993; Szota *et al.* 2007). Lateral roots explore the soil profile but remain concentrated near the soil surface. They are more superficially distributed in the restored than in the native forest sites where they could occupy much of the available space.

The abundances of medium (2–5 mm diameter) and coarse (> 5 mm diameter) roots were assessed on the walls of a series of excavated soil pits; roots were counted within areas of 2 m long in 0.10 m depth increments down the soil profile for sites ranging in restoration age from four to 26 years. These were compared to root abundances in a native forest site (Fig. 9.11). In the early years, the finer roots developed rapidly: first in the upper profile and subsequently at depth. The proliferation of fine root density at three years (Fig. 9.8) is reflected in the superficial layers in the counts of medium and coarse roots present at four years (Fig. 9.11). By year 20, a wide distribution of root sizes was observed at depth and in the more superficial layers, and at levels that greatly outnumbered the native forest in all but the coarsest of roots. By year 26 the root distribution had begun to settle into a pattern more representative of the native forest, where a wide distribution of roots occurs throughout the profile and coarser roots dominate the surface layers. This observation is in keeping with a previous study conducted in natural soils (Eamus *et al.* 2002).

The progression of root development shown here is different to others reported in landfill or mine sites in temperate zones (Handel *et al.* 1997) and where restrictions exist due to chemically hostile conditions (Gwenzi *et al.* 2011) or the presence of capping materials (Handel *et al.* 1997). In tropical forests such as these, maximum root depth might be expected to extend for many metres below the extent of our observations (Canadell *et al.* 1996; Jackson *et al.* 1996), where deep rooting is particularly important during the dry season to allow access to water stored deep below the profile.

Figure 9.10: The tap root of *E. tetrodonta* at a site restored for a period of eight years (note the lateral deflection of the root at the mine floor).

Mycorrhizas

Mycorrhizas are symbioses between plant roots and fungi in which a compatible exchange of

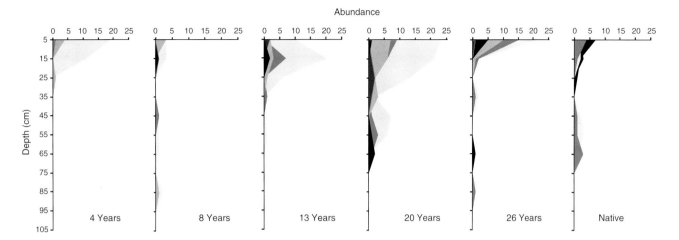

Figure 9.11: Depth distributions of medium (2–5 mm diameter) and coarse roots (> 5 mm diameter) enumerated from 0.10 m depth increments of a 2 m length of pit wall over a series of sites restored from 4–26 years of age compared to a native forest site. Darker shading represents coarser roots and lighter shading represents roots of finer diameters: ▫ = 2–5 mm, ▨ = 5–10 mm, ▦ = 10–20 mm, ▪ = 20–30 mm, ■ > 30 mm.

solutes occurs between the host plant and certain partner fungi (Harley 1989). The symbiosis may benefit both partners, but this may not always be the case and the symbiosis exists in a range of states from mutuality to parasitism (Johnson *et al.* 1997). Nonetheless they are considered important and are generally considered to be beneficial to the restoration of eucalypt forests (Jasper 2007).

Mycorrhizal symbioses occur in almost all plant species growing in the area of the Rio Tinto Alcan Gove mine and in most other tropical Australian post-mining environments (Reddell and Milnes 1992; Reddell *et al.* 1999). The fungi penetrate the fine roots, either surrounding but not penetrating roots cells (ectomycorrhizas or ECM) or internal to the cell walls, though still external to the plasmalemma (endomycorrhizas and, in particular, arbuscular mycorrhizas (AM) in this study). In ectomycorrhizal symbiosis, the fungal hyphae typically sheath the fine root tips of woody species with mycelium forming a mantle, and hyphae grow between the epidermal and cortical cells of the roots forming a covering called a Hartig net. This process usually leads to bifurcation, shortening and swelling of the root tips, or sometimes to tuberculate clumps of root tips. The morphology of these mycorrhizal structures formed in this way may be used to identify taxa. The AM association is the most common and widespread root symbiosis (Veresoglou *et al.* 2012) and in this symbiosis the hyphae grow between and within the root cells and form intracellular organs, known as arbuscules, which lend their name to this symbiosis. They form spores in the roots and soil (sometimes in large clusters on the soil surface) that may be used to identify taxa. In both kinds of mycorrhizas, the fungi produce large quantities of hyphae that ramify through the soil effectively increasing the absorptive area of the roots and consequently enhancing nutrient element and water uptake (Cairney 2012; Smith and Read 2008). Both forms of mycorrhizal association are common in the restored environments at the RTA Gove mine and in the unmined native forests. Much current knowledge of mycorrhizas at the Rio Tinto Alcan Gove restored sites is based on one previous study (Reddell *et al.* 1999).

It is likely that the chances of normal growth and reproductive success for plant species not possessing mycorrhizas, or other appropriate specialisation, would be small. These symbioses are responsible for much of the uptake of nutrient elements, particularly P, by the host plant (Smith and Read 2008). Mycorrhizal fungi can also extract water from soil with pores that are too fine for roots to exploit, or too dry for root uptake, thus being very important in enhancing plant survival under dry conditions (Allen 2007). There is now also evidence that mycorrhizas may be necessary for the long-term sustainability of restored ecosystems due to their interactions with other components of the soil biological community and through their roles in the maintenance of soil structure and plant community structure and diversity (Bever *et al.* 2010; Koide and Mosse 2004). Indeed, it has been recently proposed that the mycorrhizal symbiosis may be a major force in structuring plant communities (Klironomos *et al.* 2011).

Ectomycorrhizas
Ectomycorrhizas are symbioses formed by mutualistic fungi with a wide range of woody plant species and, in Australia, particularly with eucalypts and *Melaleuca* species, although some may also be colonised by AMF. This mycorrhizal type is important in plant N and P nutrition (e.g. Tibbett and Sanders 2002) and the fungal communities taking part are thought to change with plant successional stages in developing forest ecosystems (e.g. Chu-Chou and Grace 1982; Frankland 1992). Previous reports indicate that these fungi are poor survivors of mining activity and it may take many years for them to return unassisted to reforested ecosystems in comparable abundance and richness (Brundrett *et al.* 1996; Gardner and Malajcuk 1988; Glen *et al.* 2008; Reddell and Milnes 1992), and that their community compositions may be quite different from those of unmined forests (Tibbett 2010).

At the Rio Tinto Alcan Gove mine, ectomycorrhizal fungi form symbioses with a range of important tree species, including *E. tetrodonta, E. miniata* and several *Acacia* species (Reddell and Milnes 1992). The development of healthy ECM fungal communities is likely to be essential for the sustainable restoration of native ecosystems and to optimise the growth and nutrition of forests (Baxter and Dighton 2001).

Ectomycorrhizal root tips were enumerated from woody roots removed from soil in 0.10 m increments up to 0.30 m depth at five sites (Fig. 9.12) throughout a 15-year chronosequence (Reddell *et al.* 1993). Colonised tips were common in the surface soils of the native forest and their presence on roots declined sharply with depth; they were not found on the roots at depths of 0.20–0.30 m, either in the native or restored sites (see Dickie *et al.* 2002). Ectomycorrhizas were first found in the three-year restored site in the surface soils and colonisation continued to increase up to 15 years but remained at 50% of the root length, around half that of the native forest site. This slow initial colonisation by ectomycorrhizal fungi has previously been reported from bauxite mines elsewhere in Australia (Gardner and Malajczuk 1988). Given the importance of the symbiosis to the nutrition of woody plants, it is possible that levels of colonisation remain suppressed, possibly due to the P fertilisation of this otherwise P-limited system.

An important functional feature of the litter layers is the development of mycelial mats (Kluber *et al.* 2010) in the lower litters of the restored sites, close to the interface with the mineral soil. These mats comprise macroscopically visible masses of fungal mycelium and have been recorded as common at sites restored for periods of five and more years (Reddell *et al.* 1993). They are also common in undisturbed native forest communities dominated by *E. tetrodonta* and *E. miniata* where they have been reported to cover up to 26–35% of the soil surface (Reddell *et al.* 1999). The mats commonly present in the litters of the Gove mine are

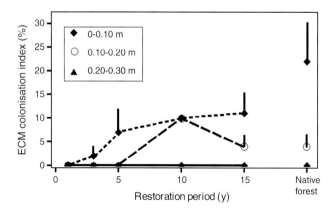

Figure 9.12: Colonisation of lateral roots by ectomycorrhizal fungi at three depths in five restored sites and in an unmined native forest site at site ($n = 5$, except for the 10-year restored site where $n = 1$; vertical bars = se).

predominantly produced by ectomycorrhizal fungi of the genera *Nothocastoreum* and *Hysterangium* although mycelia of saprotrophic fungi are also present. The roles of these mycorrhizal mats are unclear although they are considered to be of importance in plant nutrition partly because of the saprotrophic capacities of the fungi that produce them (Cairney 2011, 2012; Chapman *et al.* 2006).

Arbuscular mycorrhizas

Arbuscular mycorrhizas are an intimate symbiosis between roots and arbuscular mycorrhizal (glomeromycotan) fungi (AMF), which are ubiquitous in most terrestrial ecosystems. Arbuscular mycorrhizas are considered important in the restoration of P-limited soils due to their ability to enhance host plant uptake of P and other nutrient elements (Cardoso and Kuyper 2006; Standish *et al.* 2007), particularly where populations have been disrupted by mining (Jasper *et al.* 1989). Arbuscular mycorrhizal fungi affect host plants in other ways which include enhancing tolerance to drought and providing a degree of disease resistance (Koide and Mosse 2004). Of the plant families examined, 80% form AM symbiosis and this symbiosis is prevalent in tropical forests (Treseder and Cross 2006).

Arbuscular mycorrhizal structures were visualised and enumerated (Reddell *et al.* 1993) by clearing the fine roots, staining with trypan blue and using standard counting techniques (Bevege 1968) (Fig. 9.13). They were most common in near-surface roots in the native forest although the percentage of root length occupied declined sharply with depth and colonisation occurred only sparsely between 0.20 and 0.30 m depth in both the native and restored sites. In the latter sites, AMF were found to have colonised roots after one year and this increased to a maximum in the upper 0.10 m at five years but declined at longer restored sites. At greater depths, root colonisation increased at sites restored for up to 15 years but still remained lower than that in the native forest site. Colonisation declined with depth but (unlike ECM) occurred regularly in the 0.20–0.30 m depth range.

Arbuscular mycorrhizal fungal spores were extracted from samples taken in 2002 (Spain *et al.*

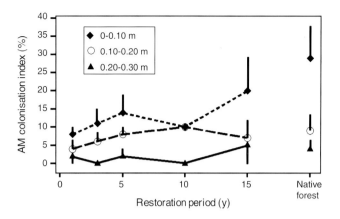

Figure 9.13: Colonisation of lateral roots by arbuscular mycorrhizal fungi at three depths in five restored sites and in an unmined native forest site ($n = 5$, except for the 10-year restored site where $n = 1$; vertical bars = se).

2009) using wet sieving and sucrose density gradient centrifugation (Daniels and Skipper 1982). The numbers of spores extracted in single 50 g soil samples are shown in Fig. 9.14 for each of the four restored sites sampled. This gives an indication of the numbers, sizes and diversities of spores extracted for sites of differing age. Figure 9.14 is linked to Table 9.3 which lists the distinct spore morphotypes found by accession number and identification, where possible. The systematic taxonomic scheme of Schüßler and Walker (2010) is followed here. Where possible, names have been given for the fungi, but doubt is indicated by use of 'cf'. Nevertheless, most of the morphotypes distinguished are likely to represent distinct AMF species. Both species' complexity and abundance of spores in the juvenile restored soil remained initially very low (after two years) but reached a maximum after 13 years. These data are based on single soil samples but the implications are intriguing: spore diversity seems to match floristic succession and the maximum diversity of spores occurred where floristic succession is in a transition phase when herbaceous perennials, re-sprouting and re-seeding trees are all strongly represented. In the 26-year restored site, spores were less abundant.

The two-year restored site produced only two spore types, each with similarity to species that are found commonly throughout the world: a small ornamented spore similar to *Acaulospora paulinae*

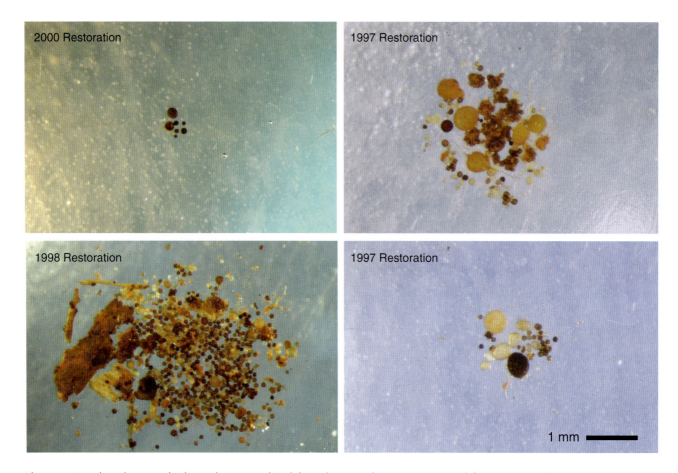

Figure 9.14: Abundances of arbuscular mycorrhizal fungal spores from sites restored for two years (2000 restoration), six years (1997 restoration), 13 years (1988 restoration) and 26 years (1975 restored) (all shown at the same magnification).

and a colourless small glomoid spore that is probably a *Paraglomus* sp., similar to *P. occultum*. Low spore abundance is in keeping with the finding of Gardner and Malajczuk (1988) who reported typically fewer than five spores per 50 g of soil in samples from the years immediately after restoration of Jarrah (*Eucalyptus marginata*) forest (in a Mediterranean climate). The six-year restored site produced eight types, primarily of one undescribed species. The abundance and species richness was higher than for the two-year restored site. For the 13-year restored site there were 18 species with high spore abundance and the species found here were not reported from the younger sites. There were several undescribed species that may be new to science. Finally, the 26-year restored site produced both common and undescribed species. The seven species found here are in keeping with the richness expected in a mature undisturbed forest (Table 9.3).

The AMF species found include several that are previously unrecorded from Australia, including some that appear not to have been found since their original descriptions as new to science. Most of the existing named species found are common and have been reported numerous times. For example, the group of organisms containing *Glomus intraradices* and *Glomus aggregatum* are amongst the most widely reported of all species.

From this preliminary work it can be hypothesised that AMF, as represented by spores, accumulate over a period of time after initial re-establishment of the topsoil, but then decline again as the large woody perennial eucalypts close the canopy and become dominant; further replicated work is required to establish if this is really

Table 9.3. Arbuscular mycorrhizal fungal spores recorded in single soil samples taken from four sites restored for different periods

Restoration year (age) Species richness	2000 (2 years) 2	1997 (6 years) 9	1988 (13 years) 18	1975 (26 years) 7
Count				
1	*Acaulospora* cf. *paulinae*	*Glomus* sp. (possibly *G. formosanum*)	*Glomus* sp. (possibly *G. formosanum*)	*Scutellospora* cf. *aurigloba*
2	*Paraglomus* cf. *occultum*	Non AM sclerotium	Undescribed species 1 with glomoid spores	cf. *Acaulospora* sp.
3	–	*Gigaspora* cf. *albida*	*Sclerocystis sinuosa*	*Glomus* sp.
4	–	cf. *Scutellospora erythropus* (?)	Possibly *Glomus magnicaulis*	*Scutellospora* sp.
5	–	Possible undescribed *Scutellospora* close to *S. cerradensis*	*Glomus clarum*	*Glomus intraradices/aggregatum* complex.
6	–	cf. *Glomus albidum*	Possibly an undescribed *Ambispora* sp.	*Acaulospora* sp.
7	–	*Acaulospora scrobiculata*	*Archaeospora trappei*	Unknown
8	–	*Paraglomus* cf. *occultum*	*Paraglomus* sp.	–
9	–	*Sclerocystis* sp. (dead spores only)	Possibly *Glomus perpusillum*	–
10	–	–	Possibly an undescribed *Ambispora* sp.	–
11	–	–	Possibly *Scutellospora alborosea*	–
12	–	–	Probably *Acaulospora tuberculata*	–
13	–	–	*Acaulospora excavate*	–
14	–	–	Undescribed species 2 with glomoid spores	–
15	–	–	–	–
16	–	–	Probably *Claroideoglomus etunicatum*	–
17	–	–	Undescribed species 3 with ornamented glomoid spores	–
18	–	–	Possibly *Funneliformis geosporus*	–

so. It is possible that non-sporulating ecotypes of AMF exist among woody perennials. If so, then the evidence from spore extractions will not represent the full spectrum of glomeromycotan species in many mature ecosystems, and molecular techniques could be used to provide a more complete evidence of their biodiversity and recovery. Further work is needed to verify species names and it is possible that some of the undescribed species could be useful in restoration of this and other mine sites.

Development and properties of the litter system

Above-ground litters are the primary source of the energy expended by organisms involved in decomposition processes and, with root litters, in many soil formation processes. Where substantial, litters represent a major source of soil organic matter, nutritionally significant stores and sources of elements critical to plant growth, soil and ecosystem development (Thompson 2011). Well developed litters may contain considerable biomasses of fine roots and complex communities of invertebrates and saprotrophic and symbiotic microorganisms. As indicated above, mycelial mats located in the litter layers and produced by ectomycorrhizal fungi are likely to play a direct role in the nutrition of the major tree species.

In conjunction with the fine roots of plants and sometimes surface cryptogams, litter layers physically protect the underlying soil from the erosive effects of raindrop impact and the entrainment of soil particles in overland flow or in wind currents. They therefore provide an important protection for the surface against accelerated erosion. Regular litter removal through firing alters the physical, chemical and biological environments of the near-surface mineral soil and exposes it to the oxidative loss of organic matter; litters thereby influence nutrient dynamics and other components of soil fertility. Firing may also modify the distributions of near-surface roots and affect their associated mycorrhizal fungi (Cairney and Bastias 2007; McMullan-Fisher et al. 2011); firing is also known to reduce the biomass (Dooley and Treseder 2012) and functional diversity of the microbial community (McMullan-Fisher et al. 2011). It may therefore be expected to influence decomposer processes and nutrient cycles in fundamental ways.

Substantial litter layers may develop in post-mining environments (Tibbett 2010) and this may be due to the higher litter fall rates consequent on fertilisation and perhaps to the limited development of decomposer communities. The latter in turn may lead to the lower decomposition rates often associated with the earlier phases of ecosystem development. As in native forest environments, the presence of substantial litters modifies physical, chemical and biological conditions within the newly restored soil materials, as it does in forests present on natural soils. They therefore contribute positively to early soil development processes, nutrient cycling and fertility maintenance in post-mining environments. Nevertheless, such benefits must be balanced against the potential fire hazard posed by the substantial litters that may form during early ecosystem development in these environments.

Studies of coverage, masses and other litter properties have been made in the restored ecosystems at the Gove mine by Brady and Noske (2010), Reddell et al. (1993), Spain et al. (2009) and by Cook (2012). These studies were conducted on sites restored for up to c. 28 years and results were compared with similar measures made in nearby native forest sites. As indicated below, litter coverage values form part of a system of restoration assessment verified at the Gove mine (Tongway and Ludwig 2011).

This section addresses the development of surface coverage by plant litter and herbaceous plants. It presents the physical composition of the litter and the development of layers that form post-canopy closure, addresses their overall mass per unit area and aspects of its chemical composition. Seasonal variation in several of these properties is also addressed. No information is available on the magnitude or dynamics of root litter, the mass of fine dead roots produced each year.

Coverage

Total coverage of the mineral soil surface by plant litter and herbaceous plants was estimated visually in both restored and native forest sites. Total ground cover and litter cover were estimated separately and herbaceous plant coverage then determined by difference (Spain *et al.* 2009). Figure 9.15 presents the mean total coverage of the soil surface based on sampling conducted in the mid to late dry season (Brady and Noske 2010; Reddell *et al.* 1993; Spain *et al.* 2009). Dry season total surface coverage increased rapidly over the chronosequence of sites from zero at sowing to *c.* 93% soon after canopy closure (Fig. 9.15), with effectively complete coverage occurring at sites restored for longer periods. In sites restored for periods of 20 years or more and in the native forest sites, variation within sites was low due to complete or near-complete coverage of the soil surface. Native forest sites had more variable coverage values, depending on both season and time since last firing (Cook 2003).

Litter is the main component of total ground cover (Fig. 9.16). In the late dry season (October 2002), mean litter coverage expressed as a percentage of total cover (litter plus herbaceous plant coverage) was 94% in the first year following rehabilitation to slightly less than 100% of total coverage in sites representing the remaining years of the chronosequence. In the post-wet season sampling (May 2002), litter coverage as a percentage of total cover increased from values of *c.* 7–21% for sites rehabilitated for two years to *c.* 90% at canopy closure. During these first two years, litter coverage was lower (*c.* 55%) although total coverage remained above 65% for sites rehabilitated for periods greater than two years, due to the increased plant cover that occurs at the end of the wet season.

Following litter closure, the soil surface is effectively protected from the direct effect of raindrop impact. Surface coverage studies conducted in pastoral environments (see, for example, McIvor in McIntyre *et al.* 2004; Roth 2004) have indicated that coverage values greater than *c.* 65–75% are associated with higher site stability and healthy soil biological functioning. At the Rio Tinto Alcan Gove mine, this highly functional level of litter cover was achieved at sites restored for approximately three to four years. As considered elsewhere in this paper, infiltration rates are extremely high in the soil materials of this mine and it is clear that acceptable levels of surface stability and biological activity are achieved much earlier in the restoration sequence and at lower surface coverage values than in many other post-mining environments.

Living herbaceous plant cover changed seasonally with higher and more variable values during and after the wet season (Fig. 9.16). Values declined as the foliage of the grasses and other annual plants died off progressively during the dry season and was converted to standing litter (Cook 2003). The greatest wet season to dry season difference occurs

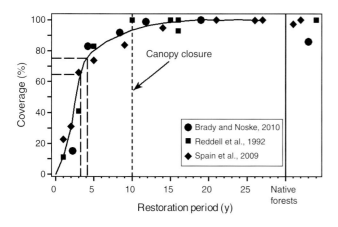

Figure 9.15: Trends in mean total dry season surface coverage values at sites restored for periods up to 26 years and in four local native forest sites (the range 65–75% of cover values at *c.* 3–4 years corresponds to the restoration period after which soil surface stability is putatively attained).

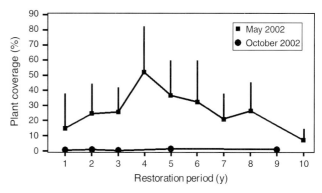

Figure 9.16: Seasonal changes in herbaceous plant cover (means and se, $n = 10$) in a chronosequence of restored sites in the post-wet season (May 2002) and in the late dry season (October 2002) when all standard errors were less than one.

at c. four years, during the period of maximum grass dominance of the herbaceous stratum.

Physical composition

As with coverage, litter composition also changed predictably with restoration period (Fig. 9.17). In the first few years following restoration, the late dry season litter formed an incomplete and diffuse layer that often exceeded 0.25 m in thickness and was dominated by standing dead grasses and forbs. A greater initial representation of acacia and later eucalypt foliage developed on restored sites with increasing restoration period. After canopy closure, litter largely comprised dead foliage originating from broad-leaved species; the proportional representation of acacia foliage subsequently declined while that of the eucalypts rose with increasing restoration period (also see George et al. 2010).

Following litter and canopy closure between 9–12 years, the litter mass increased progressively in depth and became differentiated into layers defined by colouration and the degree of leaf fracturing and decomposition. Litter depths develop to maxima of more than 0.14 m in the longer restored sites. At one site restored for 20 years, mean depths for three recognisable layers measured from the top downwards were 4, 2 and 2 cm respectively, representing progressively more decomposed materials with increasing depth.

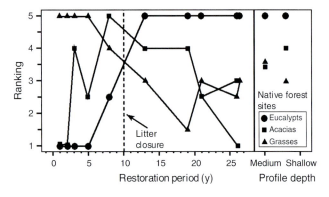

Figure 9.17: Temporal changes in the compositions of the litter layers (median ranks for foliage of eucalypts, acacias and grasses) with time at restored sites and in two native forests differing in soil depth.

Woody litter forms a very minor component of total litter in these immature ecosystems. In the first few years following restoration, woody litter was almost entirely confined to small twigs and the petioles of the shrubs and young trees. With increasing restoration period, a greater woody litter component was present in the form of fruits derived from the acacias; dead stem wood also started to be represented as the shorter-lived acacias commenced dying (see also Brady and Noske 2010). As the sites developed post-canopy closure, proportionately more woody litter derived from the eucalypts was found although this was still largely limited to twigs, small branches and woody fruits at sites restored for periods longer than eight to 11 years. Fine woody materials covered less than an estimated 1% of surface area up to the time of canopy closure but increased slightly to c. 2% at longer restored sites. The smaller branch wood of the dominant eucalypts became a more common and increasing component of the forest floor with time. However, few eucalypts died within the time frame of the present studies and, excluding substantial mortality, large branch and stem wood of these species is expected to contribute little to the mass of dead woody litter materials for an unknown period into the future, perhaps not until the first generation of eucalypts begins to senesce and die.

Fine litter mass estimates

Litter mass per unit area increased from effectively zero at sowing to estimated oven dry masses of c. 3.6 kg m^{-2} late in the dry season at sites restored for 26 to 28 years (Fig. 9.18a) (Cook 2012; Spain et al. 2009). Seasonal variation was apparent with lower masses occurring during the early post-wet season (Fig. 9.18b), as occurs in local native forests (Cook 2003). Litter accumulation may eventually approach an asymptote although more information is needed from longer-restored sites to confirm this.

The estimated litter masses post-canopy closure were substantially greater than (to more than three times) those found in local native forests (Figs. 9.18a, b). This is probably due to the higher litterfall rates (4–8 t ha^{-1} y^{-1}) recorded in unburned restored

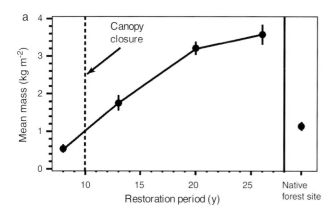

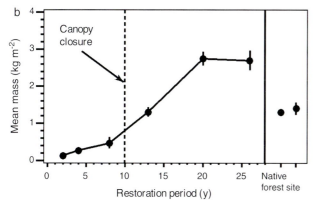

Figure 9.18: Litter mass (oven dry, means, se, $n=10$) accumulated in a chronosequence of sites restored after bauxite mining and in native forest sites in (a) late dry season (September 2001); and (b) in the post-wet season (May 2002).

sites in comparison with 2–4 t ha^{-1} y^{-1} in burned sites (Cook 2012). It is also due to the lower rates of decomposition: estimated values of the decomposition constant for the unburned restored sites were less than a third of those estimated for the unmined sites (Cook 2012). Such litter masses were also higher than the putative equilibrium value of 0.8–1.0 kg m^{-2} estimated for northern Australian savannas (Cook 2003; Williams *et al.* 2002) although substantially less than the value of 5.3 kg m^{-2} reported for a 20-year restored site at the Jarrahdale bauxite mine at 32° 20'S in West Australia (Collins 1996).

Litter chemical properties

Litter materials were very strongly acid, with mean litter pH (1:5, water) at three sites ranging from 3.9–4.9 (overall mean of 4.3) over sites restored from 8 to 26 years; there was no clear trend in pH with restoration period. At the one native forest site where litter pH was determined, mean litter pH was ranged from 3.8 to 4.2.

In the younger sites before canopy closure, total litter C concentrations (Fig. 9.19a) were high with values ranging from 320 to *c.* 480 g kg^{-1} C, the latter figure indicating minimal admixture of litter with soil materials. After canopy closure, concentrations generally declined with restoration period and ranged from 150 g kg^{-1} C in the lower layers of the litters of a 26-year-old site to 461 g kg^{-1} C in the upper layer of a 13-year-old site. Carbon concentrations in the upper litter layers were greater than in those underlying and differed little throughout the post-canopy-closure period. The lower litters also changed little until the last part of the current chronosequence of sites when considerable variability became apparent, perhaps due to patchy soil animal activity, the differing amounts of fine soil acquired as dust at each site and the greater capacity of the deeper litter layer materials to retain it. Carbon concentrations in the native forest sites were also variable, probably for the same reasons.

Nitrogen concentrations in the litters ranged from 2.7 to 11.2 g kg^{-1} over all samples (Fig. 9.19b). Before canopy closure, mean values in the grass-dominated litters of the restored sites were low (4.5 g kg^{-1}) for the first five years but increased to *c.* 9.0 g kg^{-1} in the site restored for eight years, probably due to the greater representation of acacia phyllodes. In sites restored for five or more years, N concentrations in the upper litter layer were slightly lower than in the combined lower layers. This may have partly resulted from the higher fungal biomass (possibly from the mycelial mats of ectomycorrhizal fungi) in the lower materials. Nitrogen concentrations in the lower layer declined overall with restoration period. Those in the upper layer were variable but altered little over the post-canopy-closure period.

Litter N concentrations in both layers of the sites restored for 20 or more years (combined mean for these sites was 6.5 g kg^{-1}) were slightly greater than

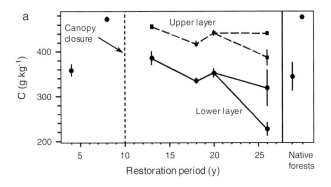
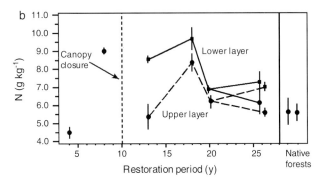
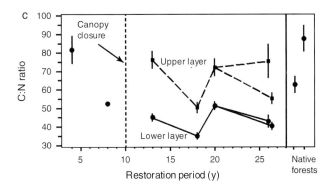

Figure 9.19: Concentrations (means, se, $n=5$ pre canopy closure, $n=10$ post closure) of (a) C; (b) N; and (c) the C:N ratio in the litter layers of a chronosequence of restored sites and at two native forest sites, late dry season (October 2002); note the differences between the samples taken from two separate sites restored for 26 years.

Excluding one sample with a very low N concentration, the C:N ratio ranged from 29 to 110 (Fig. 9.19c). The litter C:N ratio was notably high (mean = 82) in the grass-dominated litter at the site restored for four years, but was lower (mean = 53) in the acacia-dominated litters of the site restored for eight years and close to canopy closure. After canopy closure, there were no clear temporal trends in the C:N ratio with restoration period although combined values for the upper layer were greater (mean = 70) than those of the combined lower layers (mean = 43), due to the higher N concentrations in the lower layers. Following canopy closure, C:N ratios in the upper litter layers were similar to those of the local native forests while values in the lower layers were lower. The C:N ratios for the upper litter layers were slightly greater than those found in other tropical forests (typically 33–50, Lavelle and Spain 2001) but were similar in the lower layers (30–59).

Total concentrations of the essential element P lay in the range 90–310 mg kg^{-1} (mean = 182, se 7, $n=47$) in the grass and acacia-dominated litters before canopy closure and were similar to the concentrations (130–349 mg kg^{-1}, mean = 222, se 8, $n=49$) found in eucalypt-dominated sites post-canopy closure. These are also similar to the range of total P concentrations (110–240 mg kg^{-1}, mean = 170, se 6, $n=30$) found in litters studied at three unmined native forest sites within the Gove mine lease area. No temporal trends were apparent within the post-canopy-closure age range and concentrations were similar to those in the unmined native forests. One exception was a site restored for 26 years, which had a higher soil concentration (mean = 301 mg kg^{-1}, se 12, $n=10$). Total P concentrations reported here were largely within the lower part of the range (180–800 mg kg^{-1}) reported for other Australian eucalypt forests (Bevege 1978) and those reported for other tropical forests (100–14 400 mg kg^{-1}) (Lavelle and Spain 2001).

Considerable soil was visually apparent in the litter of the restored sites. This was confirmed by their Fe and Al concentrations with values commonly up to 1.0% and 1.5%, respectively (and a few

those of the unlayered litters of the two unmined native forests, mean 5.8 g kg^{-1} N. The N concentrations found at these longer-restored sites were close to the mean values reported by Bevege (1978) for Australian eucalypt forests (7.2 g kg^{-1} and were at the lower end of the range of values (7.0–22.0 g kg^{-1}) reported for other world forests (Lavelle and Spain 2001).

higher yet). These values were similar to those of local native forest sites but are greater than those reported by Bevege (1978) for native forests growing on natural soils elsewhere in Australia. The high Fe and Al concentrations in most litters are probably due to the retention of soil materials in the litter layers. This probably results from such processes as rain splash in the early years, the soil-transporting activities of termites (and possibly earthworms), deposition of dust transported onto the mined areas (Reed *et al.* 2011) or generated during mining operations, including soil replacement and cultivation procedures. Dust is also generated by wind erosion of the largely bare soil surfaces exposed for periods following site restoration.

Soil properties and development

Although this differs between soil orders, formation of fully developed soil profiles is usually considered to require periods of thousands to hundreds of thousands of years (see, for example, Birkeland 1984). The ancient soil materials of the unmined palaeosols of the Gove mining lease area have been subject to extended weathering and soil formation conditions over periods towards the upper end of the stated timescale. Such extended weathering has lead to high concentrations of residual elements such as iron and aluminium and, for the latter element, formation of the economically important indurated bauxitic horizons. Because of their highly-weathered status, these soils are low in weatherable minerals and in a wide selection of essential biological nutrient elements, particularly P (see, for example, Reed *et al.* 2011).

Despite the long-term overall nature of some aspects of soil development, early soil development is a basic part of ecosystem development and substantial changes may occur within pedologically short time periods in the artificial soil profiles successfully restored after mining, deposition of industrial materials and following major disturbances. While such changes have often been considered to be limited to the near-surface accumulation of organic matter, short-term soil development after restoration at the Gove mine has also included substantial changes to surface properties, ongoing reorganisation of the incipient profiles and the formation of soil structure, including both voids and aggregates. As flagged above, rapid and substantial increases in organic matter concentration certainly occur close to the surface although changes to the properties of most soil mineral components are likely to require long periods.

Previous information on soil development following bauxite mining in tropical Australian environments is limited to the 16- and 26-year chronosequences of sites at the Gove mine site described by Reddell *et al.* (1992, 1993) and Spain *et al.* (2009). Schwenke *et al.* (2000a, b) and Short *et al.* (2000) conducted similar chronosequence-based studies of soils at Rio Tinto Alcan's Weipa bauxite mining operation on Cape York Peninsula at sites developing under native forest ecosystems, pastures and a monoculture tree crop. Koch (Chapter 7) presents a range of soil properties from contrasting sites at the Alcoa bauxite mines in southern West Australia.

The studies discussed in this section examined the progress of soil development based on a 16-year chronosequence studied by Reddell *et al.* (1993) and seven sites studied within the 26-year chronosequence of sites available in 2001 (Spain *et al.* 2009). These studies included observations of profile morphology, field measurement of saturated infiltration and respiration rates, analysis of sampled soils and related materials for selected physical, chemical and biological properties (Spain *et al.* 2009). For many properties, comparisons were made with unmined native forest soils studied within the Gove mine lease area.

Surface properties

During the restoration process, ripping produces a pronounced ridge: hollow surface microrelief with a vertical elevation of *c.* 0.15 m and a horizontal interval of *c.* 1.7 m (O'Keefe 1992). A discontinuous surface layer (lag) of rounded pisoliths occurs at the surface variably covering up to an estimated 90% of the soil surface area. These structures may be mobile or partially embedded in the discontinuous weak vesicular surface crust (c. 0.5 mm and sometimes thicker) that forms on many exposed surfaces. Some biotic microrelief structures are present and other microrelief results from the laterite fragments that protrude through the surface after partial disinterment during the ripping process.

In sites restored for up to two to three years, and before the surface is stabilised by vegetation, pisoliths, fines, organic materials (and perhaps fertiliser) erode from the unvegetated ridges into adjacent riplines reducing the relief. Ward (2000) also found that concentrations of a putative measure of 'plant-available' P (bicarbonate-extractable inorganic P) were greater in the riplines of soils from a restored south-western Australian bauxite mine. This process leads to higher EC values and higher concentrations of C and N in these locations; subsequently initial plant colonisation may occur preferentially in the riplines. As indicated above, the soil surface is almost completely stabilised by litter and herbaceous plants beyond this period. No evidence of a surface crust was found in the sites restored for periods of 20–26 years although a weak surface crust was present at exposed soil surfaces in an undisturbed native forest site.

Several faunally created microrelief features were present at the surfaces of the restored soils. The most common of these were termite mounds, ant nests and vertical burrows.

Termite mounds were constructed by wood-feeding and grass and litter-harvesting species. The presence of populations of these structures implies a profound modification of the profile underlying and immediately adjacent to the mounds (Holt and Lepage 2000) and the construction of networks of subsurface galleries and other voids in the inter-mound soils (Lavelle and Spain 2001). At a smaller scale, termites build temporary structures comprising weakly cemented fine soil materials on the soil surface, under laterite fragments, over plant materials and on the stems of trees. These may occur as linear structures or occasionally broader sheetings and provide shelter for the foraging termites. Some types of these structures have a substantial organic component.

As indicated above, soil-dwelling ants rapidly colonise the restored areas of the Gove mine (Majer 1984a). Many nest in the soil and entry holes to their nests are a common feature of the soil surfaces. Some species nest near the surface producing characteristic areas of 'decoration' with small pebbles and multiple entry holes, seed-feeding species often discard husks around their nest entry holes. Vertical burrows c. 40 mm diameter and surrounded by debris cones up to 0.12 m in diameter were distributed more or less commonly across all but the most recently restored sites and are the surface expression of burrows dug by the 'King cricket' *Apterogryllus pedestris* (family Gryllidae) (Spain et al. 2009).

Profile properties

Profile depths in the seven restored soils examined ranged from 0.48 to 1.07 m (mean 0.82 m, se 0.09) above the variably indurated mine floor; this compares to a depth of 0.66 m to the bauxitic layer in the unmined native forest site used for comparative purposes. In the unmined lease area of the Gove mine, profile depths over the bauxitic layer ranged from 0–5.0 m although the average profile depth for the soils of the Gove mine lease area is 0.70 m (O'Keefe 1992).

In sites representing the early years of the chronosequence, the restored soil profiles were red to dark red (Munsell colouration mostly 2.5YR 3/6, moist soil) throughout. However, over time, darker colours developed in the upper profiles due to their increasing organic matter concentrations. At the 26-year restored site, soil colour in the upper 0.12 m of the profile was dark reddish-brown (Munsell colouration mostly 2.5YR 3/4), with a gradual transition again to dark red (Munsell colouration mostly 2.5YR 3/6) below this depth.

Except for the developing surface layer, all restored soils had massive structure throughout most of the profiles. Minor shrinkage cracking was occasionally observed in some profiles. Structural aggregates, including predominantly inorganic aggregates, rounded faunal aggregates and organo-mineral structures were found in the lower litters and in the near-surface layer at the longer-restored sites.

Below the pisolitic surface lag and at sites restored for more than about eight years, the upper layer of the mineral soil becomes spongy and loosely coherent due to the dense mass of fine roots that forms, reducing the bulk density. It comprised a layer *c.* 20 mm thick in profiles restored for eight years and by 13 years was up to 30 mm thick and gradually extended to greater depths at the longer-restored sites. This layer was rich in aggregates, especially close to the surface of the predominantly mineral soil. The soil underlying this layer showed considerable organic staining, which diminished with increasing depth. A similar concentration of fine roots *c.* 25 mm thick also occurred near the soil surface at the undisturbed forest site with a medium profile depth. As discussed above, coarser roots were concentrated close to the surface in the longer-restored sites, as they are in the unmined native forest site.

Within the first year following restoration, voids and organic structures attributable to termite and ant activities were occasionally found under surface laterite fragments and beneath wood transported onto the site during profile restoration (Spain *et al.* 2010). By the end of the second year, termite galleries were associated with grass roots at 0.22 m in the mineral soil and an active termite nest was found in buried wood at 0.23 m. Near-surface galleries at this stage were largely associated with decomposing logs but by four years following restoration they were regularly present to a depth of 0.12 m. At sites restored for eight or more years, networks of sub-horizontal voids (largely termite galleries) were ubiquitous (Reddell *et al.* 1993; Spain *et al.* 2010). Near-surface galleries were most common in the upper 2.0 cm of the mineral soil but extended regularly to 5.0 cm and, at the longer-restored sites, they frequently extended to depths of at least 0.14 m. Termites (and therefore their galleries), were found at a maximum depth of 0.40–0.50 m in the mineral soil in a site restored for 16 years (Spain *et al.* 2010). Ant populations and communities were not studied formally during the period 1991–2002 although the presence of their nests (indicated below) and their occurrence within near-surface galleries in the native forest sites indicates that they are likely to make some contribution to pedogenesis in these soils.

Buried wood and charcoal fragments were sparsely distributed mostly throughout the upper parts of the restored soil profiles and wood fragments sometimes protruded through the surfaces. Such fragments were generally small since most of the larger wood is taken off the site before removal of the topsoil materials that are to be transported to sites currently undergoing restoration.

Physical properties

Figure 9.20 presents the lower gravel (materials > 2 mm equivalent cylindrical diameter, or ECD) contents of the soil materials from the upper profiles of four native forest sites with deeper sola and contrasts these with materials from a solum with a medium profile depth, the most common type. The profiles of most restored sites had high percentages of gravel although some sites with lower gravel sections also occur, depending on the origins of the materials used to reconstruct the two major layers of the soil profiles. Figure 9.21 presents the frequency distribution of gravel concentrations in 61 samples taken throughout the restored soil profiles

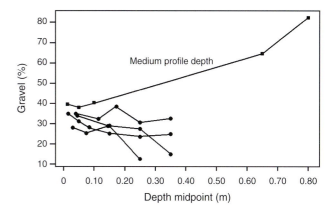

Figure 9.20: Depth distributions of gravels in four deep and one medium depth profile in the unmined native forest soils.

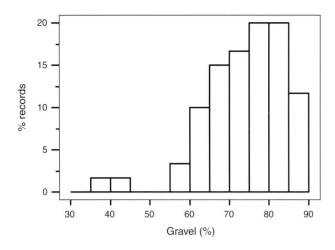

Figure 9.21: Frequency distribution of gravel percentages ($n = 61$) in the restored soils.

(Reddell *et al.* 1993; Spain *et al.* 2009). The major mode (median 75%) represents the frequency distribution of 59 samples representing the most common soil materials on the site, while the minor mode (median 41%) reflects the lower gravel concentrations originating from the smaller areas of deeper and finer-textured soils; materials from both sources are evident in some profiles.

Figure 9.22 presents boxplots of the different size classes of gravels (expressed as a percentage of sample mass) from restored soils sampled in 2002. The horizontal line in the centre of the box is the median while its upper and lower margins indicate the upper and lower quartiles; the limits delineated by the vertical lines indicate sample values lying within 1.5 times the interquartile range beyond the upper and lower quartiles; values located beyond these are outliers. Pisoliths largely fell within the sieve size range 2.0 to 4.75 mm ECD and were the predominant gravel-sized materials; the median value of these materials, 43% of sample mass, in the restored soil materials contrasts with 32% in the unmined native forest soil with a medium profile depth. The remaining gravels were largely fractured laterite fragments and medians for all larger gravel size fractions ranged from 1 to 6% of sample mass.

Schwenke *et al.* (2000a) also found high gravel contents (median 33%) in soils restored after mining at Rio Tinto Alcan's Weipa bauxite mine. As in soils studied at the Gove mine, gravel concentrations in the restored soil sites at Weipa were greater than those found in the upper horizons of the unmined sites.

Figure 9.23 presents the particle size distributions of the fine earth (< 2 mm ECD) fractions of 32 materials from the restored and native forest sites (Spain *et al.* 2009). Two soil materials from the site restored for four years appear as high outliers in the clay (< 2 μm) category and low outliers in the coarse sand (> 1 mm) category. The high outliers in the coarse sand fraction in Fig. 9.23 are five mine

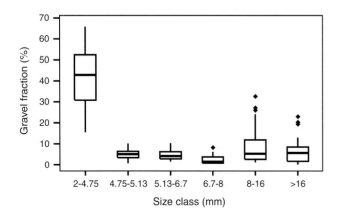

Figure 9.22: Boxplots of six size classes of gravels (as % of sample mass) in restored soils ($n = 31$) (see text for explanation of boxplots).

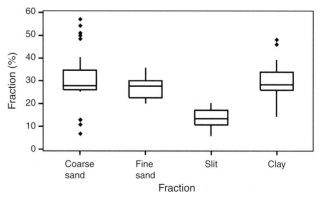

Figure 9.23: Boxplots of the components of the fine earth fraction ($n = 31$) (outlying values are indicated by solid diamonds).

floor materials. With the exception of these materials, the particle size distribution of the fine earth fraction differed only slightly among most restored soil materials.

The median texture of the fine earth (< 2 mm ECD) component of 61 restored soils reported by Reddell et al. (1993) and Spain et al. (2009) was a clay loam (62%). However, 23% of the materials were clays, 13% were loams and the remaining 2% of materials were sandy clay loams. The fine earth materials from the upper part of the variably indurated mine floor had higher coarse sand contents and were clay loams, sandy clay loams or sandy loams.

In the surface soils (0–0.10 m) of sites restored for five or more years, there is a significant negative correlation between the percentage clay ($r = -0.425$, $P = 0.017$, $n = 31$) and site restoration period. This indicates that clay is being lost from this layer either by illuviation or erosion (Reddell et al. 1993). The very high infiltration rates and low relief of these landscapes suggests that this is likely to occur through the former process and is part of the ongoing profile reorganisation that occurs at these sites.

The predominantly kaolinitic clays that occur in these soils (Reddell et al. 1993) are likely to occur as micro-aggregated compound structures 5–300 μm in size, cemented strongly with – and perhaps coated by – iron oxides (Cornell and Schwertmann 2003). This effectively produces a fine sandy structure, which contributes to the very high infiltration rates characteristic of these soils (Soil Survey Staff 1999).

Near-surface bulk density values of the restored soils studied at the Gove mine were c. 1.7 mg m^{-3} near the surface and ranged from 1.4 to 2.4 mg m^{-3}; values were higher at greater depths in the profile (Spain et al. 2009). There is some indication in the data of a small reduction in near-surface values at longer-restored sites and this may result from bioturbation due to termites and other animals and to the turnover of fine roots. Bulk density was generally greater in the restored profiles than in the native forest soils at both the Gove mine and at Rio Tinto Alcan's Weipa bauxite mine (Reddell et al. 1993; Schwenke et al. 2000a).

Infiltration rates

Saturated infiltration rates were measured using a constant head disc permeameter. Studies were carried out at eight sites ranging in restoration period from one to 26 years and in an unmined native forest soil of medium profile depth (Fig. 9.24) (Tongway 2010; Tongway and Ludwig 2011).

The major feature of these data was the extremely high saturated infiltration rates that had developed over the initial 20 years of soil formation ($r_{\text{age: infiltration rate}} = 0.790$, $P < 0.001$, $n = 60$). Although only minor net change occurred in the first four years, infiltration rates were still high, in the order of some metres per day. The higher infiltration rates reported for a few sites restored for one and two years are associated with site differentiation between bank and trough microsites and the lower infiltration rates measured at the four-year restored site are probably a consequence of its finer-textured fine earth materials and low gravel content. After c. four years, a phase of rapid and near-linear increase in infiltration rates leads to a possible asymptote at ~20 years. As indicated in Fig. 9.24, this apparent asymptote is somewhat higher than values

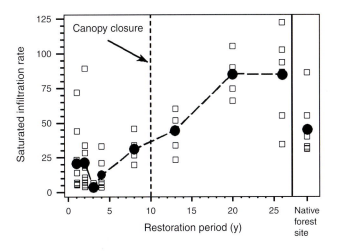

Figure 9.24: Saturated infiltration rates (means and individual observations, m day^{-1}) measured at eight restored sites and in a native forest site with a medium profile depth (Tongway 2010).

measured in the undisturbed native forest site with the medium profile depth.

While the soils at all sites studied were extremely permeable and normally drained rapidly following even the heaviest rainfall events. Nonetheless, ponding was observed at a limited number of restored sites after heavy rainfall events and it is likely that a hydraulic 'choke' layer was present or had formed at such locations.

In the restored soil profiles of the Gove mine, the highly elevated infiltration rates are considered to result from a combination of the substantial gravel contents of most materials (Poesen and Lavee, 1994), micro-aggregation (Cornell and Schwertmann 2003) and probably the development of preferred drainage pathways (Beven and Germann 1982). These pathways result from the effects of root penetration and invertebrate activity although the deep ripping process also creates preferred pathways, which contribute to the development of rapid drainage to depth (Kew et al. 2007).

Beven and Germann (1982) considered that substantial macropore systems may form in natural soils within one to two years. By the end of the latter period, the grasses, shrubs and trees of the restored sites are well established, with rapidly developing root systems and substantial populations of termites (Spain et al. 2010) and ants (Majer 1984a).

Chemical and related properties

As indicated above, extended weathering has led to high concentrations of the residual elements Fe and Al in these soils. Over 89 soil materials analysed from restored and unmined environments at the Gove mine (Reddell et al. 1993; Spain et al. 2009), median Fe and Al concentrations were, respectively, 109 g kg^{-1} and 160 g kg^{-1}; both concentrations are more than three times the respective average concentrations estimated for world soils (Kabata-Pendias 2011).

The restored Gove soils were slightly acid to neutral in reaction. Over the 79 materials analysed by Reddell et al. (1993) and Spain et al. (2009), pH measured in a 1:5 suspension of restored soil in 0.01M CaCl$_2$ (pH CaCl$_2$) ranged from 4.89 to 6.15 with a median value of 5.53. Samples from the upper part of the mine floor were in the higher part of this range (5.68 to 6.15, median 6.12). These values are comparable with the range of 5.23 to 6.20 found in three local undisturbed native forest soils. All materials sampled were within the benign range for plant growth.

The pH CaCl$_2$ is dependent on the soil organic matter concentration of the restored soils (Fig. 9.25) and was very highly significantly correlated with the C concentration (r = −0.398, $P < 0.001$, $n = 116$). While the native forest soils from the site with a medium profile depth followed the same relationship, most soils from the unmined site with the deeper profile (Fig. 9.25) had a higher pH CaCl$_2$ than their organic matter concentrations would indicate (Reddell et al. 1993).

Both the restored soils above the mine floor layer and the native forest soils are non-saline throughout. Over 79 samples, electrical conductivity measured in a 1:5 suspension of soil in water (EC 1:5) ranged from 0.01 to 0.13 dS m^{-1} (Reddell et al. 1993; Spain et al. 2009). Based on the EC 1:5 and adjusted for the dominant clay-loam, clay or loam texture (Marcar and Crawford 2004), the maximum estimated paste EC was 1.56 dS m^{-1}; this low salinity value is unlikely to cause stress to even

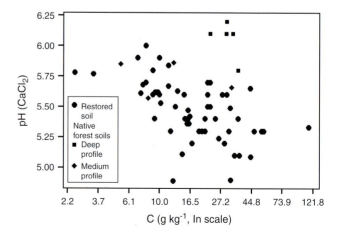

Figure 9.25: Relationship between pH CaCl$_2$ and total C concentration in soils of the restored sites and in two unmined native forest soils.

the most salt-susceptible plants (Marcar and Crawford 2004).

Consistent with their kaolinite and gibbsite-dominant clay mineralogy (Reddell *et al.* 1993), cation exchange capacity (CEC) is low in these materials (Spain *et al.* 2009). The presence of Fe and Al oxide minerals indicates that these soils have a substantial variable charge component. Further, the charge associated with the soil organic matter is entirely variable and the overall net charge on the soil is expected to vary with the pH, the soil solution strength and the soil organic matter concentration.

Over all materials from the restored sites (Spain *et al.* 2009), CEC ranged from 0.47 cmol kg^{-1} in the mine floor layer of the site restored for one year to 30 cmol kg^{-1} in the highly organic (114 g kg^{-1} C) near-surface layer (0–1.5 cm) of the site restored for 26 years. Values for near-surface soil (0–0.10 m) materials ranged from 2.2 to 30 cmol kg^{-1} and were positively correlated with restoration period (r = 0.665, P = 0.007, n = 16). Much of the CEC increase in the 0–0.10 m range occurred close to the surface, notably between sites representing restoration periods of 20 and 26 years.

For the soil materials at depths greater than 0.10 m, the median CEC value was 3.67 cmol kg^{-1} and for the mine floor materials was 1.8 cmol kg^{-1}. For these two groups of materials, the CEC values were not significantly correlated with restoration period (respectively, r = –0.219, P = 0.571, n = 9; r = –0.117, P = 0.802, n = 7). The overall median CEC for soil materials at depths greater than 0.10 m was 3.40 cmol kg^{-1} while that for the mine floor materials was 1.8 cmol kg^{-1}. Over all soil materials above the mine floor, the CEC was strongly positively correlated with the C concentration (r = 0.897, P < 0.001, n = 24).

As illustrated by the high correlation between CEC and C concentration for all soil materials above the mine floor, at very low soil organic matter concentrations soil CEC is largely governed by clay and oxide mineral concentrations but as soil organic matter concentrations rise with restoration period, this increasingly controls the CEC.

The ratio of exchangeable Ca^{2+} to exchangeable Mg^{2+} was slightly greater than unity in the restored soils and slightly lower than this value in the mine floor layer. Exchangeable acidity was low throughout all profiles, consistent with their pH status.

Soils are considered sodic if the exchangeable Na^+ percentage (ESP, expressed as a percentage of the effective CEC) is greater than six (Hazleton and Murphy 2007). Values for 24 soil materials sampled throughout the restored profiles (Spain *et al.* 2009) ranged from 0.7 to 6.96 (median 2.66) and all these materials except one are non-sodic. In contrast, ESP values for the mine floor materials ranged from 2.58–20.00 (median 8.75) and these materials were sodic at five of the seven restored sites; ESP in the mine floor increased significantly (r = 0.901, P = 0.006, n = 7) with restoration period although ESP values of the soil materials were not significantly (r = 0.020, P = 0.927, n = 24) related to restoration period.

Phosphorus is a key nutrient element that widely constrains plant growth and microbial processes (Elser *et al.* 2007) and may control net primary production (Cleveland *et al.* 2011). It is supplied to plants and their symbiotic mycorrhizal organisms almost entirely from the soil and through nutrient cycling processes associated with organic matter decomposition. Total P concentrations are low in Australian soils compared with those occurring in younger landscapes (Norrish and Rosser 1983; Wild 1958) and commonly limit plant productivity in both agricultural and natural environments. Norrish and Rosser (1983) estimated a typical concentration of total P in Australian topsoils to be 550 mg kg^{-1} and 380 mg kg^{-1} in the subsoils. It seems likely that continuing productivity in these highly weathered soils may result, at least partially, through the acquisition of dust (Reed *et al.* 2011).

At the Gove mine, total P concentrations in the soil materials of the native forest sites were generally lower than indicated above, commensurate with their highly weathered status (Reed *et al.* 2011; Vitousek *et al.* 2010). Concentrations of total P in the native forest soil with a medium profile depth declined from 220 mg kg^{-1} at the surface (0–2.5 cm) to 179 mg kg^{-1} in the upper part of the bauxitic layer (Spain *et al.* 2009). Concentrations in the surface 0–0.10 m layer of an unmined site with a deep soil profile ranged from 220 mg kg^{-1} to 260 mg kg^{-1} total P (Reddell *et al.* 1993). They were slightly greater in the 0–5 cm layer of an unmined site with a shallow profile where they ranged from 340 mg kg^{-1} to 480 mg kg^{-1} (Spain *et al.* 2009).

Total P concentrations were determined on 31 materials sampled throughout the seven profiles of the 26-year chronosequence of restored soils and in the upper part of the underlying mine floor (Spain *et al.* 2009). Combined with the materials studied by Reddell *et al.* (1993), total P concentrations in 41 materials from the 0–0.10 m layer of the soils in the restored soils ranged from 218 to 511 mg kg^{-1} (median 310). Concentrations of total P in this depth range increased with restoration period (r = 0.413, P = 0.008, n = 40), excluding one outlying value. This increasing P concentration is likely to result from incorporation of litter and decomposed litter products and from the turnover of near-surface fine roots and microbial biomass (Jobbágy and Jackson 2004). Concentrations in the 11 soil materials located below 0.10 m but above the mine floor were slightly lower and ranged from 140 to 230 mg kg^{-1} (median 210). Total P concentrations in the seven upper mine floor samples were slightly lower again and ranged from 170 to 240 mg kg^{-1}, with a median value of 200 mg kg^{-1}. The total P concentrations of 24 soils sampled throughout the restored profiles from sites restored for 1 to 26 years (Spain *et al.* 2009) were highly correlated (r = 0.809, P < 0.001) with the natural logarithm of total C concentration, indicating the presence of substantial organic P.

Due to the strong and specific adsorption of P on iron oxides and their slow and partly irreversible desorption, (Cornell and Schwertmann 2003), very little of the total P present in these iron and aluminium-rich soils is likely to be available to plants and microorganisms (Cross and Schlesinger 1995; Yang and Post 2011). As indicated above, a widely used measure of 'plant-available' P is the inorganic P extracted by 0.5M Na$_2$HCO$_3$ (bicarbonate Pi) (Tiessen and Moir 2008). In the 26-year chronosequence of restored soils and in the upper part of the underlying mine floor, concentrations of the sum of resin-extractable P and bicarbonate Pi ranged from less than detectable to 19 mg kg^{-1} in the highly organic surface (0–1.5 cm) layer of a site restored for 26 years. However, 61% of 31 restored soil materials sampled throughout the reconstructed profiles contained less than 1 mg kg^{-1}; values in the remaining materials were sporadically higher and were found mostly in the upper part of the longer-restored soils. Additionally, a slightly elevated concentration was also found in the upper part of the ripline trough of the one-year restored site perhaps indicating concentration of fertilizer P in this location. In the native forest soil with a medium profile depth, the bicarbonate Pi concentration in the upper 2.5 cm was 1.2 mg kg^{-1} with values less than detectable below this depth interval. The bicarbonate Pi values presented above are very low in comparison with those considered necessary for agricultural production (Moody and Bolland 1999).

Properties reflecting soil organic matter concentrations

In these highly weathered soil materials, soil organic matter is particularly important to soil development processes both as the dominant source of biologically important nutrient elements and through its substantial negative charge, which may largely enable cation retention, particularly in the upper soil profile. It is also a component of soil stability through its contributions to the formation of structure and of organo-mineral complexes. Further, it is the predominant source of energy for many of the soil maintenance and development processes mediated by the soil biota.

In contrast to that of fully developed soils, much of the organic matter in the upper layers of these restored soils is young and the soils contain minimal char. Soil C concentration at sites one year after restoration was c. 10 g kg^{-1} C throughout the profile and the difference between that concentration and those of the older sites indicates that most of the C present in the profiles must therefore be less than the age of the longest-restored site, 26 years. The substantial quantities (of the order of 14 t ha^{-1}) of C present as char near the surface in Australian forest soils (Graetz and Skjemstad 2003) have been diluted throughout the upper profiles of the restored soils by mixing during profile reconstruction, except where accidental burning of restored sites has occurred.

Over studies of the restored soils (Reddell *et al.* 1993; Spain *et al.* 2009), total C concentrations ranged from *c.* 7 g kg^{-1} in the early years of the chronosequence of sites in the upper 0.10 m of the restored profiles (Fig. 9.26a) to 114 g kg^{-1} in a narrow-interval (0–1.5 cm) sample from the upper part of the mineral soil at the 26-year restored site. Similarly, N concentrations over the 0–0.10 m depth interval (Fig. 9.26b) ranged from 0.30 to 5.01 g kg^{-1}. The C:N ratio (Fig. 9.26c) ranged from 13 to 43; it showed a slight decline from a mean value of 27 in the sites with restoration periods less than 10 years (canopy closure) to 23 in the subsequent years and values were lower than those found in the native forest soils. No depth variation was apparent in the restored soils below 0.10 m although values in the underlying mine floor layer ranged from 23 to 40, with a mean value of 32. In the native forest sites, the C:N ratio ranged (from 23 to 47, exceptionally to 150) in the surface 0–0.10 m interval but declined with increasing depth to 5 to 17 deeper in the profile.

Both total C and total N concentrations (Figs. 9.26a, b) in the 0–0.10 m surface depth interval increased significantly with restoration period although showing wide variation within sites ($n = 50$):

$$\ln(\text{total C}) = 2.490 + 0.048.\text{Years}, R^2 = 33.3\%$$
$$(\text{se } 0.010)$$

$$\ln(\text{total N}) = -0.811 + 0.062.\text{Years}, R^2 = 49.2\%$$
$$(\text{se } 0.009)$$

In calculating the above equations, values included for the sampling conducted at sites

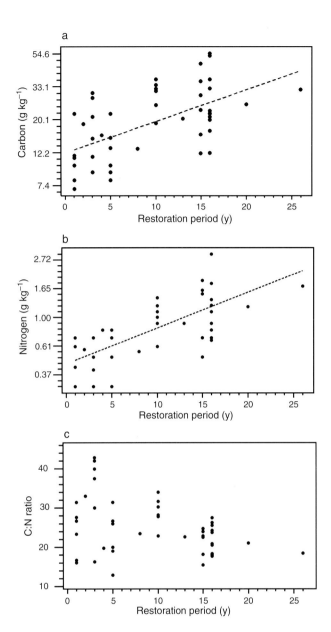

Figure 9.26: Changes in the concentrations of (a) C; and (b) N with restoration period in the surface (0–0.10 m) layer of the restored soils ($n = 50$) (C and N concentrations as natural logarithms) and (c) the C:N ratio in the same depth range.

restored for more than 16 years were the geometric means of the concentrations sampled within the 0–0.10 m depth range.

Using the above equations, estimated C and N concentrations for the sites at 26 years are 42.0 g kg^{-1} C and 2.23 g kg^{-1} N in the 0–0.10 m depth range.

Over this depth interval, these concentrations are shown to be clearly greater than the estimated values of the native forest sites: 30.4 g kg^{-1} C and 0.83 g kg^{-1} N for the native forest site with the deep profile and 19.9 g kg^{-1} C and 0.70 g kg^{-1} N for a site with a profile of medium depth.

Across the chronosequence of sites, limited increases in organic matter concentrations occurred below 0.10 m in the profile. Total C concentrations remained at or less than c. 10 g kg^{-1} except for a few higher values in the upper part of this depth range within profiles at the longer-restored sites. Nitrogen concentrations followed a similar pattern, with average concentrations not exceeding 0.5 g kg^{-1} up to 26 years. However, as with C, a gradual increase in N concentration was discernible in the upper part of this depth range in sites restored for periods of more than c. 10 years.

The concentrations of organic matter in the 0–0.10 m interval of soils of the restored sites increased to values similar to those of the adjacent native forest areas soon after canopy closure and subsequently continued to increase. In these environments organic matter concentrations diminished sharply with increasing depth in the soil profiles, perhaps partly due to illuviation and leaching in these highly permeable soil materials.

Soil respiration rates

Figure 9.27 presents soil respiration rates (measured as CO_2 emission rates) taken at the surface of the mineral soil after removal of any litter present. These measurements were made on the basis of daily rates of CO_2 emission for nine sites of differing restoration period and for two native forest sites, one recently burned (D Tongway unpublished). Emission rates were determined only once in the late dry season (September 2001). Values rose steeply from the initial rate of 3.5 g CO_2 per m^{-2} day^{-1} for bare soil to a maximum value of 14.4 g CO_2 per m^{-2} day^{-1} at 13 years, shortly after canopy closure. Values subsequently declined slightly to c. 11.4 g CO_2 per m^{-2} day^{-1} at sites restored for 26 years and were intermediate between emission rates recorded for the two native forest sites.

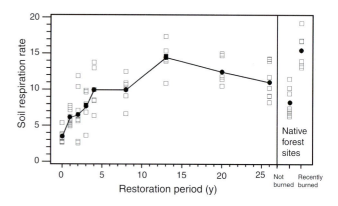

Figure 9.27: Individual observations and mean daily respiration rates (g CO_2 m^{-2} day^{-1}) from surface soils in nine restored and two native forest soils in the late dry season, September 2001.

Stable C isotope values (δ^{13}C values) of the near-surface soils

The δ^{13}C value of the near-surface soil provides information on the dominant C sources contributing C to the soil organic matter pool. The C from most tropical grasses (dominantly C4 plants) has δ^{13}C values of approximately –14.0‰ while that of broadleaved plants (C3 plants) has values of approximately –28.0‰ (Marshall et al. 2007). In the restored forests of the Gove mine, temporal changes in such values reflect the changing origins of the C acquired from the vegetation (Fig. 9.28) although

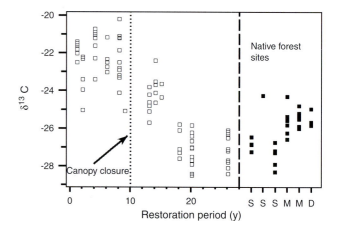

Figure 9.28: The stable C isotope values (δ^{13}C values) of the near-surface soils (0–5 cm) of restored and unmined native forest sites (Reddell et al. 1993; Spain et al. 2006).

the greater susceptibility to decomposition of the C from C4 species may lead to under-representation of the contribution of C derived from the latter source (Wynn and Bird 2007).

The $\delta^{13}C$ values of the near-surface (0–5 cm) topsoil placed on the site is commensurate with the open-forest origin and post-clearing treatment of these materials and reflects the mixed C3:C4 origin of the C present. The $\delta^{13}C$ values of the soil organic matter changed little during the period up to canopy closure reflecting the continuing C4 C inputs from the sown grasses. Subsequently, $\delta^{13}C$ values declined steeply under the influence of increasing C inputs from the dominant acacias and the now dominant eucalypts until at c. 20 years after restoration, values were similar to or more negative than those of the unmined native forest. The values at restoration periods of longer than 10 years also reflect the declining influence of the grasses beyond this point. Considerable variation occurs among the native forest sites of different soil profile depths with the shallower soils possessing lower C4 C inputs from the grasses relative to those derived from trees and shrubs.

Integration

With increasing restoration period, the dominant structural formation class of the restored plant community progresses through a series of overlapping stages from a grassland to a shrubland, to a closed forest and, at the 20- and 26-year restored sites, to an open forest structurally similar to that of the surrounding unmined forest. This pattern is accompanied by increasing above and belowground plant biomass and an emerging diversity, notably in the herbaceous layer. As development proceeds, the structure of the mature plant community emerges and by 26 years is fully established.

The trajectory of overall development of the restored ecosystems is largely determined by the successful establishment and growth of the future major structural components: the keystone eucalypt species. Through this, the developing eucalypts, notably *E. tetrodonta* and *E. miniata*, come to dominate the plant community soon after canopy closure and thereafter largely control many aspects of the physical and chemical environment of the soil and the near-surface environment aboveground. Their satisfactory establishment and continuing growth is therefore seen as critical to achieving long-term success of the restoration works.

However, for a balanced understanding of the development and functioning of these restored sites, their ecosystems must be characterised on more than the major features of the plant community. A broader appraisal leads to a multi-component characterisation based on diverse biological and other criteria, including the establishment, abundance, community development and diversity of selected key groups of microorganisms and animals. The progressive changes that occur in such components over time provide supporting evidence that the underlying ecosystem processes they mediate are also functional. Monitoring of litter properties and soil development ensures that the processes of organic matter decomposition and the development of fertility in its widest sense are progressing in a way conducive to ecosystem maturation, consistent with both theoretical knowledge and the required restoration outcomes. Final outcomes from such an appraisal are likely to be substantially more robust and have greater predictive value that those based on the plant community alone (Tongway and Ludwig 2011) and this prospect is enhanced by the considerable interdependence that occurs in the development of the different components.

Once normal patterns of ecosystem development have been established with outcomes tending towards the required restoration objectives, such patterns may be used to monitor developmental progress in newly established ecosystems in the same environment. Conversely, departures from the established patterns may reflect differences in important environmental factors such as soil type or a change in drainage status or the development of a form of stasis. Alternatively, they may also be

Table 9.4. Summary of major changes associated with the early stages of ecosystem development defined by plant community structure and occurring in the absence of fire

Stage	Age range (years)	Vegetation features	Litter types and properties	Other near-surface phenomena	Soil profile
1	0–3	Initial establishment. Grasses achieving early dominance. Total surface cover increasing. Some ruderal species present. Very sparse shrub and small tree cover.	Initially sparse grass litter but coverage increasing rapidly, seasonally variable. Minor foliar litter of broadleaved species, largely in locations close to developing shrubs and trees. Decomposing eucalypt branch and some smaller stem wood transported with soil scattered sparsely on surface.	Soil surface pisolitic, unstable; localised surface erosion occurring into rip-lines and other depressions; development of partially lagged surface. Infiltration rate variable, low for these soils. Initial colonisation by termites; grass and litter-feeding termites sporadically present; above-ground mounds from 2nd year.	Soil fabric massive. C concentration minimal; C:N ratio declining. C derived from mixed C3 and C4 sources. Microbial biomass C increasing rapidly. Grass roots dominate upper profile, extend to > 1.25 m after one year. Early colonisation of grasses and other hosts by AMF. Decomposing wood distributed, largely in upper soil profile.
2	3–9	Grasses dominant but relative representation declining. Increasing shrub and tree cover; acacias increasing to transient dominance late in this stage.	Decreasing grass litter coverage. Acacia phyllode litter increasing. Eucalypt foliar litter increasing from c. 5 years. Fine woody litter slowly increasing. First low-density stem wood becoming available from short-lived, acacias, grevilleas, etc. Eucalypt wood largely limited to petioles, twigs and fruits. Surface coverage increasing, effectively stabilising the surface soil. Mycelial mats present beyond c. 5 years.	Soil surface stabilising. Infiltration rate increasing rapidly. Termite populations broadly active from 3 years. Grass and litter-feeding termites declining late in this stage. Termite workings on litters, at soil surface and under rocks from c. 4 years. Galleries and other faunal soil voids becoming common and ubiquitous by 9 years. Very sparse near-surface earthworm populations first noted.	Upper soil profile showing first signs of aggregated structure. Increasing near-surface C concentration; C:N ratio static or declining slightly. C derived from a mixture of C4 and C3 sources. Proportion of grass roots decreasing after maximum at c. 3 years, increasing development of near-surface shrub and tree fine roots. ECM fungi established from c. 3 years.

Table 9.4. (Continued)

Stage	Age range (years)	Vegetation features	Litter types and properties	Other near-surface phenomena	Soil profile
3A	9–20	Canopy closure occurring between 9 and 12 years. Sparse grass tussocks and herbaceous plants. Acacias declining. Variable development of more diverse lower and shrubby mid-strata. Increasing dominance by eucalypts throughout stage.	Fine litter coverage nearly complete by canopy closure. Foliar litter mass and depth increasing; initiation of layering. Relative representation of grass and acacia phyllode litter declining; increasing dominance by eucalypt leaves. Fine woody litter from dying acacias, grevilleas becoming increasingly common. Eucalypt logs transported with soil largely decomposed by 13 years. Eucalypt woody litter sparse, limited to small branches, twigs and fruits.	Soil surface stable. Infiltration rate increasing rapidly. Development of a fine, organic matter-rich transitional zone between soils and litters. Increasing development of soil-litter fine root and mycelial mats. Broad earthen sheetings of termite workings on eucalypt stems from c. 16 years.	Increasing aggregate structure in upper profile. Increasing C and N concentrations in upper profile; C:N ratio declining; soil C source changing rapidly from C4 to C3 dominance. Increasing sizes and near-surface concentrations of coarser roots.
3B	20–26	Grasses moderately sparse, patchy. Acacias minor patchy component. Continuing variable development of shrubby mid-stratum. Eucalypts clearly dominant. Vines use eucalypts for support and climbing high above the forest floor.	Possible declining rate of increase in litter mass. Relative representation of grass and acacia phyllode litter static or decreasing; eucalypt foliage dominant. Fine litter comprising increasingly decomposed with depth. Increasing quantities of larger eucalypt branch litter.	Infiltration rate possibly stabilising at very high rates. Further development of fine organic-matter rich transitional zone between soils and litters.	Soil C concentration possibly developing towards an equilibrium in the absence of fire; C:N ratio declining. C now dominantly derived from C3 plants, δ 13C values similar to native forest soils. Increasing humification of organic matter in the upper 2 cm of the mineral profile. High concentrations of near-surface large roots.

Table 9.5. Possible future changes in development of the restored ecosystems, given various management options

Management	Vegetation	Litter	Near-surface features	Soils
Indefinite near-future beyond 26 y; continued exclusion of burning	Continuing maturation leading eventually to initial senescence of first cohort of eucalypts. Grasses, shrubs trees possibly achieving quasi-equilibrium densities with high density mid-storey vegetation.	Continuing but minor increases in fine litter mass, possibly approaching equilibrium. Continuing small increase in fine woody litter mass. Increasing dead eucalypt branchwood and (later) stem wood.	Increasing mass of larger woody materials on soil surface. Colonisation of sites by 'humus-feeding' termites.	Continuing smaller increases in soil C concentrations, C:N ratio static. Eventual quasi-equilibrium attained at higher C concentration than unmined native forests. Increasing biomass of dead woody eucalypt roots.
Indefinite near-future beyond 26 y; lease relinquishment and/or semi-annual firing regime instigated	Increasing senescence and then death of first cohort of eucalypts. Grasses possibly achieving post-firing equilibrium densities commensurate with firing cycle. Reduction in woody stem populations, particularly in the shrubby mid-stratum. Some upper canopy tree mortality possible on initial burning.	Sharp reduction in litter mass on first firing. Litter thereafter increasingly similar in mass and composition to unmined forests, varying commensurately with firing cycle. Increasing presence of larger eucalypt branch litter to time of senescence and death of first cohort of eucalypts when first dead eucalypt stem and root wood becomes available in large quantities.	Increasing presence of char and semi-charred plant materials at interface with mineral soil. Increasing incidence of termite attack on eucalypts leading to piping of stems. Colonisation of sites by 'humus-feeding' termites. Eventual approximate convergence of fauna with local unmined reference sites.	Declining C concentrations at interface between litter and near-surface mineral soil. Longer-term decline in near-surface C concentrations to values similar to unmined reference forests. Increasing amounts of stable soil organic materials and char in upper soil layers. Initial development of pyromorphic humus type.
Longer period; future generations of keystone eucalypts.	Possible self-thinning of eucalypts.	Surface woody debris, abundance and composition similar to local unmined reference sites.	Abundances of termite mounds (and other termite constructs) and ant nests similar to local unmined reference sites.	Soil C concentrations and quality indicators similar to values in reference sites Increasing amounts of stable soil organic materials. Humus largely of pyromorphic form. Ongoing redistribution of fine earth materials in profile.

the consequence of a perturbing event such as a disease, pest outbreak, fire or the close passage of a severe cyclone.

Based on the major stages of structural development of the plant community, Table 9.4 summarises selected features of development that have occurred in four major ecosystem components and other features at the study sites. Where sufficient information is available, similar compendia could be assembled for many other ecosystem components and particular communities of organisms. The approximate durations of the stages may differ locally between sites due to variation in soil properties, drainage status and other influences. Similarly, within each stage, some variation must be expected in the timing of the phenomena recorded.

Table 9.5 presents some of the future changes that may be expected to occur in the restored ecosystems under some of the most likely management scenarios. Many aspects of future development in these restored ecosystems will depend on when these lands are returned to their traditional owners and when regular firing is introduced. It is certain that semi-regular firing will be introduced at some stage of development although the time frame for this is currently unknown. It is also possible that the current policy of excluding fire may be maintained for some period and Table 9.5 presents some of the conditions likely to pertain during this time. Nonetheless, it is clear that firing, when and how introduced, will effect substantial changes to the restored ecosystems; depending on its seasonality and therefore severity, it may affect the canopy trees and will probably alter the compositions of the ground layer and middle stratum towards a more open-forest environment (Scott *et al.* 2012) and have consequences for all sections of the biota and for soil development.

Provided that damaging influences fall with the resilience limits of the ecosystems, it seems likely that longer-term ecosystem developmental changes will increasingly be towards a state of approximate convergence of many properties with those of the frequently burned unmined forest reference sites.

Some processes will require an unknown but considerable period for this quasi-equilibrium condition to be established. Periods necessary for the biomass and functional characteristics of the ecosystem to approximate those of the local unmined native forests may be long and require the passage of some further generations of the dominant eucalypts. Soil development, in particular, is likely to take much longer and soils are expected to remain different in some respects from those of the surrounding native forest reference sites for long periods since they are now in the initial stages of an extended period of reorganisation and development. The changes involved are, in part, qualitative since the restored sites are now positioned lower in the landscape and the materials comprising the restored profiles differ in several respects from those of the unmined profiles. As alluded to above, changes to the quality of the soil organic matter present are expected with the introduction of firing (González-Pérez *et al.* 2004).

Conclusions

Following from the requirements of the traditional owners, the objectives of the restoration process broadly necessitate the return of forests that will have a similar structure to the local native forests and possess the same dominants. From information gained to date, these requirements have clearly been met using the system of restoration described here and, overall, the present system appears to have so far realised these objectives arguably more closely than the systems of restoration used at many other mines (see, e.g. Gould 2012). Nonetheless, a substantially longer time period is needed for the full development of mature ecosystems.

Retrospectively, the setting of such a generalised aim for the rehabilitation process was perhaps reasonable in that attempts to closely replicate a specific idealised analogue or target ecosystem in restoration practice are unlikely to achieve success at any close level of detail (Society for Ecological Restoration International Science & Policy Working Group 2004). When they mature, the restored ecosystems of the Gove mine are therefore likely to differ in at least some respects from local native forest ecosystems and will thus be to some degree novel (Aerts and Honnay 2011; Hobbs *et al.* 2009). These differences are likely to result from alterations in elevation, landform particularly the degree of soil development, and the possible long-term effects of climate change (Banning *et al.* 2011; Harris *et al.* 2006). Despite some probably minor differences, it is likely that the communities of organisms of most restored sites will eventually develop to fall largely within or be close to the broad spectrum of local ecosystem variation. Further, it is expected that these ecosystems will be capable of resisting damaging influences and resilient where this does occur. A particular requirement is that, in their future management, they should be capable of being integrated with that of the rigorous and disturbance-prone local environment.

Mining-restoration system

The consistent success of the rehabilitation system developed and used at the Gove mine has been due to a range of factors. In particular, it has been in part due to supportive management which has ensured from an early stage both progressive restoration and the close integration of mining works with restoration practice (O'Keefe 1992). Management and other factors favourably influencing restoration success include the early adoption of a successful system of restoration, close observation of the restored sites (especially during the critical establishment stage) and the on-site retention of knowledge of the restored ecosystems gained over long periods. Other features of the Gove environment conducive to successful ecosystem development include the slightly elevated location, low-relief landforms and the generally free-draining, non-extreme (although chemically infertile) soil materials.

Early in mine site development, the mining-restoration system used at the Gove mine introduced many of the major elements of best practice now employed in the more successful systems of post-mining restoration elsewhere. Such elements include the now widely used system of removing, handling and replacing the valuable topsoil resource separately from the less-fertile underlying soil ('double stripping') during profile reconstruction, no stockpiling of topsoil and minimal stockpiling of subsoil, deep ripping of the reconstructed soil profile to reduce mechanical strength and promote drainage, limited fertiliser application, viability seed testing for the keystone eucalypts and mandating the use of local-provenance plant species likely to tolerate the effects of local pests and diseases. Careful observation of the developing ecosystems, particularly during the critical stages of establishment and early growth has allowed occasional deficiencies in establishment or early growth stages to be rapidly redressed with no subsequent intervention deemed necessary. One important indicator of the success of this system is the successful early establishment of the dominant eucalypts. Without this, establishment of the dominant eucalypts is likely to have been extremely slow or non-existent leading to a plant community whose structure and composition would be unlikely to resemble that of the surrounding native forests in any meaningful timescale (see also Gould 2012; Norman *et al.* 2006).

Fertilisation is likely to have had a major influence on the rapid early development of these restored systems, particularly by enhancing P supply during the critical stages of establishment and early growth. However, little detailed information is available on the responses of most major species to fertiliser application or the possible effects that such application may have on the differential nutrition of the other functional groups of plants present, including the nature of subsequent recruitment processes. This is clearly an area that needs to be addressed since inappropriate attempts to increase or maximise tree growth rates in these environments may have unintended results on subsequent plant community composition (Standish *et al.* 2008; Standish *et al.* 2010).

A major difference between the mining-restoration system at the Gove mine and such systems elsewhere is the inclusion of a one- to three-year fallow period between clearing the original native forest vegetation cover and soil removal. While this entails a substantial loss of organic matter from the surface soils, it may return the microbial community (notably the mycorrhizal fungi and associated organisms) to an earlier stage of ecosystem development, perhaps selecting for a community of soil organisms more conducive to rapid early development of the plant community. Other differences include the effective early stabilisation of the soil surface and early stimulation of nutrient cycling consequent on the planting of a sterile short-lived sorghum cultivar.

Post-restoration community development

The system of profile reconstruction and forest restoration used at the Gove mine is considered to be successful in that it has lead after 26 years of development to ecosystems which have stable soils and forests that are generally similar to the surrounding unmined native forests. However, even the longest-restored of these forests are still only part way through the first generation of the dominant tree species of the upper-stratum. It is expected that all major components of these extremely juvenile ecosystems will continue to change over many years before their major characteristics approach their equivalents in the local mature unmined native forest ecosystems.

Insofar as can be ascertained at this extremely early stage of ecosystem development, it is clear that the system of ecological restoration now employed at the Gove mine for more than 30 years has lead to the rapid development of a plant community that is structurally similar to those of the surrounding open forests, possesses the same dominant eucalypt species (including the iconic species *E. tetrodonta* as a major plant community component) and, with some caveats, is trending towards the general composition of the unmined reference sites (Brady and Noske 2010; Cook 2012; Reddell *et al.* 1993; Spain *et al.* 2009). The initially dominant tropical pasture grasses planted to control erosion and to provide the first organic matter inputs to the developing soils have been largely replaced by native perennial species. Although acacias subsequently dominated restored sites for a few years, based on basal area, stem density and height measures, eucalypts have emerged to become dominant following canopy closure at sites restored for longer than *c.* 10 years. Ongoing colonisation of the post-restoration landscape by a taxonomically diverse range of plant species has resulted in an increasingly diverse plant community although this has largely been in the herbaceous and middle strata. From the 39 species sown, the establishment on the restored sites of more than 170 additional species distributed over 133 genera confirms the increasing taxonomic and functional diversity of the restored plant communities; it almost certainly implies a similar or greater increase in associated animal species.

Despite their critical importance to ecosystem functional processes (Schlesinger 1997; van der Heijden *et al.* 2008) in these infertile soils, little is known of the communities of microorganisms colonising the sites, their functional diversity and patterns of development. Preliminary studies indicate substantial diversity and clear patterns of change with restoration period in the community structures of both arbuscular and ectomycorrhizal fungi. Direct acquisition of P and perhaps N from decomposing litter and the more-available fractions of soil organic matter in the upper profiles is likely to be mediated by both the ectomycorrhizal fungi (including mycelial mat forming species) and arbuscular mycorrhizal fungi (see, for example, Cairney 2011; Inselbacher and Näsholm 2012; Plassard and Dell 2010; Turner 2008). Almost nothing is known of the communities of saprotrophic microorganisms present, or of their activities. The importance of saprotrophic capacity should not be underrated since much of the N and P used by the plant community is likely to be absorbed in organic form (see Inselsbacher and Näsholm 2012).

Both structured studies and casual observations have indicated colonisation of the restored sites by a diverse range of animal groups, including regular use of ecosystem resources by such widely mobile groups as birds and mammals. The three groups most studied to date (ants, termites and birds) have clear functional roles and, together with the mammals, possess important indicator or conservation value. However, despite their provision of a range of essential ecosystem services, only limited studies have been published of the many animal groups that colonise the rapidly developing ecosystems of restored mine sites (Cristescu *et al.* 2012).

Initial colonisation of the restored environments by the more mobile animal groups has been rapid and both species richness and functional diversity have increased with restoration period in the three groups most studied to date. Both past and future colonisation is clearly related to environmental conditions, the resources available at particular stages of development of the plant community and, for certain groups, the stage of soil development. Communities of these three groups and other animals will in turn influence the further development of both the biotic environment and soils through feedback effects.

Termites are particularly important to soil development in the restored areas because of the lack of pedogenically effective earthworm populations in this seasonally dry environment and for their roles in the decomposition of dead plant materials. Ants have major functional roles as predators and seed spreaders and through their complex interactions with plants; where populous, they are likely to have substantial pedogenic effects (Paton *et al.* 2003). In contrast, the important invasive ant species, *Anoplolepis gracilipes*, reported from the restored sites, is known from elsewhere to have potentially large negative effects on the biota (see, for example, Holway *et al.* 2002). Together with the mammals, birds play major roles in providing a range of ecosystem services, including seed dispersal (Whelan *et al.* 2008) and are therefore important in promoting plant biodiversity beyond the sown and wind-dispersed species.

Published information on most other animal groups is largely limited to casual records of presence, with little known of their detailed functional roles in the restored environments. Due to the close proximity of the local unmined native forests, colonisation of the restored sites by the more mobile animal groups is unlikely to be limited by dispersal distance. Nonetheless, the development of sparse earthworm populations on the longer-restored sites indicates that even putatively less mobile groups may successfully colonise these restored sites.

Future development of the plant communities will continue to be reflected in the characteristics of and interactions with the animal communities. Both of these will clearly depend on future management decisions, most notably those relating to the use of fire.

Litter layer development

A protective cover of fine litter and herbaceous plants rapidly (within three to four years) stabilises the surfaces of the newly restored soil materials and subsequent litter accumulation leads to the complete coverage of the soil soon after canopy closure. Continuing forest development after this stage has lead to rapidly increasing fine litter masses and the formation of multiple layers of progressively decomposed litter at the longer-restored sites.

The substantial litter layers that have developed at the latter sites are an important feature of such areas. They appear to be partly due to fertiliser application, which effects higher initial plant growth rates and leads to increased rates of litterfall. Also contributing to the build up of these layers, is the lower estimated rate of litter decomposition (Cook 2012) that pertains during development in these immature ecosystems. Litter masses at the longer-restored sites are substantial, as much as three times those that occur in the frequently burned local native forest environments, and represent considerable stores of nutrient elements and energy.

Litter physical composition followed a pattern of serially changing dominance paralleling that of the plant community: from the initial grasses, to acacias, to eucalypts. The litters at the longer-restored sites are unusual for eucalypt forests in their very small quantities of woody materials and in the absence of charred and partly charred plant materials. The amounts of wood from the shorter lived acacias and other woody species present increase with restoration period although it is not until the first cohort of eucalypts matures and senesces that the denser wood of eucalypts is likely to occur in any quantity in the restored landscapes. This increase in coarse woody debris that occurs with time may help to expedite animal colonisation by providing habitat and food materials for a wide range of species, invertebrate and vertebrate. In some mines, logs and other woody debris have been consciously introduced to restored sites for these purposes (Cristescu *et al.* 2012).

Over the course of development, the chemical properties of the litters altered predictably, with N concentrations increasing from low values in the initial grassy litters to higher values during the period of acacia dominance to lower values again in the eucalypt-dominant litters occurring following canopy closure. Nitrogen concentrations at the longer-restored sites were of similar magnitude to those found in native eucalypt forests (Bevege 1978). Litter P concentrations did not alter appreciably during development and were at the lower end of those found in native eucalypt forests (Bevege 1978). Soil materials become increasingly apparent in the litters with development period and this has resulted in higher Fe and Al concentrations at the longer-restored sites. This accumulation was most notable in the lower layers and appears due to bioturbation, particularly through termite activities, and the accumulation of dust in this seasonally dry environment.

These litters provide valuable ecosystem services, notably in promoting surface stability and soil development. Further, they contribute directly to the development of the dominant eucalypts in that they provide habitat for the mycelial mats of the mycorrhizal fungi so important to their nutrition in these infertile soils. However, their role in provision of these and other services must be balanced against the degree of risk of damage posed by uncontrolled firing (Cook 2012).

Post-restoration soil development

As restored following the mining-restoration sequence described here, the soil materials at the Gove mine differ substantially from those of their parent soils. This is initially due to the loss of organic matter associated with fallowing and subsequently to the soil mixing associated with handling during mining and profile reconstruction. This leads to the dilution of the remaining organic matter present with the mineral soil components and, through the subsequent ripping process, further coarse material (as fractured laterite) is introduced into the profile.

The highly weathered soil materials that comprise the reconstructed profiles have unusually high gravel contents; this occurs mostly as pisoliths but to a lesser degree as fractured lateritic materials. While pisoliths form a large portion of the gravels they were not studied here although it is possible that they may contribute to the P and perhaps N nutrition of organisms growing in these soils (see, for example, Abekoe and Tiessen 1998). The fine earth fraction is dominantly of clay loam and clay texture. The major clay-sized materials are kaolinite and oxide minerals of Fe and Al. A particular feature of these soils is the strong micro-aggregation of their fine particles, leading to a functionally coarser texture than is indicated by their particle size distributions.

At the time of soil replacement, the chemical environment that these largely mineral materials provide for plant roots and their associated mycorrhizal organisms is one of medium pH and low salinity; the soil materials above the mine floor are non-sodic. However, because of their highly weathered status, they possess high concentrations of Fe and Al and low concentrations of the nutrient elements required for plant growth. Phosphorus is a key element limiting the production of both plants and microorganisms in highly weathered soils (Reed *et al.* 2011). Even in comparison with Australia's low soil P concentrations, P is present in low total concentrations (*c.* 210 mg kg^{-1}) and much of that present is unavailable to plants and microorganisms in these oxidic soils (Reed *et al.* 2011; Yang and Post 2011). Putatively plant-available P concentrations are extremely low despite pre-planting P fertiliser application and this element is likely to be the most limiting element for the growth of unadapted plants (see, for example, Grant *et al.* 2007; Jasper and Davy 1993; Short *et al.* 2000, 2007) and microbial activity (Cleveland *et al.* 2002; Cleveland *et al.* 2006). Because of the dominantly kaolinitic clays present, permanent charge on the soil colloids is low. Exchangeable base status is also low in these soils. The overall net negative charge is largely variable and is dependent on the oxide minerals present and on soil organic matter concentrations. Therefore, at least initially, such soil materials have little capacity to retain nutrient cations.

Substantial changes emerged in the physical properties of the newly created soil profiles during the 26 years following profile reconstruction. The bulk density of the newly restored profiles is high although near-surface values appear to decline with restoration period due to biological activity. A major feature of the restored soils is their development of extremely high infiltration and drainage rates. This is attributed to the combined effects of their high gravel concentrations, micro-aggregation of their clays and structural development, including the development of preferred pathways. Such high permeability implies a well aerated soil environment over all but a very limited area of the restored sites.

Evidence of rapid profile reorganisation over what are pedogenically short time periods includes the probable eluviation of clays from the surface layer of the restored soils and the large increases in the infiltration rate referred to above. Biological features of the latter are likely to be partly associated with structural development (both as voids and structural aggregates) associated with termite and ant bioturbation and the turnover of plant roots that takes place following profile reconstruction. Longer-term reorganisation is inherent in the ongoing transport of fine soil materials to the surface by termites and ants (Tardy and Roquin 1992).

Organic matter concentrations are directly relevant to CEC and nutrient supply in these highly weathered soil materials. In the restored soils, the char and partly charred plant material originally present in the transported topsoil is diluted throughout the upper soil layer by partial mixing and freshly created char is absent from the soil surface. Concentrations of total C and total N in the upper (0–0.1 m) profiles of the restored soils increased from initial values of less than 10 g kg^{-1} C and 0.46 g kg^{-1} N in the recently restored soils to values similar to those of the surrounding unmined forests by ~20 years for C and at about the time of canopy closure for N. Carbon and N concentrations in the near-surface materials both continued to increase thereafter and, after 26 years of soil development, their concentrations substantially exceeded those in the soils of the surrounding unmined forests. The build up of soil organic matter occurring at the Gove mine substantially exceeds that normally occurring in the reafforested natural soils of tropical areas (Li *et al.* 2012). Despite the rapid build up of soil organic matter in the upper profile, only minor increases in C and N concentrations occurred below the upper 0.10 m layer. The C:N ratios of the restored Gove soils ranged from *c.* 23 to 27 and such values are lower than those found in the unmined native forest soils.

Organic matter concentrations attained at the Gove mine were similar to those reported for sites restored to native forest species at the Weipa bauxite mining operation for sites of similar restoration period (Schwenke 1992; Schwenke *et al.* 2000a, b). In contrast, at Alcoa's Darling Plateau sites in Western Australia, total N concentration in the upper 0.10 m after 8.5 years was much greater than at the present study sites. This difference is partly attributable to the high concentration of understorey legumes and to the greater amounts of fertiliser P applied at these sites (Grant *et al.* 2007).

In contrast to the unmined native forests virtually all of the C present in the sites studied here is young, less than 26 years, and this implies considerable chemical differences. Further, the absence of large quantities of char and partly charred plant materials from the restored sites contrasts with that in the native forest reference sites although the eventual introduction of firing will convert the humus form of these juvenile soils to the pyromorphic form (González-Pérez *et al.* 2004).

Soil development following soil replacement and ecosystem establishment is rapid. Over the course of the current studies, it has lead to the early development of surface stability, ongoing upper profile reorganisation, increasing structural development and the near-surface accumulation of substantial quantities of soil organic matter.

Final remarks

Basic functionality has been confirmed for many features of the developing ecosystems through previous studies (Brady and Noske 2010; Cook 2012; Hinz 1992; Reddell et al. 1992, 1993; Spain et al. 2006, 2009, 2010; Woinarski et al. 2009). Studies (Tongway 2010; Tongway and Ludwig 2011) conducted on the same sites and in conjunction with those described by Spain et al. (2009) have confirmed the broad success of the current rehabilitation system using 'Ecosystem Function Analysis', a widely used system of indicators of restoration success.

In addition to confirming basic functionality of the restored ecosystems, the developmental trajectory under current conditions has become largely predictable due to the highly interdependent developmental processes evident among components. Nonetheless, much still needs to be learned about their detailed functioning and interactions. Such knowledge would allow a greater certainty in management during development and probable ecosystem dynamics following eventual lease relinquishment. Basic understanding of the factors influencing the nutrition of the plant community as it progresses through the different successional stages is currently deficient and should be redressed since it has consequences at all levels of ecosystem organisation and limits the capacity to generalise findings to other mine sites. This further necessitates a greater understanding of soil development in relation to litter and soil organic matter dynamics, fertilisation and microbial activity (notably that relating to the roles of mycorrhizal fungi and N-fixing organisms).

In addition to the interdependence between components discussed above, it is expected that time lags of various lengths will occur between components although these should be relatively consistent among processes between sites of similar restoration periods. This indicates that formal multivariate analyses of the similarities occurring within and between ecosystem components over the course of development would be profitable. Such analyses could be profitably used to explore the relationships, *inter alia*, between and within different sections of the developing biological communities and of these with relevant litter and soil properties: physical, chemical and biological. Outcomes from successful application of such an approach would lead towards a more robust definition of post-mining ecosystem development and the derivation of parsimonious systems of indicators.

The trajectory of future development in these ecosystems will be substantially influenced by management decisions relating to burning, particularly in sites at the earlier stages of development. The basal areas of the dominant eucalypts at the longer-restored sites have similar magnitudes to those of the native forests and the older trees are sufficiently tall to be largely tolerant of the immediate effects of fire. Nonetheless, as indicated above, firing is likely to substantially reduce their growth rates and to increase mortality both directly (Cook 2012) and through interactions with fungal colonisation and termite activity (Perry et al. 1985; Werner and Prior 2007; Werner et al. 2008).

Future management decisions also require greater knowledge of the effects of burning on ecosystem functioning at different developmental stages to determine the most appropriate firing regimes. The effects of fires of differing severity on a range of ecosystem processes across all development stages are important since these will almost certainly be dissimilar to those occurring in the mature native forest ecosystems. Protecting the restored forests from firing for several years has promoted rapid early tree growth and ecosystem development although the substantial litter masses that develop after canopy closure pose some degree of fire risk for the developing community dominants and other woody species (see, for example, Reddell et al. 1993; Cook 2012; Werner 2012).

Beyond the effects of fire, potential threats to the sustainability of the restored communities include cyclones, tornadoes (Franklin *et al.* 2010), invasive plant species, plant pathogens and pest animal species in addition to the longer-term effects of climate change. While little evidence exists to indicate that they will seriously threaten the existence of the restored ecosystems, the resistance and resilience characteristics of the restored systems largely remain to be tested against such potential threats. It is clear that caution should be exercised through appropriate monitoring and the development of remediation plans, should significant threats emerge.

Acknowledgements

We thank Christopher Walker (Royal Botanic Garden, Edinburgh and Cranfield University) for his assistance with the identification and nomenclature of the arbuscular mycorrhizal fungi; C.R. Dunlop (Herbarium, Darwin), Bob Harwood (Herbarium, Darwin), Bernie Hyland (CSIRO, Atherton) and J.R. Maconochie (Alice Springs Herbarium) for determination of plant species. D. Hinz would particularly like to thank his many Yolngu Aboriginal friends and field assistants; over a period of 30 years during the restoration of local native land ecosystems after mining, they freely shared their invaluable knowledge of ethnobiology.

References

Abekoe MK, Tiessen H (1998) Phosphorus forms, lateritic nodules and soil properties along a hillslope in northern Ghana. *Catena* **33**, 1–15. doi:10.1016/S0341-8162(98)00063-0

Aerts R, Honnay O (2011) Forest restoration, biodiversity and ecosystem functioning. *BMC Ecology* **11**, 29. doi:10.1186/1472-6785-11-29

Ali FH (2010) Use of vegetation for slope protection: root mechanical properties of some tropical plants. *International Journal of Physical Sciences* **5**, 496–506.

Allen MF (2007) Mycorrhizal fungi: highways for water and nutrients in arid soils. *Vadose Zone Journal* **6**, 291–297. doi:10.2136/vzj2006.0068

Andersen AN, Lonsdale WM (1990) Herbivory by insects in Australian tropical savannas: a review. *Journal of Biogeography* **17**, 433–444. doi:10.2307/2845374

Andersen AN, Majer JD (2004) Ants show the way down under: invertebrates as indicators in land management. *Frontiers in Ecology and the Environment* **2**, 291–298. doi:10.1890/1540-9295(2004)002[0292:ASTWDU]2.0.CO;2

Andersen AN, Morrison S, Belbin L, Ashwath N, Brennan K (1998) 'The role of ants in minesite restoration in the Kakadu region of Australia's Northern Territory, with particular reference to their use as bioindicators'. Department of Sustainability, Environment, Water, Population and Communities, Supervising Scientist Report No. 130, Darwin.

Andersen AN, Fisher A, Hoffmann BD, Read JL, Richards R (2004) Use of terrestrial invertebrates for biodiversity monitoring in Australian rangelands, with particular reference to ants. *Austral Ecology* **29**, 87–92. doi:10.1111/j.1442-9993.2004.01362.x

Andersen AN, Woinarski JCZ, Parr CL (2012) Savanna burning for biodiversity: fire management for faunal conservation in Australian tropical savannas. *Austral Ecology* doi:10.1111/j.1442-9993.2011.02334.x

Anonymous (2011) Yirrkala bark petitions 1963. http://foundingdocs.gov.au/items-did-100.Html#description

Atlas of Living Australia (2012a) *Eucalyptus tetrodonta*. http://spatial.ala.org.au/?q=lsid:urn:lsid:biodiversity.org.au:apni.taxon:252554

Atlas of Living Australia (2012b) *Eucalyptus miniata*. http://spatial.ala.org.au/?q=lsid:urn:lsid:biodiversity.org.au:apni.taxon:82

Australian Plant Census. http://www.anbg.gov.au/chah/apc/about-APC.html

Australian Plant Name index. http://www.anbg.gov.au/cgi-bin/apni

Banning NC, Lalor BM, Grigg AH, Philips IR, Colquhoun IJ, Jones DL, Murphy DV (2011) Rehabilitated mine-site management, soil health and climate change. In *Soil Health and Climate Change*. (Eds BP Singh, AL Cowie and KY Chan) pp. 287–315. Springer, Berlin.

Baxter JW, Dighton J (2001) Ectomycorrhizal diversity alters growth and nutrient acquisition of grey birch (*Betula populifolia*) seedlings in host-symbiont culture conditions. *New Phytologist* **152**, 139–149. doi:10.1046/j.0028-646x.2001.00245.x

Beardsell DV, O'Brien SP, Williams EG, Knox RB, Calder DM (1993) Reproductive biology of Australian Myrtaceae. *Australian Journal of Botany* **41**, 511–526. doi:10.1071/BT9930511

Bell LC (2001) Establishment of native ecosystems after mining – Australian experience across diverse biogeographic zones. *Ecological Engineering* **17**(2–3), 179–186. doi:10.1016/S0925-8574(00)00157-9

Bevege DI (1968) A rapid technique for clearing tannins and staining intact roots for detection of mycorrhizas caused by *Endogone* spp. and some records of infection in Australasian plants. *Transactions of the British Mycological Society* **51**, 808–810. doi:10.1016/S0007-1536(68)80106-8

Bevege DI (1978) 'Biomass and nutrient distribution in indigenous forest ecosystems'. Queensland Department of Forestry, Technical Paper No. 6.

Beven K, Germann P (1982) Macropores and water flow in soils. *Water Resources Research* **18**(5), 1311–1325. doi:10.1029/WR018i005p01311

Bever JD, Dickie IA, Facelli E, Facelli J, Klironomos J, Moora M, Rillig MC, Stock WD, Tibbett M, Zobel M (2010) Rooting theories of plant community ecology in soil microbial interactions. *Trends in Ecology & Evolution* **25**, 468–478. doi:10.1016/j.tree.2010.05.004

Bignell DE (2006) Termites as soil engineers and soil processors. In *Intestinal Microorganisms of Termites and Other Invertebrates*. (Eds H König and A Varma) pp. 183–220. Springer-Verlag, Berlin.

Birkeland PW (1984) *Soils and Geomorphology*. Oxford University Press, NY.

Bowen ME, McAlpine CA, House APN, Smith GC (2007) Regrowth forests on abandoned agricultural land: a review of their habitat values for recovering forest fauna. *Biological Conservation* **140**, 273–296. doi:10.1016/j.biocon.2007.08.012

Bowman DMJS, Brown GK, Braby MF, Brown JR, Cook LG, Crisp MD, Ford F, Haberle S, Hughes J, Isagi Y, Joseph L, McBride J, Nelson G, Ladiges PY (2010) Biogeography of the Australian monsoon tropics. *Journal of Biogeography* **37**, 201–216. doi:10.1111/j.1365-2699.2009.02210.x

Brady CJ (2008) Male–male conflict and breeding of the Australian bustard *Ardeotis australis* in rehabilitated mine land in Arnhem Land, Northern Australia. *Australian Field Ornithology* **25**, 203–206.

Brady CJ, Noske RA (2010) Succession in bird and plant communities over a 24-year chronosequence of mine rehabilitation in the Australian monsoon tropics. *Restoration Ecology* **18**(6), 855–864. doi:10.1111/j.1526-100X.2008.00511.x

Braithwaite RW, Miller L, Wood JT (1988) The structure of termite communities in the Australian tropics. *Australian Journal of Ecology* **18**, 375–391.

Brooksbank K, Veneklaas EJ, White DA, Carter JL (2011) The fate of hydraulically redistributed water in a semi-arid zone eucalyptus species. *Tree Physiology* **31**, 649–658. doi:10.1093/treephys/tpr052

Brundrett MC, Ashwath N, Jasper DA (1996) Mycorrhizas in the Kakadu region of tropical Australia I. Propagules of mycorrhizal fungi and soil properties in natural habitats. *Plant and Soil* **184**, 159–171. doi:10.1007/BF00029285

Bureau of Meteorology (2011) Climate statistics for Australian locations, monthly climate statistics, summary statistics Gove airport. http://reg.bom.gov.au/climate/averages/tables/cw_014508.shtml

Cairney JWG (2011) Ectomycorrhizal fungi: the symbiotic route to the root for phosphorus in forest soils. *Plant and Soil* **344**, 51–71. doi:10.1007/s11104-011-0731-0

Cairney JWG (2012) Extramatrical mycelia of ectomycorrhizal fungi as moderators of carbon dynamics in forest soil. *Soil Biology & Biochemistry* **47**, 198–208. doi:10.1016/j.soilbio.2011.12.029

Cairney JWG, Bastias BA (2007) Influences of fire on forest soil fungal communities. *Canadian Journal of Forest Research* **37**, 207–215. doi:10.1139/x06-190

Canadell J, Jackson RB, Ehleringer JR, Mooney HA, Sala OE, Schulze ED (1996) Maximum rooting depth of vegetation types at the global scale. *Oecologia* **108**, 583–595. doi:10.1007/BF00329030

Cardoso IM, Kuyper TW (2006) Mycorrhizas and tropical soil fertility. *Agriculture, Ecosystems & Environment* **116**, 72–84. doi:10.1016/j.agee.2006.03.011

Chapman SK, Langely JA, Hart SC, Koch GW (2006) Plants actively control nitrogen cycling: uncorking the microbial bottleneck. *New Phytologist* **169**, 27–34. doi:10.1111/j.1469-8137.2005.01571.x

Chen X, Eamus D, Hutley LB (2002) Seasonal patterns of soil carbon dioxide efflux from a wet–dry tropical savanna of northern Australia. *Australian Journal of Botany* **50**, 43–51. doi:10.1071/BT01049

Chen X, Eamus D, Hutley LB (2004) Seasonal patterns of fine-root productivity and turnover in a tropical savanna of northern Australia. *Journal of Tropical Ecology* **20**, 221–224. doi:10.1017/S0266467403001135

Chu-Chou M, Grace LJ (1982) Mycorrhizal fungi of *Eucalyptus* in the North Island of New Zealand. *Soil Biology & Biochemistry* **14**, 133–137. doi:10.1016/0038-0717(82)90056-6

Cleveland CC, Townsend AR, Schmidt SK (2002) Phosphorus limitation of microbial processes in moist tropical forests: evidence from short-term laboratory incubations and field studies. *Ecosystems* **5**, 680–691.

Cleveland CC, Townsend AR, Schmidt SK (2006) Nutrient additions to a tropical rain forest drive substantial soil carbon dioxide losses to the atmosphere. *Proceedings of the National Academy of Sciences of the United States of America* **103**(27), 10316–10321. doi:10.1073/pnas.0600989103

Cleveland CC, Townsend AR, Taylor P, Alvarez-Clare S, Bustamente MMC, Chuyong G, Dobrowski SZ, Grierson P, Harms KE, Houlton BZ, Marklein A, Parton W, Porder S, Reed SC, Sierra CA, Silver WL, Tanner EVJ, Wiede WR (2011) Relationships among net primary productivity, nutrients and climate in tropical rain forest: a pan-tropical analysis. *Ecology Letters* **14**(9), 939–947. doi:10.1111/j.1461-0248.2011.01658.x

Collins S (1996) Fuel loads, composition and distribution in rehabilitated bauxite mines in Western Australia. In *Third International and 21st Annual Minerals Council of Australia Environmental Workshop*. 14–18 October, Canberra. pp. 213–222. Minerals Council of Australia, Dickson, ACT.

Cook GD (2003) Fuel dynamics, nutrients and atmospheric chemistry. In *Fire in Tropical Savannas: The Kapalga Experiment*. (Eds AN Andersen, GD Cook and RJ Williams) pp. 47–58. Springer, NY.

Cook GD (2012) Fire management and minesite rehabilitation in a frequently burnt tropical savanna. *Austral Ecology* **37**, 686–692. doi:10.1111/j.1442-9993.2012.02375.x.

Cornell RM, Schwertmann U (2003) *The Iron Oxides - Structure, Properties, Reactions, Occurrences and Uses.* 2nd edn, Wiley-VCH, Weinheim.

Cornwell WK, Cornelissen JHC, Allison SD, Bauhus J, Eggleton P, Preston CM, Scarf F, Weedon JT, Wirth C, Zanne AE (2009) Plant traits and wood fates across the globe: rotted, burned, or consumed? *Global Change Biology* **15**, 2431–2449. doi:10.1111/j.1365-2486.2009.01916.x

Cowie RH, Logan JWM, Wood TG (1991) Termite damage and control in tropical forestry. In *Advances in Management and Conservation of Soil Fauna*. (Eds GK Vereesh, D Rajagopal and CA Viraktamath) pp. 161–167. Oxford and IBH Publishing, New Delhi.

Cristescu RH, Frère C, Banks PB (2012) A review of fauna in mine rehabilitation in Australia: current state and future directions. *Biological Conservation* **149**, 60–72. doi:10.1016/j.biocon.2012.02.003

Cross AF, Schlesinger WH (1995) A literature review and evaluation of the Hedley fractionation: applications to the biogeochemical cycle of soil phosphorus in natural ecosystems. *Geoderma* **64**, 197–214. doi:10.1016/0016-7061(94)00023-4

Croton JT, Ainsworth GL (2007) Development of a winged tine to relieve mining related soil compaction. *Restoration Ecology* **15**, S48–S53. doi:10.1111/j.1526-100X.2007.00292.x

Daniels BA, Skipper HD (1982) Methods for the recovery and quantitative estimation of propagules from soil. In *Methods and Principles of Mycorrhizal Research*. (Ed NC Schenck) pp. 29–36. The American Phytopathological Society, St. Paul, Minnesota.

DeJong JT, Tibbett M, Fourie A (2015) Geotechnical systems that evolve with ecological processes. *Environmental & Earth Science* **73**, 1067–1082. doi:10.1007/s12665-014-3460-x

Department of Sustainability, Environment, Water, Population and Communities (2008) *Rangelands – taking the pulse, Arnhem coast bioregion.* http://www.environment.gov.au/land/publications/acris/pubs/bioregion-arnhem-coast.pdf

Department of Sustainability, Environment, Water, Population and Communities (2011) *Australia's ecoregions.* http://www.environment.gov.au/parks/nrs/science/bioregion-framework/terrestrial-habitats.html

Dickie IA, Xu B, Koide RT (2002) Vertical niche differentiation of ectomycorrhizal hyphae in soil as shown by T-RFLP analysis. *New Phytologist* **156**, 527–535. doi:10.1046/j.1469-8137.2002.00535.x

Dooley SR, Treseder KK (2012) The effect of fire on microbial biomass: a meta-analysis of field studies. *Biogeochemistry* **109**, 49–61. doi:10.1007/s10533-011-9633-8

Eamus D, Chen X, Kelley G, Hutley LB (2002) Root biomass and root fractal analyses of an open *Eucalyptus* forest in a savanna of north Australia. *Australian Journal of Botany* **50**, 31–41. doi:10.1071/BT01054

Elser J, Bracken MES, Cleland EE, Gruner DS, Harpole WS, Hillebrand H, Ngai JT, Seabloom EW, Shurin JB, Smith JE (2007) Global analysis of nitrogen and phosphorus limitation of primary producers in freshwater, marine and terrestrial ecosystems. *Ecology Letters* **10**(12), 1135–1142. doi:10.1111/j.1461-0248.2007.01113.x

Fattet M, Fu Y, Ghestem M, Ma W, Foulonneau M, Nespoulous J, Le Bissonnais Y, Stokes A (2011) Effects of vegetation type on soil resistance to erosion: Relationship between aggregate stability and shear strength. *Catena* **87**, 60–69. doi:10.1016/j.catena.2011.05.006

February EC, Cook GD, Richards AE (2012) Root dynamics influence tree–grass coexistence in an Australian savanna. *Austral Ecology* doi:10.1111/j.1442-9993.2012.02376.x

Federal Law Report (1971) 'Milirrpum v Nabalco Pty Ltd and the Commonwealth of Australia', Supreme Court of the Northern Territory.

Ford HA, Paton DC, Forde N (1979) Birds as pollinators of Australian plants. *New Zealand Journal of Botany* **17**, 509–519. doi:10.1080/0028825X.1979.10432566

Fox ID, Neldner VJ, Wilson GW, Bannink PJ (2001) *The Vegetation of the Australian Tropical Savannas.* Environmental Protection Agency, Queensland Government, Brisbane.

Frankland JC (1992) Mechanisms in fungal succession. In *The Fungal Community – Its Organization and Role in the Ecosystem, 2nd edition.* (Eds JC Carrol and DT Wicklow) pp. 383–402. CRC Press, Baton Rouge.

Franklin DC, Gunton RM, Schatz J, Lawes MJ (2010) Resprouting responses of trees in a fire-prone tropical savanna following severe tornado damage. *Austral Ecology* **35**, 685–694. doi:10.1111/j.1442-9993.2009.02075.x

Gardner JH, Malajczuk N (1988) Recolonisation of rehabilitated bauxite mine sites in Western Australia by mycorrhizal fungi. *Forest Ecology and Management* **24**, 27–42. doi:10.1016/0378-1127(88)90022-9

George SJ, Kelly RN, Greenwood PF, Tibbett M (2010) Soil carbon and litter development along a reconstructed biodiverse forest chronosequence of South-Western Australia. *Biogeochemistry* **101**, 197–209. doi:10.1007/s10533-010-9519-1

Glen M, Bougher NL, Colquhoun IJ, Vlahos S, Loneragan WA, O'Brien PA, Hardy GEStJ (2008)

Ectomycorrhizal species in fungal communities of rehabilitated bauxite mine sites in the Jarrah forest of Western Australia. *Forest Ecology and Management* **255**, 214–225. doi:10.1016/j.foreco.2007.09.007

Gómez C, Espadaler X (1998) Myrmecochorous dispersal distances: a world survey. *Journal of Biogeography* **25**, 573–580. doi:10.1046/j.1365-2699.1998.2530573.x

González-Pérez JA, González-Vila FJ, Almendros G, Knicker H (2004) The effect of fire on soil organic matter–a review. *Environment International* **30**, 855–870. doi:10.1016/j.envint.2004.02.003

Gould S (2012) Comparison of post-mining rehabilitation with reference ecosystems in monsoonal eucalypt woodlands, Northern Australia. *Restoration Ecology* **20**, 250–259. doi:10.1111/j.1526-100X.2010.00757.x

Graetz RD, Skjemstad JO (2003) The charcoal sink of biomass burning on the Australian continent. CSIRO Atmospheric Research Technical Paper No. 64, CSIRO Australia, Aspendale.

Grant CD, Ward SC, Morley SC (2007) Return of function to restored bauxite mines in Western Australia. *Restoration Ecology* **15**(4), S94–S103. doi:10.1111/j.1526-100X.2007.00297.x

Gray DH, Leiser AT (1982) *Biotechnical Slope Protection and Erosion Control.* Van Nostrand Reinhold Co., NY.

Gwenzi W, Veneklaas EJ, Holmes KW, Bleby TM, Phillips IR, Hinz C (2011) Spatial analysis of fine root distribution on a recently constructed ecosystem in a water-limited environment. *Plant and Soil* **344**, 255–272. doi:10.1007/s11104-011-0744-8

Handel SN, Robinson GR, Parsons WFJ, Mattei JH (1997) Restoration of woody plants to capped landfills: root dynamics in an engineered soil. *Restoration Ecology* **5**, 178–186. doi:10.1046/j.1526-100X.1997.09721.x

Harley JL (1989) The significance of mycorrhiza. *Mycological Research* **92**, 129–139. doi:10.1016/S0953-7562(89)80001-2

Harris JA, Hobbs RJ, Higgs E, Aronson J (2006) Ecological restoration and global climate change. *Restoration Ecology* **14**, 170–176. doi:10.1111/j.1526-100X.2006.00136.x

Hazleton P, Murphy B (2007) *Interpreting Soil Test Results: What Do All the Numbers Mean?* CSIRO Publishing, Collingwood, Victoria.

Hill GF (1942) *Termites (Isoptera) From The Australian Region.* CSIR, Melbourne.

Hinz DA (1979) Rehabilitation of mined land and bauxite residue at Gove, Northern Territory, In *Mining Rehabilitation–1979*, pp. 11–16. Australian Mining Industry Council, Canberra.

Hinz DA (1981) Returning land to eucalypt forest after bauxite mining at Gove, N.T. In *Proceedings North Australian Mine Rehabilitation Workshop.* Gove NT, pp. 16–34. Nabalco Pty Ltd, Gove.

Hinz DA (1992) Bauxite mining and Walyamiri, the return of the living environment – paper two. In *Proceedings of the Seventeenth Annual Environmental Workshop.* Yeppoon, Queensland. pp. 100–114. Australian Mining Industry Council, Dickson, ACT.

Hinz DA (1997) Termites in land reclamation. *Mining Environmental Management* **September 1997**, 1215

Hinz DA (1999) Legend of the firebird. *Down to Earth, March 1999*, 4–6. Northern Territory Minerals Council.

Hinz DA (2000) Termites as ecological indicators of mine-land rehabilitation in tropical Australia. In *Workshop on the Rehabilitation of Narbalek.* 18–19 April, Darwin. (Ed. DA Klessa) pp. 84–89. Supervising Scientist Report 160, Supervising Scientist, Darwin.

Hinz DA (2007) Restoration of Wäna (the land) after bauxite mining on the Gove Peninsula, Northern Territory. In *Sustainable Development Conference, SD07.* 29 October–2 November 2007, Cairns. 9 pp. Minerals Council of Australia, Kingston ACT.

Hobbs RJ, Higgs E, Harris JA (2009) Novel ecosystems: implications for conservation and restoration. *Trends in Ecology & Evolution* **24**, 599–605. doi:10.1016/j.tree.2009.05.012

Hoffmann BD (2007) *North east Arnhem Land Yellow crazy ant eradication project.* http://www.issg.

org/database/species/references.asp?si=110&fr=1&sts=&lang=EN

Hoffmann BD (2010) Project summaries 28.11. Yellow Crazy Ant, *Anoplolepis gracilipes*, eradication in NE Arnhem Land. *Ecological Management & Restoration* **11**(1), 82–83.

Hoffmann BD (2011) Eradication of populations of an invasive ant in northern Australia: successes, failures and lessons for management. *Biodiversity and Conservation* **20**, 3267–3278. doi:10.1007/s10531-011-0106-0

Hoffmann BD, Saul WC (2010) Yellow crazy ant (*Anoplolepis gracilipes*) invasions within undisturbed mainland Australian habitats: no support for biotic resistance hypothesis. *Biological Invasions* **12**, 3093–3108. doi:10.1007/s10530-010-9701-3

Hölldobler B, Wilson EO (1990) *The Ants*. Belknap Press, Harvard University Press, Cambridge, Massachusetts.

Holt JA, Lepage M (2000) Termites and soil properties. In *Termites: Evolution, Sociality, Symbioses, Ecology.* (Eds T Abe, DE Bignell and M Higashi) pp. 389– 407. Kluwer Academic Publications, Dordrecht.

Holway DA, Lach L, Suarez AV, Tsutsui ND, Case TJ (2002) The causes and consequences of ant invasions. *Annual Review of Ecology and Systematics* **33**, 181–233. doi:10.1146/annurev.ecolsys.33.010802.150444

Huggett RJ (1998) Soil chronosequences, soil development, and soil evolution: a critical review. *Catena* **32**, 155–172. doi:10.1016/S0341-8162(98)00053-8

Inselsbacher E, Näsholm T (2012) The belowground perspective of forest plants: soil provides mainly organic nitrogen for plants and mycorrhizal fungi. *New Phytologist* **195**, 329–334. doi:10.1111/j.1469-8137.2012.04169.x

Isbell RF (2002) *The Australian Soil Classification*. Revised edition. CSIRO Publishing, Collingwood, Victoria.

IUSS Working Group WRB (2007) *World Reference Base for Soil Resources 2006, first update 2007*, World Soil Resources Reports No. 103. FAO, Rome.

Jackson RB, Canadell J, Ehleringer JR, Mooney HA, Sala OE, Schulze ED (1996) A global analysis of root distributions for terrestial biomes. *Oecologia* **108**, 389–411. doi:10.1007/BF00333714

Janos DP, Scott J, Bowman DMJS (2008) Temporal and spatial variation of fine roots in a northern Australian *Eucalyptus tetrodonta* savanna. *Journal of Tropical Ecology* **24**(2), 177–188. doi:10.1017/S0266467408004860

Jasper DA (2007) Beneficial soil microorganisms of the Jarrah forest and their recovery in bauxite mine restoration in Southwestern Australia. *Restoration Ecology* **15**, S74–S84. doi:10.1111/j.1526-100X.2007.00295.x

Jasper DA, Davy JA (1993) Root characteristics of native plant species in relation to the benefits of mycorrhizal colonisation for phosphorus uptake. *Plant and Soil* **155–156**, 281–284. doi:10.1007/BF00025037

Jasper DA, Robson AD, Abbott LK (1987) The effect of surface mining on the infectivity of vesicular-arbuscular mycorrhizal fungi. *Australian Journal of Botany* **35**, 641–652. doi:10.1071/BT9870641

Jasper DA, Abbott LK, Robson AD (1989) The loss of VA mycorrhizal infectivity during bauxite mining may limit the growth of *Acacia pulchella* R.Br. *Australian Journal of Botany* **37**, 33–42. doi:10.1071/BT9890033

Jobbágy EG, Jackson RL (2004) The uplift of soil nutrients by plants: biogeochemical consequences across scales. *Ecology* **85**(9), 2380–2389. doi:10.1890/03-0245

Johnson EA, Miyanishi K (2008) Testing the assumptions of chronosequences in succession. *Ecology Letters* **11**(5), 419–431. doi:10.1111/j.1461-0248.2008.01173.x

Johnson DB, Williamson JC, Bailey AJ (1991) Microbiology of soils at opencast coal sites. 1. Short-term and long-term transformations in stockpiled soils. *Journal of Soil Science* **42**, 1–8. doi:10.1111/j.1365-2389.1991.tb00085.x

Johnson NC, Graham JH, Smith FA (1997) Functioning of mycorrhizal associations along the mutualism-parasitism continuum. *New Phytolo-*

gist **135**, 575–585. doi:10.1046/j.1469-8137.1997.00729.x

Jones DL, Hodge A, Kuzyakov Y (2004) Plant and mycorrhizal regulation of rhizodeposition. *New Phytologist* **163**, 459–480. doi:10.1111/j.1469-8137.2004.01130.x

Kabata-Pendias A (2011) *Trace Elements in Soils and Plants*. 4th edn. CRC Press, Boca Raton.

Kauffman P (1998) *Wik, Mining and Aborigines*. Allen and Unwin, Sydney.

Kew GA, Mengler FC, Gilkes RJ (2007) Regolith strength, water retention, and implications for ripping and plant root growth in bauxite mine restoration. *Restoration Ecology* **15**(4), S54–S64. doi:10.1111/j.1526-100X.2007.00293.x

Klironomos J, Zobel M, Tibbett M, Stock WD, Rillig MC, Parrent JL, Moora M, Koch AM, Facelli JM, Facelli E, Dickie IA, Bever JD (2011) Forces that structure plant communities: quantifying the importance of mycorrhizal symbiosis. *New Phytologist* **189**, 366–370. doi:10.1111/j.1469-8137.2010.03550.x

Kluber LA, Tinnes and KM, Caldwell BA, Dunham SM, Yarwood RR, Bottomley PJ, Myrold DD (2010) Ectomycorrhizal mats alter forest soil biogeochemistry. *Soil Biology & Biochemistry* **42**, 1607–1613. doi:10.1016/j.soilbio.2010.06.001

Koch JM (2007) Alcoa's mining and restoration process in south Western Australia. *Restoration Ecology* **15**(4), S11–S16. doi:10.1111/j.1526-100X.2007.00288.x

Koide RT, Mosse B (2004) A history of research on arbuscular mycorrhiza. *Mycorrhiza* **14**, 145–163. doi:10.1007/s00572-004-0307-4

Lach L, Hoffmann BD (2011) Are invasive ants better plant-defense mutualists? A comparison of foliage patrolling and herbivory in sites with invasive yellow crazy ants and native weaver ants. *Oikos* **120**, 9–16. doi:10.1111/j.1600-0706.2010.18803.x

Lardner T, Worthington TR, Baimbridge MF, Vlahos S, Tibbett M (2011) Optimising soil physical properties for rehabilitation of mined land – effects of tine type on soil strength and root proliferation. In *Mine Closure 2011, Volume 1: Mine Site Reclamation*. Lake Louise, Alberta. (Eds A Fourie, M Tibbett and A Beersing), pp. 153–162. Australian Centre for Geomechanics, Perth.

Lavelle P, Spain AV (2001) *Soil Ecology*, Kluwer Academic, Dordrecht.

Leavitt KJ, Fernandez GCJ, Nowak RS (2000) Plant establishment on angle of repose mine waste dumps. *Journal of Range Management* **53**, 442–452. doi:10.2307/4003758

Lee KE, Wood TG (1971) *Termites and Soils*. ACADEMIC PRESS, London.

Lenz M (1994) Food resources, colony growth and caste development in wood-feeding termites, In *Nourishment and Evolution in Insect Societies*. (Eds JH Hunt and CA Nalepa) pp. 159–209. Westview Press, Boulder.

Li D, Niu S, Luo Y (2012) Global patterns of the dynamics of soil carbon and nitrogen stocks following afforestation: a meta-analysis. *New Phytologist* **195**, 172–181. doi:10.1111/j.1469-8137.2012.04150.x

Livesley SJ, Grover S, Hutley LB, Jamali H, Butterbach-Bahl K, Fest B, Beringer J, Arndt SK (2011) Seasonal variation and fire effects on CH_4, N_2O and CO_2 exchange in savanna soils of northern Australia. *Agricultural and Forest Meteorology* **151**, 1440–1452. doi:10.1016/j.agrformet.2011.02.001

Lowe S, Browne M, Boudjelas S, De Porter M (2000) *A selection of the world's worst invasive alien species, A selection from the global invasive species database*. Invasive Species Specialist Group, Species Survival Commission, IUCN, 12 pp.

Mack RN, Lonsdale M (2001) Humans as global plant dispersers: getting more than we bargained for? *Bioscience* **51**(2), 95–102. doi:10.1641/0006-3568(2001)051[0095:HAGPDG]2.0.CO;2

Majer JD (1984a) Recolonisation by ants in rehabilitated open-cut mines in northern Australia. *Reclamation and Revegetation Research* **2**, 279–298.

Majer JD (1984b) The influence of ants on seeding operations in northern Australian mined areas. *Reclamation and Revegetation Research* **2**, 299–313.

Majer JD, Brennan KEC, Moir ML (2007) Invertebrates and the restoration of a forest ecosystem: 30 years of research following bauxite mining in Western Australia. *Restoration Ecology* **15**, S104–S115. doi:10.1111/j.1526-100X.2007.00298.x

Marcar NE, Crawford DF (2004) *Trees for Saline Landscapes.* Rural Industries Research and Development Corporation, Barton, ACT.

Marshall JD, Brooks JR, Lajtha K (2007) Sources of variation in the stable isotopic composition of plants. In *Stable Isotopes in Ecology and Environmental Science*, 2nd Edition, (Eds R Michener and K Lajtha), pp. 22–60, Blackwell Publishing, Carlton, Victoria.

McIntyre S, McIvor JG, Heard KM (Eds) (2004) *Managing and Conserving Grassy Woodlands.* CSIRO Publishing, Collingwood, Victoria.

McKenzie N, Jacquier D, Isbell RF, Brown K (2004) *Australian Soils and Landscapes: An Illustrated Compendium.* CSIRO Publishing, Collingwood, Victoria.

McMullan-Fisher SJM, May TW, Robinson RM, Bell TL, Lebel T, Catcheside P, York A (2011) Fungi and fire in Australian ecosystems: a review of current knowledge, management implications and future directions. *Australian Journal of Botany* **59**, 70–90. doi:10.1071/BT10059

Miller LR, Hoschke M, Neal M (1996) Depth of foraging of subterranean termites in tropical Australia. International Research Group on Wood Protection, *Proceedings of Conference 96–05–19/21*, Document IRG/WP 96–10141, 6 pp. (only abstract seen).

Moody PW, Bolland MDA (1999) Phosphorus. In *Soil Analysis, an Interpretation Manual.* (Eds KI Peverill, LA Sparrow and DJ Reuter) pp. 187–220. CSIRO Publishing, Collingwood, Victoria.

Mulligan D, Scougall J, Williams DJ, Nichols O, Tibbett M, Lindbeck K, Loch R, Allan J, Grun W, Farrell C, Benbow R (2006) Leading Practice Sustainable Development for the Mining Industry: Mine Rehabilitation. Department of Industry Tourism and Resources, Canberra.

Nichols OG, Carbon BA, Colquhoun IJ, Croton JT, Murray NJ (1985) Rehabilitation after bauxite mining in south-western Australia. *Landscape Planning* **12**, 75–92. doi:10.1016/0304-3924(85)90071-1

Norman MA, Koch JM, Grant CD, Morald TK, Ward SC (2006) Vegetation succession after bauxite mining in Western Australia. *Restoration Ecology* **14**, 278–288. doi:10.1111/j.1526-100X.2006.00130.x

Norrish K, Rosser H (1983) Mineral phosphate. In *Soils an Australian Viewpoint*, pp. 335–361. CSIRO Division of Soils, Melbourne, Academic Press, London.

O'Grady AP, Chen X, Eamus D, Hutley LB (2000) Composition, leaf area index and standing biomass of eucalypt open forests near Darwin in the Northern Territory, Australia. *Australian Journal of Botany* **48**, 629–638. doi:10.1071/BT99022

O'Keefe FD (1992) Bauxite mining and Walyamiri, the mining operation – paper one. In *Proceedings of the Seventeenth Annual Environmental Workshop*, 5–9 October, Yeppoon, Queensland. pp. 88–99, Australian Mining Industry Council, Dickson, ACT.

Pang J, Wang Y, Lambers H, Tibbett M, Siddique K, Ryan M (2013) Commensalism in an agroecosystem: hydraulic redistribution by deep-rooted legumes improves survival of a droughted shallow-rooted legume companion. *Physiologia Plantarum* **149**, 79–90. doi:10.1111/ppl.12020

Parr CL, Andersen AN, Chastagnol C, DuVaud C (2007) Savanna fires increase rates and distances of seed dispersal by ants. *Oecologia* **151**, 33–41. doi:10.1007/s00442-006-0570-5

Paton TR, Humphreys GS, Mitchell PB (2003) *Soils: A New Global View.* Routledge, London.

Perry DH, Lenz M, Watson JAL (1985) Relationships between fire, fungal rots and termite damage in Australian forest trees. *Australian Forestry* **48**, 46–53.

Pinno BD, Belanger N (2008) Ecosystem carbon gains from afforestation in the boreal transition zone of Saskatchewan (Canada) are coupled with the devolution of Black Chernozems. *Agriculture, Ecosystems & Environment* **123**, 56–62. doi:10.1016/j.agee.2007.04.002

Plassard C, Dell B (2010) Phosphorus nutrition of mycorrhizal trees. *Tree Physiology* **30**(9), 1129–1139. doi:10.1093/treephys/tpq063

Poesen J, Lavee H (1994) Rock fragments in topsoils: significance and processes. *Catena* **23**, 1–28. doi:10.1016/0341-8162(94)90050-7

Prieto I, Armas C, Pugnaire FI (2012) Water release through plant roots: new insights into its consequences at the plant and ecosystem level. *New Phytologist* **193**(4), 830–841. doi:10.1111/j.1469-8137.2011.04039.x

Prior LD, Eamus D, Bowman DMJS (2004) Tree growth rates in northern Australian savanna habitats: seasonal patterns and correlations with leaf attributes. *Australian Journal of Botany* **52**, 303–314. doi:10.1071/BT03119

Reddell P, Milnes AR (1992) Mycorrhizas and other specialised nutrient-acquisition strategies: their occurrence in woodland plants from Kakadu and their role in rehabilitation of waste rock dumps at a local uranium mine. *Australian Journal of Botany* **40**, 223–242. doi:10.1071/BT9920223

Reddell P, Spain AV, Milnes AR, Hopkins M, Hignett CT, Joyce S, Playfair LA (1992) Bauxite mining and Walyamiri, indicators of ecosystem recovery – paper three. In *Proceedings of the Seventeenth Annual Environmental Workshop.* 5–9 October, Yeppoon, Queensland. pp. 115–127. Australian Mining Industry Council, Dickson, ACT.

Reddell P, Spain AV, Milnes AR, Hopkins M, Hignett CT, Joyce S, Playfair LA (1993) 'Indicators of ecosystem recovery in rehabilitated areas of the open strip bauxite mine, Gove, Northern Territory'. CSIRO Minesite Rehabilitation Research Program report to Nabalco Pty Ltd Gove Joint Venture, Adelaide.

Reddell P, Spain AV, Hopkins M (1997) Dispersal of spores of mycorrhizal fungi in scats of native mammals in tropical forests of northeastern Australia. *Biotropica* **29**, 184–192. doi:10.1111/j.1744-7429.1997.tb00023.x

Reddell P, Gordon V, Hopkins MS (1999) Ectomycorrhizas in *Eucalyptus tetrodonta* and *E. miniata* forest communities in tropical northern Australia and their role in the rehabilitation of these forests following mining. *Australian Journal of Botany* **47**(6), 881–907. doi:10.1071/BT97126

Reed SC, Townsend AR, Taylor PG, Cleveland CC (2011) Phosphorus cycling in tropical forests growing on highly weathered soils. In *Phosphorus in Action.* (Eds EK Bünneman, A Oberson and E Frossard) pp. 339–369. Springer, Heidelberg.

Richards RJ, Calder GJ (1978) 'Rehabilitation at Gove, a bauxite mine in northern Australia'. Land Conservation Unit, Darwin, NT.

Rio Tinto Alcan (2009) *Gove 2009 sustainable development review.* http://www.riotinto.com/documents/ReportsPublications/2009_SD_Gove.pdf>

Roth CH (2004) A framework relating soil surface condition to infiltration and sediment and nutrient mobilization in grazed rangelands of north-eastern Queensland, Australia. *Earth Surface Processes and Landforms* **29**, 1093–1104. doi:10.1002/esp.1104

Schäffer B, Eggenschwiler L, Suter B, Vogt L, Buchter B, Pfister H, Schulin R (2007) Influence of temporary stockpiling on the initial development of restored topsoils. *Journal of Plant Nutrition and Soil Science* **170**, 669–681.

Schäffer B, Schulin R, Boivin P (2008) Changes in shrinkage of restored soil caused by compaction beneath heavy agricultural machinery. *European Journal of Soil Science* **59**, 771–783. doi:10.1111/j.1365-2389.2008.01024.x

Schlesinger WH (1997) *Biogeochemistry: An Analysis of Global Change.* Academic Press, San Diego.

Schüßler A, Walker C (2010) *Phylogeny and taxonomy of Glomeromycota ('arbuscular mycorrhizal (AM) and related fungi').* Arthur Schüßler and Christopher Walker, Gloucester. http://www.amf-phylogeny.com

Schwenke GD (1992) Organic matter dynamics in the post-mining landscape at Weipa, North Queensland. In *Proceedings of the Seventeenth Annual Environmental Workshop.* 5–9 October, Yeppoon, Queensland. pp. 155–167. Australian Mining Industry Council, Dickson, ACT.

Schwenke GD, Mulligan DR, Bell LC (2000a) Soil stripping and replacement for the rehabilitation of bauxite-mined land at Weipa. I. Initial changes to soil organic matter and related parameters. *Australian Journal of Soil Research* **38**, 345–369. doi:10.1071/SR99043

Schwenke GD, Ayre L, Mulligan DR, Bell LC (2000b) Soil stripping and replacement for the rehabilitation of bauxite-mined land at Weipa. II. Soil organic matter dynamics in mine soil chronosequences. *Australian Journal of Soil Research* **38**, 371–393. doi:10.1071/SR99044

Scott K, Setterfield S, Douglas M, Andersen A (2010) Soil seed banks confer resilience to savanna grass-layer plants during seasonal disturbance. *Acta Oecologica* **36**, 202–210. doi:10.1016/j.actao.2009.12.007

Scott K, Setterfield SA, Douglas MM, Parr CL, Schatz J, Andersen AN (2012) Does long-term fire exclusion in an Australian tropical savanna result in a biome shift? A test using the reintroduction of fire. *Austral Ecology* doi:10.1111/j.1442-9993.2012.02379.x

Sea WB (2012) Self-thinning and tree competition in savannas. *Biotropica* **44**, 189–196. doi:10.1111/j.1744-7429.2011.00789.x

Setterfield SA, Williams RJ (1996) Patterns of flowering and seed production in *Eucalyptus miniata* and *E. tetrodonta* in a tropical savanna woodland, Northern Australia. *Australian Journal of Botany* **44**, 107–122. doi:10.1071/BT9960107

Short TA, Menzies NW, Mulligan DR (2000) Mining disturbance alters phosphorus fractions in northern Australian soils. *Australian Journal of Soil Research* **38**, 411–421. doi:10.1071/SR99033

Short TA, Kopittke PM, Mulligan DR, Menzies NW (2007) Growth of *Eucalyptus* species in a brown kandosol, and changes in soil phosphorus fractionation following fertilisation. *Australian Journal of Soil Research* **45**, 190–198. doi:10.1071/SR06147

Smith SE, Read DJ (2008) *Mycorrhizal Symbiosis*, 3rd edition. Elsevier, Netherlands.

Society for Ecological Restoration International Science & Policy Working Group 2004. *The SER International Primer on Ecological Restoration*. http://www.ser.org/content/ecological_restoration primer.asp

Soil Survey Staff (1999) *Soil Taxonomy: A Basic System of Soil Classification for Making and Interpreting Soil Surveys*. 2nd edn. U.S. Department of Agriculture, Soil Conservation Service, Washington D.C.

Spain AV, Hinz DA, Ludwig J, Tibbett M, Tongway D (2006) Mine closure and ecosystem development: Alcan Gove bauxite mine, NT, Australia. In *Mine Closure 2006, Proceedings of the First International Seminar on Mine Closure*. 13–15 September 2006, Perth. (Ed A Fourie and M Tibbett) pp. 299–308. Australian Centre for Geomechanics, Perth.

Spain AV, Ludwig J, Tibbett M, Tongway D (2009) '*Ecological and minesoil development studies at the Rio Tinto Alcan Gove Mine Site, Gove, NT, Final Report, Volume One – Text*'. Centre for Land Rehabilitation, University of Western Australia, Perth.

Spain AV, Hinz DA, Tibbett M (2010) Colonisation of rehabilitated lands by termites (Dictyoptera), Rio Tinto Alcan Gove bauxite mine, Northern Territory, Australia. In *Mine Closure 2010, Proceedings of the Fifth International Seminar on Mine Closure*. 23–26 November, 2010, Viña del Mar. (Eds A Fourie, M Tibbett and J Wiertz) pp. 437–448. Australian Centre for Geomechanics, Perth.

Stace HCT, Hubble GD, Brewer R, Northcote KH, Sleeman JR, Mulcahy MJ, Hallsworth EG (1968) *A Handbook of Australian Soils*. Rellim Technical Publications, Glenside, South Australia.

Standish RJ, Stokes BA, Tibbett M, Hobbs RJ (2007) Seedling response to phosphate addition and inoculation with arbuscular mycorrhizas and the implications for old-field restoration in Western Australia. *Environmental and Experimental Botany* **61**, 58–65. doi:10.1016/j.envexpbot.2007.03.004

Standish RJ, Morald TK, Koch JM, Hobbs RJ, Tibbett M (2008) Restoring Jarrah forest after bauxite mining in Western Australia —The effect of fertilizer on floristic diversity and composition.

In *Mine Closure 2008, Proceedings of the Third International Seminar on Mine Closure*. 14–17 October 2008, Johannesburg, South Africa. (Eds AB Fourie, M Tibbett, IM Weiersbye and PJ Dye) pp. 717–725. Australian Centre for Geomechanics, Perth.

Standish RJ, Tibbett M, Vlahos S, Stokes BA, Hobbs RJ (2010) Effect of fertiliser on early-successional Jarrah forest restored after bauxite mining in south-western Australia. In *Mine Closure 2010, Proceedings of the Fifth International Seminar on Mine Closure*. 23–26 November 2010, Viña del Mar, Chile. (Eds AB Fourie, M Tibbett and J Wiertz) pp. 387–396, Australian Centre for Geomechanics, Perth.

Szota C, Veneklaas EJ, Koch JM, Lambers H (2007) Root architecture of jarrah (*Eucalyptus marginata*) trees in relation to post-mining deep ripping in Western Australia. *Restoration Ecology* **15**. (Supplement S65–S73).

Tacey WH, Glossop BL (1980) Assessment of topsoil handling techniques of sites mined for bauxite within the Jarrah forest of Western Australia. *Journal of Applied Ecology* **17**, 195–201. doi:10.2307/2402974

Tardy Y, Roquin C (1992) Geochemistry and evolution of lateritic landscapes. In *Weathering, Soils and Paleaosols*. (Eds IP Martini and W Chesworth) pp. 407–443. Elsevier, Amsterdam.

The Plant List. http://www.theplantlist.org/

Thompson K (2011) Life after death: the role of litter in ecosystems, introduction. *Functional Ecology*, Virtual issue. http://www.functionalecology.org/view/0/virtualIssues/lifeafterdeathvirtualissue.html

Tibbett M (2008) Carbon accumulation in soils during reforestation–the Australian experience after bauxite mining. In *Proceedings of the Third International Seminar on Mine Closure. Johannesburg, South Africa*. (Eds A Fourie, M Tibbett, I Weiersbye and P Dye) pp. 3–11. Australian Centre for Geomechanics, Perth.

Tibbett M (2010) Large scale mine site restoration of Australian eucalypt forests after bauxite mining: soil management and ecosystem development. In *Ecology of Industrial Pollution*. (Eds LC Batty and KB Hallberg), pp. 309–326. Cambridge University Press, Cambridge.

Tibbett M, Sanders FE (2002) Ectomycorrhizal symbiosis can enhance plant nutrition through improved access to discrete organic nutrient patches of high resource quality. *Annals of Botany* **89**, 783–789. doi:10.1093/aob/mcf129

Tiessen H, Moir JO (2008) Characterization of available P by sequential extraction. In *Soil Sampling and Methods of Analysis*. 2nd edn. (Eds MR Carter and EG Gregorich) pp. 293–306. CRC Press, Boca Raton.

Tongway DJ (2010) Development of physical, chemical and biological soil properties at the Rio Tinto Alcan Gove bauxite mine, Northern Territory, Australia. In *Mine Closure 2010, Proceedings of the Fifth International Seminar on Mine Closure*. 23–26 November 2010, Viña del Mar, Chile. (Eds A Fourie, M Tibbett and J Wiertz) pp. 417–435. Australian Centre for Geomechanics, Perth.

Tongway DJ, Ludwig JA (2011) *Restoring Disturbed Landscapes, Putting Principles into Practice*. Island Press, Washington.

Treseder KK, Cross A (2006) Global distributions of arbuscular mycorrhizal fungi. *Ecosystems* **9**, 305–316. doi:10.1007/s10021-005-0110-x

Turner BL (2008) Resource partitioning for soil phosphorus: a hypothesis. *Journal of Ecology* **96**, 698–702. doi:10.1111/j.1365-2745.2008.01384.x

van der Heijden GA, Bardgett RD, van Straalen NM (2008) The unseen majority: soil microbes as drivers of plant diversity and productivity in terrestrial ecosystems. *Ecology Letters* **11**, 296–310. doi:10.1111/j.1461-0248.2007.01139.x

Veresoglou SD, Chen B, Rillig MC (2012) Arbuscular mycorrhizas and soil nitrogen cycling. *Soil Biology & Biochemistry* **46**, 53–62. doi:10.1016/j.soilbio.2011.11.018

Vitousek PM, Porder S, Houlton BZ, Chadwick OA (2010) Terrestrial phosphorus limitation: mechanisms, implications, and nitrogen-phosphorus interactions. *Ecological Applications* **20**, 5–15. doi:10.1890/08-0127.1

Vlahos S, Bastow BB, Rayner GA (1999) Bauxite mining rehabilitation in the northern Jarrah forest. In *Proceedings of Fifth International Alumina Quality Workshop, Vol. 2*. Bunbury, Western Australia, pp. 559–569.

Walker LR, Wardle DA, Bardgett RD, Clarkson BD (2010) The use of chronosequences in studies of ecological succession and soil development. *Journal of Ecology* **98**, 725–736. doi:10.1111/j.1365-2745.2010.01664.x

Ward SC (2000) Soil development on rehabilitated bauxite mines in south-west Australia. *Australian Journal of Soil Research* **38**, 453–464. doi:10.1071/SR99032

Watson JAL, Gay FJ (1991) Isoptera. In *The Insects of Australia: a Textbook for Students and Research Workers*. 2nd edn. (Ed ID Naumann) pp. 330–347. CSIRO Division of Entomology, Melbourne University Press, Carlton South, Victoria.

Wedd RF (2002) Vegetation compositional and structural changes over 27 years on the revegetated minesite, Nhulunbuy, N.T., Masters in Tropical Environmental Management thesis, Charles Darwin University, Australia.

Werner PA (2012) Growth of juvenile and sapling trees differs with both fire-season and understorey type: trade-offs and transitions out of the fire-trap in an Australian savanna. *Austral Ecology* **37**, 644–657. doi:10.1111/j.1442-9993.2011.02333.x

Werner PA, Prior LD (2007) Tree-piping termites and growth and survival of host trees in savanna woodland of north Australia. *Journal of Tropical Ecology* **23**, 611–622. doi:10.1017/S0266467407004476

Werner PA, Prior LD, Forner J (2008) Growth and survival of termite-piped *Eucalyptus tetrodonta* and *E. miniata* in northern Australia: implications for harvest of trees for didgeridoos. *Forest Ecology and Management* **256**, 328–334. doi:10.1016/j.foreco.2008.04.027

Whelan CJ, Wenny DG, Marquis RJ (2008) Ecosystem services provided by birds. *Annals of the New York Academy of Sciences* **1134**, 25–60. doi:10.1196/annals.1439.003

Wild A (1958) The phosphate content of Australian soils *Australian Journal of Agricultural Research* **9**(2), 193–204. doi:10.1071/AR9580193

Williams RJ, Griffiths AD, Allan GE (2002) Fire regimes and biodiversity in the savannas of northern Australia. In *Flammable Australia: The Fire Regimes and Biodiversity of a Continent*. (Eds RA Bradstock, JE Williams and MA Gill) pp. 282–304. Cambridge University Press, Cambridge.

Wilson BA, Brocklehurst PS, Clark MJ, Dickinson KJM (1990) 'Vegetation survey of the Northern Territory, Australia. Explanatory notes to accompany 1:1,000,000 Map Sheets'. Technical Report No. 49, Conservation Commission of the Northern Territory, Darwin.

Woinarski J, Mackey B, Nix H, Traill B (2007) *The Nature of Northern Australia: Natural Values, Ecological Processes and Future Prospects*. ANU E-PRESS, Australian National University, Canberra, ACT. http://epress.anu.edu.au/nature_na_citation.html

Woinarski JCZ, Rankmore B, Hill B, Griffiths AD, Stewart A, Grace B (2009) Fauna assemblages in regrowth vegetation in tropical open forests of the Northern Territory, Australia. *Wildlife Research* **36**(8), 675–690. doi:10.1071/WR08128

Wynn JG, Bird MI (2007) C4-derived soil organic carbon decomposes faster than its C3 counterpart in mixed C3/C4 soils. *Global Change Biology* **13**(10), 2206–2217. doi:10.1111/j.1365-2486.2007.01435.x

Yang X, Post WM (2011) Phosphorus transformations as a function of pedogenesis: a synthesis of soil phosphorus data using Hedley fractionation method. *Biogeosciences Discussions* **8**, 5907–5934. doi:10.5194/bgd-8-5907-2011

Yunupingu B, Yunupingu-Marika L, Marika D, Marika B, Marika B, Marika R, Wightman G (1995) 'Rirratjingu ethnobotany: Aboriginal plant use from Yirrkala, Arnhem Land, Australia'. Northern Territory Botanical Bulletin No. 21, Parks and Wildlife Commission of the Northern Territory.

Appendix 9.1. Listing of plant species recorded from restored sites at the Gove mine

The species listed were reported by Richards and Calder (1978), Hinz (1992), Reddell *et al.* (1993), Hinz (1992, 1999, 2007 and personal communication 2011) and by J Ludwig in Spain *et al.* (2009). Nomenclature follows: 'Australian Plant Name index', 'Australian Plant Census' and 'The Plant List'.

Species	Family
Acacia gonocarpa	Fabaceae
Alloteropsis semialata	Poaceae
Alysicarpus rugosus	Fabaceae
Alysicarpus vaginalis	Fabaceae
Alyxia spicata	Apocynaceae
Ampelocissus acetosa	Vitaceae
Andrographis paniculata	Acanthaceae
Anthobolus filifolius	Santalaceae
Aristida browniana	Poaceae
Aristida holathera	Poaceae
Asparagus racemosus	Asparagaceae
Austrodolichos errabundus	Fabaceae
Blumea sp.	Asteraceae
Bonamia brevifolia	Convolvulaceae
Borreria carpentariae	Rubiaceae
Borreria exserta	Rubiaceae
Bossiaea bossiaeoides	Fabaceae
Brachiaria decumbens	Poaceae
Brachiaria holosericea	Poaceae
Brachychiton diversifolius	Sterculiaceae
Brachychiton megaphyllus	Sterculiaceae
Brachychiton paradoxus	Sterculiaceae
Breynia sp.	Euphorbiaceae
Bridelia tomentosa	Euphorbiaceae
Cajanus marmoratus	Fabaceae
Callitris intratropica	Cupressaceae
Calopogonium mucunoides	Fabaceae
Canarium australianum	Burseraceae
Capillipedium parviflorum	Poaceae
Carallia brachiata	Rhizophoraceae

Appendix 9.1. (Continued)

Species	Family
Cassia absus	Caesalpiniaceae
Cassia mimosoides	Caesalpiniaceae
Cassytha sp.	Lauraceae
Cayratia trifolia	Vitaceae
Celtis philippinensis	Cannabaceae
Cenchrus ciliaris	Poaceae
Cenchrus setigerus	Poaceae
Chrysopogon fallax	Poaceae
Clerodendrum floribundum	Verbenaceae
Colubrina asiatica	Rhamnaceae
Commelina ensifolia	Commelinaceae
Crinum uniflorum	Amaryllidaceae
Crotalaria trifoliastrum	Fabaceae
Crotalaria brevis	Fabaceae
Crotalaria calycina	Fabaceae
Crotalaria lanceolata	Fabaceae
Crotalaria linifolia	Fabaceae
Crotalaria medicaginea	Fabaceae
Cupaniopsis anacardioides	Sapindaceae
Cymbopogon procerus	Poaceae
Cynodon arcuatus	Poaceae
Cyperus sp.	Cyperaceae
Dactyloctenium aegyptiacum	Poaceae
Desmodium flagellare	Fabaceae
Digitaria bicornis	Poaceae
Digitaria decumbens	Poaceae
Dioscorea bulbifera	Dioscoreaceae
Dioscorea transversa	Dioscoreaceae
Dipodium punctatum	Orchidaceae
Distichostemon hispidulus	Sapindaceae
Dodonaea sp.	Sapindaceae

(Continued)

Appendix 9.1. (Continued)

Species	Family
Drypetes australasica	Euphorbiaceae
Drypetes lasiogyna	Euphorbiaceae
Ectrosia leporina	Poaceae
Eleusine indica	Poaceae
Eragrostis cumingii	Poaceae
Eriosema chinense	Fabaceae
Erythrina variegata	Fabaceae
Euphorbia mitchelliana	Euphorbiaceae
Euphorbia schultzii	Euphorbiaceae
Evolvulus alsinoides	Convolvulaceae
Exocarpos latifolius	Santalaceae
Ficus opposita	Moraceae
Ficus superba	Moraceae
Fimbristylis xyridis	Cyperaceae
Fimbrystylis densa	Cyperaceae
Flueggea virosa	Euphorbiaceae
Galactia muelleri	Fabaceae
Galactia tenuiflora	Fabaceae
Galactia striata	Fabaceae
Ganophyllum falcatum	Sapindaceae
Gardenia sp.	Rubiaceae
Geodorum pictum	Orchidaceae
Glochidion xerocarpum	Euphorbiaceae
Glycine tomentella	Fabaceae
Gompholobium subulatum	Fabaceae
Gonocarpus leptothecus	Haloragaceae
Goodenia hispida	Goodeniaceae
Grevillea dryandri	Proteaceae
Grevillea goodii	Proteaceae
Grewia retusifolia	Tiliaceae
Habenaria sp.	Orchidaceae
Helicteres cana	Sterculiaceae

Appendix 9.1. (Continued)

Species	Family
Heliotropium ventricosum	Boraginaceae
Heteropogon contortus	Poaceae
Heteropogon triticeus	Poaceae
Hibbertia dealbata	Dilleniaceae
Hibbertia lepidota	Dilleniaceae
Hibiscus meraukensis	Malvaceae
Hybanthus enneaspermus	Violaceae
Hyptis suaveolens	Lamiaceae
Indigofera linifolia	Fabaceae
Ipomoea gracilis	Convolvulaceae
Ipomoea pes-tigridis	Convolvulaceae
Jacksonia sp.	Fabaceae
Jacquemontia browniana	Convolvulaceae
Jacquemontia paniculata	Convolvulaceae
Leucaena leucocephala	Fabaceae
Lomandra tropica	Dasypogonaceae
Macaranga tenarius	Euphorbiaceae
Marsdenia viridiflora	Apocynaceae
Melinis minutiflora	Poaceae
Melinis repens	Poaceae
Merremia hastata	Convolvulaceae
Merremia quinata	Convolvulaceae
Mnesithea formosa	Poaceae
Myristica insipida	Myristicaceae
Neptunia gracilis	Fabaceae
Olearia sp.	Asteraceae
Panicum maximum	Poaceae
Paspalum plicatulum	Poaceae
Passiflora sp.	Passifloraceae
Pennisetum polystachion	Poaceae
Petalostigma pubescens	Euphorbiaceae
Petalostigma quadriloculare	Euphorbiaceae

(Continued)

Appendix 9.1. (Continued)

Species	Family
Phyllanthus sp.	Euphorbiaceae
Pimelea punicea	Thymelaeaceae
Planchonella sp.	Sapotaceae
Planchonia careya	Lecythideae
Polygala orbicularis	Polygalaceae
Pycnospora lutescens	Fabaceae
Pseudopogonatherum contortum	Poaceae
Rhynchospora heterochaeta	Cyperaceae
Rhynchospora sp.	Cyperaceae
Schizachyrium fragile	Poaceae
Scleria sp.	Cyperaceae
Sebastiania chamaelea	Euphorbiaceae
Setaria apiculata	Poaceae
Sida cordifolia	Malvaceae
Sorghum brevicallosum	Poaceae
Sorghum plumosum	Poaceae
Stackhousia viminea	Stackhousiaceae
Stenocarpus cunninghamii	Proteaceae
Striga curviflora	Scrophulariaceae
Stylosanthes guianensis var. *gracilis*	Fabaceae
Stylosanthes hamata	Fabaceae
Stylosanthes humilis	Fabaceae
Tacca leontopetaloides	Taccaceae
Tephrosia juncea	Fabaceae
Terminalia ferdinandiana	Combretaceae
Terminalia latipes	Combretaceae
Thaumastochloa major	Poaceae
Thecanthes punicea	Thymelaeaceae
Themeda arguens	Poaceae
Themeda quadrivalvis	Poaceae
Thysanotus banksii	Asparagaceae
Trema aspera	Ulmaceae

Appendix 9.1. (Continued)

Species	Family
Trema carpentariae	Ulmaceae
Trema orientalis	Ulmaceae
Tridax procumbens	Asteraceae
Uraria lagopodioides	Fabaceae
Urochloa mosambicensis	Poaceae
Verticordia sp.	Myrtaceae
Vigna lanceolata	Fabaceae
Vitex acuminata	Verbenaceae
Vitex glabrata	Verbenaceae
Vittadinia brachycomoides	Asteraceae
Waltheria indica	Sterculiaceae
Whiteochloa airoides	Poaceae
Ziziphus oenopolia	Rhamnaceae

WAYS FORWARD

10

Key values of metallophytes for the minerals industry in Australasia

Antony van der Ent

Summary

The unrealised resource of plant species naturally occurring on metalliferous soils ('metallophytes') could provide the minerals industry with an economic and sustainable means of mine site rehabilitation and effective progression towards mine closure. Rather than viewing mining solely as an industry that contributes to the destruction of metallophyte habitats and subsequent biodiversity loss, offsetting could be applied through the use of metallophyte species in the rehabilitation to not only be an effective way to mitigate the loss of habitat, but also to secure the survival of the metallophyte species themselves. Of the range of available phytotechnologies, phytomining offers the potential to capitalise on sub-economic ore bodies, particularly ultramafic regoliths. This technology could be complementary to conventional strip-mining methods as part of the subsequent rehabilitation strategy and presents a potential use for hyperaccumulator species in commercial "metal farming".

Introduction

Most plant species native to Australasia are unable to survive and reproduce on metalliferous soils that impose metal toxicity (Baker *et al.* 2010). A minority of plant species however have the unique ability to thrive under metal toxicity and are called metallophytes (Baker 1987; Baker and Whiting 2008; Brooks 1987). Some of these metallophytes are ecotypes of more common species that have evolved metal-tolerance (facultative or pseudo-metallophytes) and represent opportunistic species with a wide ecological range. An example is *Equisetum ramosissimum* (Equisetaceae) which can grow in copper-rich acidic mine drainage (Fig. 10.1).

Other metallophytes are solely restricted to metalliferous soils (obligate or strict metallophytes). Globally, distinct plant communities are known to exist on metalliferous soils. The most extensive metalliferous soils are those derived from ultramafic outcrops (ferromagnesian-rich mantle rocks with anomalous nickel, chromium and cobalt concentrations), which are represented on all continents. Other examples of plant communities on metalliferous soils include those on calamine (zinc, lead and cadmium-enriched soils) in Europe and China; and those on copper and cobalt-enriched soils in the DR Congo and Zambia.

Figure 10.1: Few edaphic conditions are too challenging for plants to adapt as these *Equisetum ramosissimum* (Equisetaceae) demonstrate by growing in highly phytotoxic copper leachates.

Strategies for metal-tolerance

Plants may exclude, indicate or hyperaccumulate metallic elements in their shoots in response to growing on metalliferous soils (Baker 1981). These responses result from a range of ecophysiological mechanisms enabling tolerance or counteracting toxicity of soil metals to which they are exposed (Baker 1981). Most plant species however lack specific metal tolerance and/or exclusion mechanisms and can therefore not survive on metalliferous soils. Ecophysiological responses differ greatly between plants and metals (Baker *et al.* 2010).

1. **Hyperaccumulators** accumulate metallic elements (such as cadmium, nickel, manganese or zinc) or metalloids (such as selenium or arsenic) in their shoots over a wide range of (available) metal concentrations in the soil, with a concentration factor generally higher >1 (e.g. concentrations in the shoots exceed those in the soil). The shoot: root ratio is >1, due to enhanced translocation and storage in the shoots (Baker 1981; Baker *et al.* 2000). Most hyperaccumulators are restricted ('obligate') to metalliferous soils, such as ultramafic soils to which most nickel hyperaccumulators are confined (Kazakou *et al.* 2008), although some hyperaccumulator species occur on metalliferous and normal soils ('facultative') while only displaying hyperaccumulator behaviour on the metalliferous soils (Pollard *et al.* 2014).

2. **Indicators** uptake and translocate metals into the shoots in a response to (available) concentrations in the soil, until toxicity occurs (Baker

1981; Baker 1987). With only limited ecophysiological tolerance or exclusion mechanisms indicator-type plants cannot tolerate high (available) metal concentrations in soils (Baker 1981), and hence such plants do not specifically occur on metalliferous soils.

3. **Excluders** maintain low metal concentrations in the shoot over a wide range of concentrations of (available) metals in the soil, up to a critical threshold after which ecophysiological mechanisms restricting transport fail (Baker 1981). The shoot:root ratio is < 1 (and often root metal concentrations are high). Excluder-type plants occur on metalliferous and normal soils, with distinct tolerant and non-tolerant ecotypes.

Accumulation of metallic and metalloid elements by plants

Hyperaccumulators share the ability to grow on metalliferous soils and to accumulate exceptional concentrations of specific metallic and metalloid elements in their shoots (Reeves 2003; Van der Ent et al. 2013a). The shoot metal concentrations which hyperaccumulators develop are greatly in excess of other plant species growing on similar metalliferous soils (Ginocchio and Baker 2004) without apparent toxicity occurring (Baker 2009). To date ~500 plant species have been identified as hyperaccumulators for arsenic, cadmium, cobalt, copper, manganese, zinc, nickel, lead, thallium and selenium (Van der Ent et al. 2013a). Considering the total number of known plant species in the world, the phenomenon is hence rare. Hyperaccumulator plants can be divided into four main groups on the basis of the accumulated metals or metalloids: (1) Arsenic or selenium hyperaccumulators, (2) Copper and/or cobalt hyperaccumulators, (3) Zinc, lead and/or cadmium hyperaccumulators, and (4) Nickel hyperaccumulators.

Hyperaccumulator plants

Hyperaccumulation in plants is operationally defined in relative rather than absolute concentrations (Brooks et al. 1998) with different notional thresholds set for different metals. Accepted definitions for hyperaccumulators are those plants which contain in their dry weight foliar tissue more than 100 µg/g cadmium or thallium, more than 1000 µg/g of nickel or lead, more than 300 µg/g of cobalt, copper or chromium, more than 10 000 µg/g of manganese, or more than 3000 µg/g of zinc, while growing in their natural habitats (Baker and Brooks 1989; Brooks et al. 1977; Brooks 1979; Reeves 2006; Van der Ent et al. 2013a). For metalloids such as arsenic and selenium the threshold is set at 1000 µg/g and 100 µg/g respectively (Meharg 2003; Reeves 2003; Van der Ent et al. 2013a) to qualify for hyperaccumulator status. Although these criteria seem arbitrary, in general the concentrations of metals or metalloids in hyperaccumulator plants are several orders of magnitude greater than the natural concentrations of these metals in non-accumulating ('normal') plants (McGrath and Zhao 2003; Raskin et al. 1994). The shoot metal concentrations of hyperaccumulators do not directly reflect those of the soil, because hyperaccumulators have highly efficient metal-uptake mechanisms (Ernst 2006; Ginocchio and Baker 2004). Some common characterises for hyperaccumulators are a bioconcentration factor of >1 (but often >50), a shoot-to-root metal content ratio >1 and extreme metal tolerance, due to effective biochemical detoxification (McGrath et al. 1993). The total number of known hyperaccumulators has increased rapidly through new discoveries and improved description. Total species numbers for nickel (450), zinc (12), manganese (10), cobalt (30), copper (32), selenium (20), lead (14), arsenic (5), and cadmium (2), hyperaccumulators have been described (Van der Ent et al. 2013a). Examples of hyperaccumulators include, *Sebertia acuminata* (now *Pycnandra acuminata* – Sapotaceae) from New Caledonia, a tree up to 10 metres tall with green latex containing 11.2% Ni on a fresh weight basis or 25.7% on a dry weight basis (Jaffré et al. 1976). Another example is *Psychotria douarrei* (now *Psychotria gabriellae* – Rubiaceae), also from New Caledonia, which contains up to 9.3% in the bark of the roots and 4.7% nickel in its leaves (Jaffré and Schmid 1974). Only a small proportion of all known metallophytes are hyperaccumulators (Rajakaruna and Baker 2006).

Some authors have suggested the term 'hypernickelophore' for hyperaccumulators with foliar nickel concentrations >1% (Jaffré and Schmid 1974), and similar terminology for zinc and manganese (Chaney et al. 1997). So far only ~35 species of such hypernickelophores have been recorded, but these plants have special significance for phytomining technology (described later). In addition, towards the lower end of accumulation, Boyd and Jaffré (2009) suggested 'hemi-accumulator' for plants with foliar nickel concentrations of 100–1000 μg/g.

Nickel hyperaccumulators and ultramafic soils

Ultramafic ('serpentine') soils are derived from ferromagnesian-rich mantle rocks and characterised by relatively high iron and magnesium concentrations, and concomitantly enrichment in nickel, chromium and cobalt while nutrient concentrations (potassium, phosphorus) are low (Brooks 1987). Often such soils have a blue-green hue from serpentinite minerals with high nickel phytoavailability (Fig. 10.2). Given that ultramafic soils represent the largest global extent of metalliferous soils most nickel hyperaccumulators are found obligate to these soils. Other natural metalliferous soils, such as those with high concentrations of copper/cobalt (DR Congo and Zambia) or zinc/lead/cadmium (Europe and China) are far smaller and highly localised. Continuous surface expression of ultramafic soils, in several instances hundreds or even thousands of km² in size, create large areas where plants have been able to colonise this habitat over millions of years, especially in tropical regions (Reeves 2006). The much higher plant diversity in the tropics also created prime evolutionary conditions for metal-tolerance and hyperaccumulation to evolve, and as a result nickel hyperaccumulators are widely distributed in different genera across at least 45 plant families (Baker et al. 1999) with a great variety of growth forms, ecophysiological characteristics and ecological requirements (Pollard et al. 2002). Nickel hyperaccumulators are well-represented in the following plant families: Asteraceae (especially South Africa), Brassicaceae (Mediterranean Region), Buxaceae (mainly Cuba), Salicaceae, Phyllanthaceae, Rubiaceae and Violaceae (New Caledonia, South-East Asia) (*sensu* Reeves 2003). Anomalously high foliar nickel concentrations are easily identified by chemical analysis of dried plant material in the laboratory or in the field by pressing fresh leaves against white test paper impregnated with the nickel specific reagent dimethylglyoxime ('DMG') (Baker et al. 1992).

Metallophytes as a key resource for the global minerals industry

The ability of metallophytes to cope with potentially toxic metal concentrations and otherwise difficult edaphic conditions suggests potential for rehabilitation of minerals wastes (Baker et al. 2010; Whiting et al. 2004). This provides the minerals industry with a potentially sustainable and economic means of site rehabilitation by introducing plant species naturally occurring on metalliferous soils. Possible applications of hyperaccumulator capabilities include phytoextraction for remediation purposes and its subset phytomining (Anderson et al. 1999a; McGrath and Zhao 2003; Pollard et al. 2002). These technologies exploit the metal accumulation characteristics in hyperaccumulator plants to extract metallic elements from contaminated soils or sub-economic ore bodies. In addition there is a wide range of metal-tolerant (excluder-type) plant species that are potentially suitable for phytostabilisation (Rajakaruna and Baker 2006; Whiting et al. 2004). As such the use of these plants can contribute to 're-greening' and reducing soil erosion while reducing entry of potentiality toxic metals into the food chain (Baker 2009). Excluder-type plants can also indirectly contribute to phytoremediation by root-action such as in *Typha angustifolia* (Typhaceae), which forms iron-rich plaques (Ye et al. 1997) (Fig. 10.3). Site-specific application of either

Figure 10.2: Ultramafic soils form the natural habitats of many plant species adapted to grow in nickel-rich and nutrient-poor edaphic conditions.

phytoextraction/phytomining or phytostabilisation may improve ecological performance at mine closure, thus improving environmental sustainability and resilience (Baker *et al.* 2010).

Using metallophytes for phytostabilisation

Over the last half a century ore grades have been declining while mined volumes are increasing (Mudd 2007). As a consequence ever-increasing amounts of minerals wastes such as tailings and overburden are generated posing a challenge to traditional minerals waste cover designs. Establishing plant cover is important to minimise impacts and metallophytes are uniquely suited for this purpose (Bradshaw and Chadwick 1980). The pioneering work of Bradshaw (1952, 1959) on the use of metal-tolerant grasses firmly launched phytostabilisation technology. If so used, metallophytes may stabilise minerals waste and thereby provide improved conditions for natural succession. For example, plants with supplementary nutrient-acquisition strategies such as carnivorous pitcher plants (Fig. 10.4) can stimulate the later establishment of other plant species. In some environments, evapotranspiration acts to prevent leachate migration towards groundwater (Keeling *et al.* 2003). Selected plant species for phytostabilisation not only need to be metal tolerant, but also drought resistant, establish on poorly structured substrates, and be able to cope with major nutrient deficiencies. The addition of soil amendments, such as organic matter (compost), nutrients (NPK) and lime, may enhance the establishment of plants on sites (Ernst 2005). Although it is clear that metallophytes with 'excluder' characteristics may find more ready application than hyperaccumulator plants, to date this group of plants have been

Figure 10.3: Some plant species are able to thrive under extreme chemical conditions such as in this pool with copper-mining leachate at pH 2–3 where *Typha angustifolia* (Typhaceae) survives because of its ecophysiological adaptations.

somewhat neglected by research (Dickinson *et al.* 2009).

Utilising hyperaccumulators for phytoextraction

The phenomenon of hyperaccumulation when discovered in the 1970s was generally regarded as a scientific oddity until Chaney (1983) and Baker and Brooks (1989) proposed the use of such plants to extract metallic elements from the soil in the new technology of phytoextraction. The benchmark papers by McGrath *et al.* (1993) and Baker *et al.* (1991, 1994) sparked these developments by proposing parameters for phytoextraction with the purpose of soil remediation. The first field trials began in 1991 in the UK and in 1993 in the US, after which several seminal papers followed (Baker and Smith 2000; Chaney *et al.* 1997; Lombi *et al.* 2001; McGrath and Zhao 2003; McGrath *et al.* 2000; Robinson *et al.* 1997a, b). Since the early 1990s phytoremediation applications have been tested on soils and substrates contaminated with cadmium, copper, cobalt, mercury, lead, nickel, thallium, arsenic, selenium and radioactive elements (Harris *et al.* 2009).

Phytoextraction technology

Metallic elements such as nickel, zinc or copper, cannot be broken down to carbon dioxide and water, but may be extracted from soils, tailings and minerals waste using hyperaccumulator plants (Chaney *et al.* 1997; McGrath *et al.* 1993). Living plants can in essence be considered as solar-driven ion-pumps, capable of transferring soil metals into their shoots (Raskin *et al.* 1994). This technology offers an economical *in-situ* approach

Figure 10.4: These carnivorous pitcher plants (*Nepenthes reinwardiana* – Nepenthaceae) are able to supplement their need of nitrogen by special organs that have evolved to attract, collect and digest insects, while being metal-tolerant to prevailing high soil copper concentrations.

to soil remediation, while reducing soil erosion. Phytoextraction may therefore offer significant benefits over soil *ex-situ* approaches, although regulators often prefer such invasive methods, because of greater confidence in the outcomes. Two different approaches may be taken in phytoextraction operations: natural phytoextraction, where selected hyperaccumulators are used; and chemically assisted phytoextraction, which uses non-specific crops (such as *Brassica juncea*) and metal-chelating agents (EDTA and the like). Although still heavily researched and promoted in some regions, chemically assisted phytoextraction must be strongly denounced because of the major (uncontrolled) soil pollution it causes by metal re-distribution in the soil and groundwater. Phytoextraction usually requires a long-term commitment and metals may not be able to be entirely removed and ultimately the success depends on the bioavailability of the target metal(s), nutrient availability, water status and plant species spatial access to the target metal (Ernst 1996). Each target metal must be considered separate because of the different soil chemistry and plant ecophysiological responses (Chaney *et al.* 1997). A summary of desirable outcomes and limitations of phytoextraction are:

Advantages of phytoextraction:
- overall costs are lower than many conventional *ex-situ* technologies
- recovery of low concentrations of metals in soils
- minimal environmental disruption, comparable to agriculture or agroforestry

- conservation of biodiversity when using native hyperaccumulators.

Limitations of phytoextraction:
- effectiveness is limited to the surface area and soil depth available to plant roots
- slow growth rates and low biomass require long-term commitment
- success rate of plant growth is significantly affected by metal concentrations, and edaphic factors such as water and nutrient availability, and structural characteristics of the soils
- shoots with high metal concentrations impose a risk for transfer of metals to higher trophic levels in the food chain.

phytoextraction application are site specific. Target properties for plants potentially used for phytoextraction include:

- high level of hyperaccumulation in harvestable shoots
- high biomass production and fast growth rate
- preferable perennial, regrowth from root base or after pruning
- tolerant to major nutrient deficiencies (retaining biomass yield)
- tolerant to other soil elements other than the target metal
- wide ecological amplitude, without being susceptive to becoming a weed.

Selecting plants for phytoextraction

The success of plant growth is profoundly affected by prevailing edaphic factors such as drought, salinity and physical characteristics of the soils/substrate being remediated. Plant species for phytoremediation therefore need to be selected on basis of their total suite of characteristics, such as their uptake efficiency (Clemens 2006), the translocation of the metals from the root to the shoot, the level of accumulation in the shoots, the growth form, and the metal tolerance of the plant species or ecotype and their associated rhizosphere microbes (Ernst 1996; Lasat 2002). Although most (temperate region) hyperaccumulators have low biomass there are fast-growing hyperaccumulators producing high biomass (McGrath *et al.* 1993), such as *Berkheya coddii* (Asteraceae) from South Africa. Several nickel hyperaccumulators from Malaysia and Indonesia have a high biomass and high metal concentrations, and also relatively fast growth rates, such as *Phyllanthus*-species (Phyllanthaceae). Soils contaminated with different metals present a significant problem for phytoextraction, because of the specificity of hyperaccumulation, although some hyperaccumulators are capable of hyperaccumulation of several metals simultaneously (for example zinc and cadmium). Generally the characteristics of a suitable plant species for a

Ecotype selection and plant breeding

Populations of hyperaccumulators can be highly variable in metal accumulation behaviour. This is the result of phenotypic plasticity, which allows plants to adapt to the edaphic conditions to which they are confronted as well as local soil chemical conditions (Lasat 2002). Ideally hyperaccumulators to be used for phytoextraction should be fast growing with high biomass and high accumulation (McGrath *et al.* 2002; Pollard *et al.* 2002). Bioengineering could potentially create such 'ideal phytoextraction crops' as transgenic plants, however, given the negative public attitude to genetically modified organisms and the technical challenges in molecular engineering, improving 'natural' hyperaccumulators by conventional plant breeding technology seems the best approach. The combination of traditional and novel methods of hybridisation to use natural genetic variability may improve phytoextraction crops substantially (Chaney *et al.* 2000; Raskin *et al.* 1994).

Phytomining

The technology of phytomining is based on phytoextraction and relies on hyperaccumulators to extract metals in biomass for economic gain rather than pollution remediation (Chaney 1983; Brooks and Robinson 1998). In this approach hyperaccumulators are grown over (spatially large) sub-economic ore bodies or ultramafic soils followed by harvesting and incineration of the biomass to produce a commercial high-grade bio-ore. The nickel hyperaccumulator, *Streptanthus polygaloides* (Brassicaceae), yielded 100 kg/ha Ni from ultramafic substrates in initial experiments (Nicks and Chambers 1995, 1998). Similar success were achieved with the nickel hyperaccumulators; *Alyssum bertolonii* (Brassicaceae) in Italy (72 kg/Ni/ha) by Robinson *et al.* (1997a) and with *Berkheya coddii* in South Africa (100 kg/Ni/ha), with a thallium hyperaccumulator; *Biscutella laevigata* (Brassicaceae) in France (8 kg/ha Tl) and with gold by induced hyperaccumulation using ammonium thiocyanate in *Brassica juncea* (Brassicaceae) (up to 57 mg/kg Au) (Anderson *et al.* 1998, 1999b, 2005). Societal pressure to reduce the environmental impacts of conventional mining, technical difficulties in economic recovery of metals from low-grade ores, and high metal prices have contributed to the sustained research in phytomining (Harris *et al.* 2009). Compared to strip-mining operations, phytomining has an environmental impact similar to agriculture or agroforestry and it does not require mine site rehabilitation at the end of life (Harris *et al.* 2009). Phytomining might also generate renewable energy by the necessary incineration of plant biomass before smelting or hydrometallurgical refining (Anderson *et al.* 1999a; Robinson *et al.* 1999a). It should be emphasised though that phytomining is unlikely to replace conventional mining at any scale, and should not been seen as a standalone technology, but rather a 'top-up' as part of mine site rehabilitation.

Limitations to phytomining

Phytomining depends on the distribution of the target metal and plant roots within the soil profile, as well as the bioavailability of the target metal. The rate and level of addition of bioavailable metal from the non-bioavailable and potentially bioavailable pools due to extraction of metals by plants in phytomining is an important factor to consider in feasibility studies. Research by Robinson *et al.* (1999b) showed that between 3 and 18 successive economic extracts could be obtained before the soil was practically depleted. However once the initial metal content of the topsoil has been depleted, the topsoil can be ploughed to bring fresh material to the surface (Anderson *et al.* 1999b). Ultimately phytomining is finite as the target metal is removed, as opposed to conventional agriculture (Anderson *et al.* 1999a).

Implementation and finances of phytomining

The range of metals suitable for phytomining are constrained by several conditions: (a) presence of known hyperaccumulators of the target metal; (b) large surface areas with mineralisation of this metal; and (c) sufficient economic value to balance the costs of crop production and metal extraction. In the first instance, potentially suitable hyperaccumulators for specific metals are limited to nickel, cobalt, zinc, thallium, cadmium and copper. While current (average June 2014) metal prices (per ton) for nickel US$18 350, cobalt US$30 200 and thallium US$60 000 are high, zinc US$2167 and copper US$6777 are relatively low. The theoretical requirements for commercial investment are therefore met by only nickel, cobalt and thallium, but of these metals, large surface areas with enrichment only exist for nickel.

The potential for nickel phytomining

Many species show potential for phytomining of nickel compared to any other metal (Chaney *et al.* 2007a) with a range of plant species accumulating at least 1–3% nickel in dry matter, which translated into 12% to >20% in the ash. This is an order of magnitude higher concentration than current lateritic ores (1–2%). Nickel bio-ore could be an addition to smelter feed stocks (Dickinson *et al.* 2009) and the relatively high price of nickel on global financial markets indicates economic viability in the future (Chaney *et al.* 2007b). Both high biomass yields and metal hyperaccumulation of the 'metal crop' are required to make phytomining efficient and commercially viable (McGrath *et al.* 1993). *Berkheya coddii* shows high potential for phytomining as it has high biomass, it grows fast, is a perennial which is able to grow after excision, it is a strong hyperaccumulator of nickel and it grows well in a variety of climatic conditions and water regimes (Robinson *et al.* 1997a). *Berkheya coddii* is able to accumulate 11 600 µg/g (1.16%) to 17 000 µg/g (1.7%) dry mass nickel (Morrey 1998; Robinson *et al.* 1997a). In experimental trials, accumulation levels up to 10 000 µg/g (1%) Ni (Robinson *et al.* 1997a), and a nickel yield of 110 kg/ha have been achieved with moderate fertilisation and adequate irrigation (Anderson *et al.* 1999a; Robinson *et al.* 1997a). While *Berkheya coddii* can withstand a wide temperature range (Robinson *et al.* 1997a) its growth performance under the prevalent semi-arid conditions in many parts of Australia is unknown and there is a major risk of this species becoming an invasive weed because of the production of large quantities of wind-borne seeds, despite its narrow ecological range in South Africa (Robinson *et al.* 1997a). The Anglo-American Platinum Corporation in South Africa has trialled phytomining with *Berkheya coddii* successfully in 1996 near a refinery site, and was able to recover nickel metal from the biomass (Anderson *et al.* 1999b). Trials at Soroako, Sulawesi by Vale INCO used *Alyssum* species from Europe, which although strong hyperaccumulators in their native temperate ranges, did not prove to be successful in tropical humid climates of Sulawesi (Van der Ent *et al.* 2013b). However, regionally a range of potentially suitable plant species for phytomining occurs, for example the shrub *Phyllanthus* cf. *securinegoides*(Fig. 10.5). Elsewhere, in Italy and the US, phytomining with *Alyssum* species has been highly successfully demonstrated in full field trials, and a patent has been registered for this technology (Chaney *et al.* 2007b).

Examples of hyperaccumulators from the region

Some of the earliest nickel hyperaccumulators to be reported from the region were species of *Rinorea*

Figure 10.5: *Phyllanthus* cf. *securinegoides* (Phyllanthaceae) from Sabah (Malaysia) may prove to be an ideal candidate for nickel phytomining. The test paper impregnated with DMG shows a strong colour reaction, indicating high concentrations of nickel in the plant tissue (1.4–1.8%).

(Violaceae), including *Rinorea bengalensis* (17 350 µg/g nickel) and *R. javanica* (2170 µg/g nickel) (Brooks and Wither 1977; Brooks 1987). On the basis of extreme nickel accumulation, noteworthy are especially species of *Phyllanthus* (Phyllanthaceae), such as *P. insulae-japen* (34 330–38 720 µg/g nickel) from Yapen Island in Indonesia (Reeves 2003), *P. securinegoides* (34 750 µg/g) from Palawan and Mindanao in the Philippines (Baker *et al.* 1992) and *P. favieri* (21 750 µg/g nickel) from New Caledonia (Amir *et al.* 2007). In Australia the strongest nickel hyperaccumulators are *Stackhousia tryonii* (Celastraceae) (3640–41 260 µg/g nickel) from Queensland (Batianoff and Specht 1992) and *Hybanthus floribundus* (Violaceae) (13 500 µg/g nickel) from Western Australia (Farago and Mahmoud 1983). In New Caledonia, besides the earlier mentioned *Pycnandra acuminata* and *Psychotria gabriellae* another strong nickel hyperaccumulator is *Hybanthus austrocaledonicus* (Violaceae) (20 540 µg/g nickel) (Amir *et al.* 2007). The most unusual hyperaccumulators in the region are several sub-species of *Dichapetalum gelonioides* (Dichapetalaceae) from the Philippines and Malaysia that hyperaccumulate nickel on ultramafic soils and zinc on non-ultramafic soils, for example *D. gelonioides* subsp. *tuberculatum* with 26 600 µg/g nickel and *D. gelonioides* subsp. *sumatranum* with 30 000 µg/g zinc (Baker *et al.* 1992). Examples of manganese hyperaccumulators include *Macadamia neurophylla* (Proteaceae) with up to 51 800 µg/g manganese from New Caledonia (Jaffré *et al.* 1979a) and *Gossia bamagensis* (Myrtaceae) with up to 21 483 µg/g manganese from Australia (Fernando *et al.* 2009). Selenium hyperaccumulators in the region include *Neptunia amplexicaulis* (Mimosaceae), which occurs in Australia and contains up to 4334 µg/g selenium (McCray and Hurwood 1963). Hyperaccumulators for other elements (cadmium, chromium, copper, lead and zinc) are extremely rare, and any reports need to be treated with great caution because contamination of samples with soil/dust particles is a major risk (Van der Ent *et al.* 2013a). Dried herbarium samples are particularly prone for such contamination (Macnair 2003; McGrath *et al.* 2002). A possible example, requiring further investigation is *Polycarpaea synandra* (Caryophyllaceae) with 6960 µg/g zinc and 1044 µg/g lead from Australia (Cole *et al.* 1968).

Target areas and species for nickel phytomining

Extensive areas of ultramafic soils (covering thousands of km^2) containing nickel in the range of 1000–6000 mg/kg are found in Australia, New Caledonia, Malaysia and Indonesia. Although large scale nickel strip-mining is conducted in some of these areas, the nickel concentrations of most ultramafic soils are generally well below concentrations required for profitable conventional mining technology (>1%). However, such sub-economic soils are perfectly adequate for nickel hyperaccumulators to attain 1–3% nickel in foliar matter. Food crops do not grow very well on these soils, but native hyperaccumulators are known in all of these environs, which could be exploited in phytomining operations. The relatively low cost of labour in developing countries such as Indonesia further benefit the economic viability of any future nickel phytomining projects.

Australia

In Australia, ultramafic areas cover roughly 3000 km^2 with several known nickel hyperaccumulators. Extensive ultramafics in the vicinity of Rockhampton, Queensland (Batianoff and Specht 1992) may be suitable for phytomining, but the only known strong native hyperaccumulator *Stackhousia tryonii* is slow-growing and has very low biomass, and hence not suitable for phytomining. In the Western Australian Greenstone Belt composed of nickel-rich soils, the native hyperaccumulator *Hybanthus floribundus* might be proposed for phytomining. This shrub accumulates up to 13 500 µg/g or 1.3% nickel in its foliage and is adapted to local semi-arid conditions (Severne and Brooks 1972), but again biomass production of this species is rather low.

New Caledonia

The island is largely composed of ultramafic outcrops (exceeding 5500 km^2) that to date harbour at least 57 described hyperaccumulator species (Jaffré 1977; Jaffré 1980; Jaffré *et al.* 1976; Jaffré *et al.* 1979a, b). Several might be potential candidates for future phytomining operations, for example *Psychotria douarrei* (*P. gabriellae*), *Hybanthus austrocaledonicus* (Violaceae) and *Geissois pruinosa* (Cunoniaceae). These woody shrubs might be regularly pruned and biomass collected.

Indonesia, Malaysia and the Philippines

Of all ultramafic areas in the world, this region is the most promising for nickel phytomining for several reasons: this region has the world's largest tracts of ultramafic outcrops, Indonesia for instance has >15 000 km^2 in Sulawesi alone (Van der Ent *et al.* 2013b), the climate is favourable for

Figure 10.6: The nickel hyperaccumulator *Phyllanthus balgooyi* (Phyllanthaceae), here in Sabah (Malaysia), can accumulate up to 18% of nickel in its phloem sap.

year-round tropical agriculture, and the region contains several strong hyperaccumulators (with >1% foliar nickel), for example, *Phyllanthus balgooyi* (Phyllanthaceae), which is known from the Philippines and Malaysia, which has up to 18% nickel in its phloem sap (Fig. 10.6) and *Rinorea bengalensis* (Violaceae), which occurs throughout South-East Asia.

Case example for nickel phytomining in Sulawesi, Indonesia

Globally nickel deposits are found in either sulphide (40% of world reserves) or lateritic (ultramafic) deposits (60% of worlds reserves) with some of the largest reserves in nickel laterites in Indonesia, Cuba, New Caledonia and Australia. The USA Geological Survey (USGS 2010a) estimates the nickel reserves at 7 100 000 for New Caledonia, 3 200 000 for Indonesia and 26 000 000 tonnes for Australia, with 2009 productions of 107 000, 189 000 and 167 000 tonnes of nickel respectively. High-grade sulphide deposits are depleting, and as a result a higher proportion of future production is expected to come from laterite deposits (Mudd 2007). Historically nickel laterites were difficult to process, but with the development of the 'high pressure acid leach' (HPAL) technology laterite ores have become profitable (Mudd 2007, 2009). Retrieving nickel from laterites is energy intensive and produces large volumes of waste materials, for instance in 2008, a total 86 000 tons of nickel was produced from 4.7 million tons of ore in New Caledonia (USGS 2010b).

Ultramafic outcrops containing nickel laterites are found mainly in Central and Eastern Sulawesi and commonly strip-mined. In 2000, Indonesia's reserve base was estimated at 13 million tons, or 8% of total world nickel reserves (Sangadji 2002). Mining operations are dominated by Vale INCO's subsidiary P.T. Vale Inco, which commenced commercial production in 1978 in Central Sulawesi with current production of ~70 ktpa. Mining for nickel is a major industry in Sulawesi, and other islands in Indonesia, with significant environmental and biodiversity impacts (Sangadji 2002). Strip-mining for nickel laterites involves complete removal of vegetation. The result of more than

30 years of extensive nickel mining around Soroako has been extensive destruction of natural vegetation communities. The true extent of biodiversity loss is unknown due to a lack of knowledge of plant diversity in these areas.

The stripped land after lateritic nickel mining in Indonesia presents a promising area for implementation of nickel phytomining (Van der Ent *et al.* 2013b). Such soils do not present significant barriers to plant growth with current rehabilitation using large plantings of local plant species. Phytomining could be applied to recover additional nickel. In this context, phytomining merely replaces currently used species with hyperaccumulators in a specialised form of agroforestry. Using native species for rehabilitation strategies holds many benefits over using non-native imported species. Native species are generally well-adapted to the local climate, diseases and the particular edaphic conditions (Fig. 10.7). As such native species can provide a resilient and reliable vegetation cover over mined land. It is conceivable that agricultural practices could be applied using several selected hyperaccumulator species with complementary growth habits and ecological requirements.

Using metallophytes to prospect for mineral deposits

Because metallophytes thrive on metalliferous soils these plants have historically been used in mineral exploration (Baker and Brooks 1988; Ginocchio and Baker 2004). For example 20th century prospectors used vegetation to identify ore deposits, such as the well known 'copper flower' *Polycarpaea spirostylis* in Queensland, Australia (Cannon 1960) Extensive use of this technique did not occur until rapid analytical techniques became available (Cole and Smith 1984). Since then many metallophytes continue to be used as 'indicators' in

Figure 10.7: Minerals wastes may be colonised by plants if these have intrinsic capabilities to cope with such stresses as metal-toxicity, salinity, lack of water and nutrient-deficiency.

prospecting for metal deposits (Baker and Brooks 1988; Cannon 1960). Ultramafic regoliths may contain metal deposits of nickel and chromium of potential economic value, and the vegetation on these soils is often highly distinctive (Brooks, 1987).

Threats to global metallophyte biodiversity

The discovery of new deposits is necessary to meet the need for metals by society (Baker and Whiting 2008). New deposits are most likely to be found in the remote and ecologically intact areas of Australasia. The restriction of metallophytes to localised metalliferous soils, which are mining targets, means that most metallophytes are both rare and threatened (Baker *et al.* 2010). With increasing demand for metals, it is inevitable that the mineral industry will continue to expand into remote areas, which have also been identified as priorities for biodiversity conservation (ICMM 2010). This creates tension between biodiversity conservation and capitalising on resources. Particularly in Indonesia and the Philippines there is a critical nexus of developing economy, globally significant biodiversity hotspots (Barthlott *et al.* 2007), and significant mining targets (IUCN 1999). Historically the result has been the mining related destruction of metallophyte habitats and subsequent biodiversity loss, particularly on nickel mining targets on ultramafics. A lack of knowledge of metallophytes in terms of their value to efficient resource extraction, site rehabilitation and mine closure process limits the potential for practical application. Addressing this knowledge gap requires stewardship of current metallophyte communities particularly on future targets for mineral extraction and adequate methods of conservation. The ICMM Sustainable Development Framework provides support for exploration of metallophyte biodiversity. In the ICMM framework it is specifically acknowledged that: 'Local extinctions can be caused by any sectoral activity, but there is one group of plants that is likely to go extinct as a result of mining activity alone... metallophytes ... mining activities can easily obliterate them, resulting in the loss of a potentially valuable resource' (MMSD 2002, p. 261). The minerals industry now recognises that biodiversity conservation is integral to the responsible exploitation of mineral resources (Baker 2009).

Outlook

Research interest in metallophytes is driven on by the increasing ecological understanding of the importance and benefits that native species provide compared to monocultures of imported species for sustainable rehabilitation of mined lands. Rather than view mining solely as an industry that contributes to the destruction of metallophyte habitats, an offsetting perspective could be applied using metallophyte species in the rehabilitation to not only effectively mitigate the loss of habitat, but also to secure the survival of the metallophyte species themselves. To ensure metallophytes are not lost 'in the process' and to benefit from their advantages in rehabilitation, this approach must be given priority in closure plans (MMSD 2002). The basic knowledge crucial to success of this approach includes systematic screening and cataloguing of metallophytes and metal-tolerant ecotypes in natural metalliferous soils before mining. Screening would also support strategies aimed at the preservation of metallophyte germplasm, which is especially critical for local ecotype populations possessing enhanced metal-tolerance or hyperaccumulation traits. Local knowledge is indispensable for the identification of metallophytes, and vital to assess priorities for those most under threat (Baker and Whiting 2008).

Acknowledgements

The author wishes to thank Prof David Mulligan and Dr Peter Erskine from the Centre for Mined Land Rehabilitation (CMLR) at the University of Queensland for providing continuous support. Special thanks to Prof Mark Tibbett for his advice and offering the opportunity to contribute this chapter. Thanks also to Dr Andrew Fletcher from the CMLR, and Prof Alan Baker from the University of Melbourne for critical comments on an earlier version of the manuscript. Antony van der Ent has been the recipient of IPRS and UQRS scholarships in Australia.

References

Amir H, Perrier N, Rigault F, Jaffré T (2007) Relationships between Ni-hyperaccumulation and mycorrhizal status of different endemic plant species from New Caledonian ultramafic soils. *Plant and Soil* **293**, 23–35. doi:10.1007/s11104-007-9238-0

Anderson CWN, Brooks RR, Stewart RB, Simcock R (1998) Harvesting a crop of gold in plants. *Nature* **395**, 553–554. doi:10.1038/26875

Anderson CWN, Brooks R, Chiarucci A, Lacoste C, Leblanc M, Robinson B, Simcock R, Stewart R (1999a) Phytomining for nickel, thallium and gold. *Journal of Geochemical Exploration* **67**, 407–415. doi:10.1016/S0375-6742(99)00055-2

Anderson CWN, Brooks R, Stewart RB, Simcock R, Robinson B (1999b) The phytoremediation and phytomining of heavy metals. *PACRIM* **99**, 1–10

Anderson CWN, Moreno F, Meech J (2005) A field demonstration of gold phytoextraction technology. *Minerals Engineering* **18**, 385–392. doi:10.1016/j.mineng.2004.07.002

Baker AJM (1981) Accumulators and excluders-strategies in the response of plants to heavy metals. *Journal of Plant Nutrition* **3**, 643–654. doi:10.1080/01904168109362867

Baker AJM (1987) Metal tolerance. *New Phytologist* **106**, 93–111. doi:10.1111/j.1469-8137.1987.tb04685.x

Baker AJM (2009) A Darwinian Approach to Mine Closure and Restoration. *Mine Closure 2009*. Perth.

Baker AJM, Brooks RR (1988) Botanical exploration for minerals in the humid tropics. *Journal of Biogeography* **15**, 221–229. doi:10.2307/2845062

Baker AJM, Brooks RR (1989) Terrestrial higher plants which hyperaccumulate metallic elements: a review of their distribution, ecology and phytochemistry. *Biorecovery* **1**, 81–126.

Baker AJM, Smith JAC (2000) Metal hyperaccumulator plants: biological resources for exploitation in the phytoextraction of metal-contaminated soils. *Conference Paper*, 1–39.

Baker AJM, Whiting SN (2008) Metallophytes – a unique biodiversity and biotechnological resource in the care of the minerals industry. In: *Proceedings of the Third International Seminar on Mine Closure, 14–17 October 2008, Johannesburg, South Africa.* Ed. by A. Fourie, M. Tibbett, I. Weiersbye and P. Dye, Australian Centre for Geomechanics, Nedlands, Western Australia, pp. 13–20.

Baker AJM, Reeves RD, McGrath SP (1991) In situ decontamination of heavy metal polluted soils using crops of metal-accumulating plants – a feasibility study. In: Hinchee, R and Olfenuttel, R (eds.) *In Situ Bioreclamation: Applications and Investigations for Hydrocarbon and Contaminated Site Remediation.* Boston/London: Butterworth-Heinemann.

Baker AJM, Proctor J, Van Balgooy MMJ, Reeves RD (1992) Hyperaccumulation of nickel by the flora of the ultramafics of Palawan, Republic of the Philippines. In: BAKER, AJM, Proctor, J and Reeves, RD (eds.) *The vegetation of ultramafic (serpentine) soils.* Andover, UK.: Intercept Ltd.

Baker AJM, McGrath SP, Sidoli CMD, Reeves RD (1994) The possibility of in situ heavy metal decontamination of polluted soils using crops of metal-accumulating plants. *Resources, Conservation and Recycling* **11**(1–4), 41–49.

Baker AJM, McGrath SP, Reeves RD, Smith JAC (1999) Metal hyperaccumulator plants: a review of the ecology and physiology of a biological resource for phytoremediation of metal-polluted soils. In: Terry, N and Banuelos, G (eds.) *Phytoremediation of contaminated soil and water.* Boca Raton, Florida, USA: Lewis Publishers.

Baker AJM, McGrath SP, Reeves RD, Smith JAC (2000) Metal hyperaccumulator plants: a review of the ecology and physiology of a biological resource for phytoremediation of metal-polluted soils. In: Terry, N. and Banuelos, G. (eds.) *Phytoremediation of contaminated soil and water.* Boca Raton, Florida, USA: Lewis Publishers.

Baker AJM, Ernst WHO, van der Ent A, Malaisse F, Ginicchio R (2010) Metallophytes: the unique biological resource, its ecology and conservational status in Europe, central Africa and Latin America. In: Batty, L and Hallberg, K (eds.) *The Ecology of Industrial Pollution.* Cambridge University Press.

Barthlott W, Hostert A, Kier G, Kuper W, Kreft H, Mutke J, Rafiqpoor MD, Sommer JH (2007) Geographic patterns of vascular plant diversity at continental to global scales. *Erdkunde* **61**, 305–315. doi:10.3112/erdkunde.2007.04.01

Batianoff GN, Specht RL (1992) Queensland (Australia) serpentinite vegetation. In: Baker, AJM, Proctor, J and Reeves, RD (eds.) *The vegetation of ultramafic (serpentine) soils.* Intercept Ltd., Andover, UK.

Boyd R, Jaffré T (2009) Elemental concentrations of eleven New Caledonian plant species from serpentine soils: Elemental correlations and leafage effects. *Northeastern Naturalist* **16**, 93–110. doi:10.1656/045.016.0508

Bradshaw AD (1952) Populations of *Agrostis tenuis* resistant to lead zinc poisoning. *Nature* **169**, 109810.1038/1691098a0

Bradshaw AD (1959) Population differentiation in *Agrostis tenuis* I. Morphological differentiation. *New Phytologist* **58**, 208–227. doi:10.1111/j.1469-8137.1959.tb05352.x

Bradshaw AD, Chadwick MJ (1980) *The restoration of land, the ecology and reclamation of derelict and degraded land.* Berkeley and Los Angelos, University of California Press.

Brooks RR (1979) Indicator plants for mineral prospecting–a critique. *Journal of Geochemical Exploration* **12**, 67–78. doi:10.1016/0375-6742(79)90064-5

Brooks RR (1987) *Serpentine and its vegetation: a multidisciplinary approach.* Dioscorides Press, Portland, Oregon, USA.

Brooks RR, Robinson B (1998) The potential use of hyperaccumulators and other plants for phytomining. In: *Plants that hyperaccumulate heavy metals: Their role in phytoremediation, microbiology, archaeology, mineral exploration and phytomining.* pp. 327–356.

Brooks RR, Wither ED (1977) Nickel accumulation by *Rinorea bengalensis* (Wall) O.K. *Journal of Geochemical Exploration* **7**, 295–300. doi:10.1016/0375-6742(77)90085-1

Brooks RR, Lee J, Reeves RD, Jaffré T (1977) Detection of nickeliferous rocks by analysis of herbarium specimens of indicator plants. *Journal of Geochemical Exploration* **7**, 49–57. doi:10.1016/0375-6742(77)90074-7

Brooks RR, Chambers MF, Nicks LJ, Robinson BH (1998) Phytomining. *Trends in Plant Science* **3**, 359–362. doi:10.1016/S1360-1385(98)01283-7

Cannon HL (1960) Botanical prospecting for ore deposits. *Science* **132**, 591–598. doi:10.1126/science.132.3427.591

Chaney RL (1983) Plant uptake of inorganic waste constituents. In: Parr JF, Marsh PB and KLA, JM (eds.) *Land Treatment of Hazadous Wastes.* Park Ridge, NJ: Noyes Data Corporation, pp. 50–76.

Chaney RL, Malik M, Li YM, Brown SL, Brewer EP, Angle JS, Baker AJM (1997) Phytoremediation of soil metals. *Current Opinion in Biotechnology* **8**, 279–284. doi:10.1016/S0958-1669(97)80004-3

Chaney RL, Li YM, Brown SL, Homer FA, Malik M, Angle JS, Baker AJM, Reeves RD, Chin M (2000) Improving metal hyperaccumulator wild plants to develop commercial phytoextraction systems: Approaches and progress. In: Terry, N. and Banuelos, G. (eds.) *Phytoremediation of contaminated soil and water.* Boca Raton, FL: CRC Press, pp. 129–158.

Chaney RL, Angle JS, Broadhurst CL, Peters CA, Tappero RV, Sparks DL (2007a) Improved understanding of hyperaccumulation yields commercial phytoextraction and phytomining technologies. *Journal of Environmental Quality* **36**, 1429. doi:10.2134/jeq2006.0514

Chaney R, Angle J, Li Y, Baker AJM (2007b) Recovering metals from soil. *United States Patent US 7,268,273 B2*, 1–12.

Clemens S (2006) Toxic metal accumulation, responses to exposure and mechanisms of tolerance in plants. *Biochimie* **88**, 1707–1719. doi:10.1016/j.biochi.2006.07.003

Cole MM, Provan DMJ, Tooms JS (1968) Geobotany, biogeochemistry and geochemistry in the Bulman-Waimuna Springs area, Northern Territory, Australia. *Transactions Institute of Mining and Metallurgy Section B* **77**, 81–104.

Dickinson NM, Baker AJM, Doronila A, Laidlaw WS, Reeves RD (2009) Phytoremediation of inorganics: realism and synergies. *International Journal of Phytoremediation* **11**, 97–114. doi:10.1080/15226510802378368

Ernst WHO (1996) Bioavailability of heavy metals and decontamination of soils by plants. *Applied Geochemistry* **11**, 163–167. doi:10.1016/0883-2927(95)00040-2

Ernst WHO (2005) Phytoextraction of mine wastes – options and impossibilities. *Chemie der Erde – Geochemistry* **65**, 29–42.

Ernst WHO (2006) Evolution of metal tolerance in higher plants. *Forest Snow and Landscape Research* **80**, 251–274.

Farago ME, Mahmoud ED (1983) Plants that accumulate metals (Part VI): further studies of an Australian nickel accumulating plant. *Minerals and the Environment* **5**(4), 113–121. doi:10.1007/BF02093332

Fernando DR, Guymer G, Reeves RD, Woodrow IE, Baker AJM, Batianoff G (2009) Foliar Mn accumulation in eastern Australian herbarium specimens: prospecting for 'new' Mn hyperaccumulators and potential applications in taxonomy. *Annals of Botany* **103**(1), 931–939. doi:10.1093/aob/mcp013

Ginocchio R, Baker AJM (2004) Metallophytes in Latin America: a remarkable biological and genetic resource scarcely known and studied in the region. *Revista Chilena de Historia Natural (Valparaiso, Chile)* **77**, 185–194.

Harris AT, Naidoo K, Nokes J, Walker T, Orton F (2009) Indicative assessment of the feasibility of Ni and Au phytomining in Australia. *Journal of Cleaner Production* **17**, 194–200. doi:10.1016/j.jclepro.2008.04.011

ICMM (2010) ICMM Biodiversity case studies 2010. pp. 1–36.

IUCN (1999) Metals from the forests. Mining and forest degradation. *Special issue of the newsletter Arborvitae.* ISBN: 2-8317-0468-5, 1–36.

Jaffré T (1977) Composition chimique élémentaire des tissus foliaires des espèces végétales colonisatrices des anciennes mines de nickel en Nouvelle Calédonie. *Cah. O.R.S.T.O.M., sér. Biol.* **XII**, 323–330.

Jaffré T (1980) Vegetation des roches ultrabasiques en Nouvelle-Caledonie. *Travaux et Documents de l'ORSTOM* 124. ORSTOM, Paris.

Jaffré T, Schmid M (1974) Accumulation du nickel par une Rubiacee de Nouvelle-Caledonie, *Pychotria douarrei* (G. Beauvisage) Däniker. *Comptes Rendus, Academie des Sciences (Paris), Series D* **278**, 1727–1730.

Jaffré T, Brooks RD, Lee J, Reeves RD (1976) *Sebertia acuminata*: a hyperaccumulator of nickel from New Caledonia. *Science* **193**, 579–580. doi:10.1126/science.193.4253.579

Jaffré T, Brooks RD, Trow J (1979a) Hyperaccumulation of nickel by Geissois species. *Plant and Soil* **51**, 157–161. doi:10.1007/BF02205937

Jaffré T, Kersten WJ, Brooks RR, Reeves RD (1979b) Nickel uptake by Flacourtiaceae of New Caledonia. *Proceedings of the Royal Society of London. Series B, Biological Sciences* **205**, 385–394. doi:10.1098/rspb.1979.0072

Kazakou E, Dimitrakopoulos PG, Baker AJM, Reeves RD, Troumbis AY (2008) Hypotheses, mechanisms and trade-offs of tolerance and adaptation to serpentine soils: from species to ecosystem level. *Biological Reviews of the Cambridge Philosophical Society* **83**(4), 495–508.

Keeling SM, Stewart RB, Anderson CWN, Robinson BH (2003) Nickel and cobalt phytoextraction by the hyperaccumulator *Berkheya coddii*: implications for polymetallic phytomining and phytoremediation. *International Journal of Phytoremediation* **5**, 235–244. doi:10.1080/713779223

Lasat MM (2002) Phytoextraction of toxic metals: a review of biological mechanisms. *Journal of Environmental Quality* **31**, 109–120. doi:10.2134/jeq2002.0109

Lombi EZ, Zhao FJ, Dunham SJ, McGrath SP (2001) Phytoremediation of heavy metal-contaminated soils: natural hyperaccumulation versus chemically enhanced phytoextraction. *Journal of Environmental Quality* **30**, 1919–1926. doi:10.2134/jeq2001.1919

Macnair MR (2003) Hyperaccumulation of metals by plants. *Advances in Botanical Research* **40**, 63–105. doi:10.1016/S0065-2296(05)40002-6

McCray CWR, Hurwood ID (1963) Selenosis in northwestern Queensland associated with a marine cretaceous formation. *Queensland Journal of Agricultural Science* **20**, 475–498.

McGrath SP, Zhao FJ (2003) Phytoextraction of metals and metalloids from contaminated soils. *Current Opinion in Biotechnology* **14**, 277–282. doi:10.1016/S0958-1669(03)00060-0

McGrath SP, Sidoli CMD, Baker AJM, Reeves RD (1993) The potential for the use of metal-accumulating plants for the in-situ decontamination of metal-polluted soils. In: Eijsackers, HJ and Hammers, T (eds.) *Integrated Soil and Sediment Research: A Basis for Proper Protection.* Dordrecht: Kluwer Academic Publ.

McGrath SP, Dunham SJ, Corell RL (2000) Potential phytoextraction of zinc and cadmium from soils using hyperaccumulator plants. In: Terry, N and Banuelos, G (ed.) *Phytoremediation of contaminated soil and water.* Lewis Publishers, pp. 110–127.

McGrath SP, Zhao FJ, Lombi E (2002) Phytoremediation of metals, metalloids, and radionuclides. *Advances in Agronomy* **75**, 1–56. doi:10.1016/S0065-2113(02)75002-5

Meharg AA (2003) Variation in arsenic accumulation: hyperaccumulation in ferns and their allies. *New Phytologist* **157**, 25–31. doi:10.1046/j.1469-8137.2003.00541.x

MMSD (2002) MMSD Breaking New Ground. *MMSD*, 1–476.

Morrey DR (1998) Using metal-tolerant plants to reclaim mining wastes. *Mining Engineering* **47**, 247–249.

Mudd GM (2007) An analysis of historic production trends in Australian base metal mining. *Ore Geology Reviews* **32**, 227–261. doi:10.1016/j.oregeorev.2006.05.005

Mudd GM (2009) Nickel Sulfide Versus Laterite: The Hard Sustainability Challenge Remains. Proceedings of the 48th Annual Conference of Metallurgists, Canadian Metallurgical Society, Sudbury, Ontario, Canada, August 2009.

Nicks LJ, Chambers MF (1995) Farming for metals. *Mining Environ. Mgt.*

Nicks LJ, Chambers MF (1998) A pioneering study of the potential of phytomining for nickel. In: Brooks, R (ed.) *Plants that Hyperaccumulate Heavy Metals.* CAB International.

Pollard AJ, Powell KD, Harper FA, Smith JAC (2002) The genetic basis of metal hyperaccumulation in plants. *Critical Reviews in Plant Sciences* **21**, 539–566. doi:10.1080/0735-260291044359

Pollard AJ, Reeves RD, Baker AJM (2014) Facultative hyperaccumulation of heavy metals and metalloids. *Plant Science* **217–218**, 8–17. doi:10.1016/j.plantsci.2013.11.011

Rajakaruna N, Baker AJM (2006) Serpentine: a model habitat for botanical research in Sri Lanka. *Ceylon Journal of Science* **32**, 1–19.

Raskin I, Kumar PBAN, Dushenkov S, Salt DE (1994) Bioconcentration of heavy metals. *Current Opinion in Biotechnology* **5**, 285–290. doi:10.1016/0958-1669(94)90030-2

Reeves RD (2003) Tropical hyperaccumulators of metals and their potential for phytoextraction. *Plant and Soil* **249**, 57–65. doi:10.1023/A:1022572517197

Reeves RD (2006) Hyperaccumulation of trace elements by plants. In: *Phytoremediation of metal-contaminated soils.* (Eds JL Morel, G Echevarria, N Goncharova) pp. 25–52.

Robinson BH, Brooks RR, Howes AW, Kirkman JH, Gregg PEH (1997a) The potential of the high-biomass nickel hyperaccumulator *Berkheya coddii* for phytoremediation and phytomining. *Journal of Geochemical Exploration* **60**, 115–126. doi:10.1016/S0375-6742(97)00036-8

Robinson BH, Chiarucci A, Brooks RR, Petit D, Kirkman JH, Gregg P, De Dominicis V (1997b)

The nickel hyperaccumulator plant *Alyssum bertolonii* as a potential agent for phytoremediation and phytomining of nickel. *Journal of Geochemical Exploration* **59**, 75–86. doi:10.1016/S0375-6742(97)00010-1

Robinson BH, Brooks RR, Clothier BE (1999a) Soil amendments affecting nickel and cobalt uptake by *Berkheya coddii*: potential use for phytomining and phytoremediation. *Annals of Botany* **84**, 689–694. doi:10.1006/anbo.1999.0970

Robinson BH, Brooks RR, Gregg PEH, Kirkman JH (1999b) The nickel phytoextraction potential of some ultramafic soils as determined by sequential extraction. *Geoderma* **87**, 293–304. doi:10.1016/S0016-7061(98)00062-7

Sangadji A (2002) Japanese involvement in Nickel mining in Indonesia. Discussion paper prepared for a meeting of Japanese NGOs in Tokyo, 18 July 2002. The meeting was organized by NINDJA (Network for Indonesian Democracy, Japan) and FoE. pp. 1–8.

Severne BC, Brooks RR (1972) A nickel-accumulating plant from Western Australia. *Planta* **103**, 91–94. doi:10.1007/BF00394610

USGS (2010a) 2008 Minerals Yearbook. The Mineral Industry of New Caledonia in 2008, by Susan Wacaster, pp. 1–4.

USGS (2010b) Minerals Information: Nickel. Prepared by Peter H. Kuck. Mineral Commodities Summaries. 1–2.

Van der Ent A, Baker AJ, Reeves RD, Pollard AJ, Schat H (2013a) Hyperaccumulators of metal and metalloid trace elements: facts and fiction.*Plant and Soil* **362**(1–2), 319–334.

Van der Ent A, Baker AJM, van Balgooy MMJ, Tjoa A (2013b) Ultramafic nickel laterites in Indonesia (Sulawesi, Halmahera): mining, nickel hyperaccumulators and opportunities for phytomining. *Journal of Geochemical Exploration* **128**, 72–79. doi:10.1016/j.gexplo.2013.01.009

Whiting SN, Reeves RD, Richards D, Johnson MS, Cooke JA, Malaisse F, Paton A, Smith JAC, Angle JS, Chaney RL, Ginocchio R, Jaffré T, Johns R, McIntyre T, Purvis OW, Salt DE, Schat H, Zhao FJ, Baker AJM (2004) Research priorities for conservation of metallophyte biodiversity and their potential for restoration and site remediation. *Restoration Ecology* **12**, 106–116. doi:10.1111/j.1061-2971.2004.00367.x

Ye ZH, Baker A, Wong MH, Willis AJ (1997) Copper and nickel uptake, accumulation and tolerance in *Typha latifolia* with and without iron plaque on the root surface. *New Phytologist* **136**(3), 481–488. doi:10.1046/j.1469-8137.1997.00758.x

11

A collaborative approach for mining, environment organisations and Traditional Owners to manage and conserve biodiversity in Australia's Great Western Woodlands

Alexander WT Watson, Simon Judd, Wayne O'Sullivan and James EM Watson

Summary

The Goldfields region of Western Australia is one of the most geologically mineralised areas in the world, being particularly rich in gold, nickel and iron ore. It also contains the largest remaining temperate and semi-arid woodland on Earth. In this paper, we describe some of the outstanding natural values of the Great Western Woodlands (GWW). We then show why contemporary conservation methods, which aim to capture specific target levels of biodiversity in a 'protected area system', will not guarantee the persistence of biodiversity across this landscape. Instead, the region's biodiversity can only be conserved by appropriately managing key ecological processes (e.g. fire, eco-hydrology, gross primary productivity, trophically important species) over the entire landscape. We argue that targeted research is required to identify what is needed to manage these processes. We describe the first phase of an initiative that works with all major stakeholders, including Traditional Owners, state and federal governments, local communities, non-government organisations and resource-based industry groups, which ultimately aims to forge a holistic long-term land use plan for the Great Western Woodlands.

Introduction

The Great Western Woodlands (GWW), located in the south-west of the continent of Australia, and entirely within the state of Western Australia, is a landscape that retains a relatively undisturbed natural vegetation (Fig. 11.1). At almost 16 million hectares, it is more than twice the size of Tasmania and similar in size to England (Fig. 11.2). Despite being ecologically distinct, the region has never had a unique name, usually referred to as part of the broader 'Goldfields' region. The name 'Great Western Woodlands' (Watson *et al.* 2008a) was selected because it best reflects its position and status as being in the west of the continent and containing the largest remaining area of temperate woodland in Australia. After only two years, the name 'Great Western Woodlands' is now used by governments (e.g. Department of Environment and Conservation 2010), Traditional Owners, local communities, mining companies and non-government organisations, and is widely accepted.

People in the landscape

The Traditional Owners of GWW have one of the longest continuous connections to Country of any culture on Earth. Archaeologists think that the earliest permanent occupation of Australia's arid areas occurred at least 22 000 years ago (Department of Environment and Conservation 2010; Watson *et al.* 2008a). Over such a long timescale, the use and management of the area by its Traditional Owners had a direct influence on many characteristics we see today. Although we cannot be sure of details, there can be little doubt that, at times, particular types of vegetation were favoured or reduced by management decisions about where, when and how to burn. Similarly, fauna populations would have responded in varying ways to fire regimes, hunting pressures and other practices. Presently there are 18 unresolved native title groups with claims that cover 95% of the Woodlands. As we discuss later in the paper, the involvement of Traditional Owners in management of the

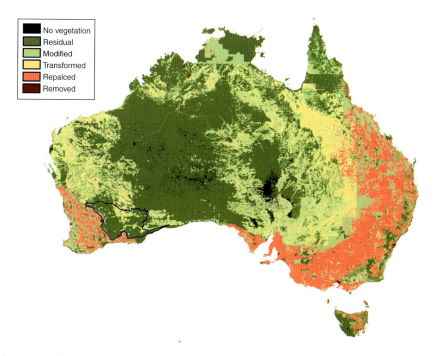

Figure 11.1: The relative condition of Australia's vegetation condition using the Vegetation Assets, States and Transitions (VAST) framework (from Thackway and Lesslie 2006). The dark green colour that predominates in the Great Western Woodlands suggests that the vegetation has not been significantly affected by post-European land use change. The black line shows the boundary of the Great Western Woodlands. The Great Western Woodlands is the largest residual (or 'intact') non-arid landscape in southern Australia.

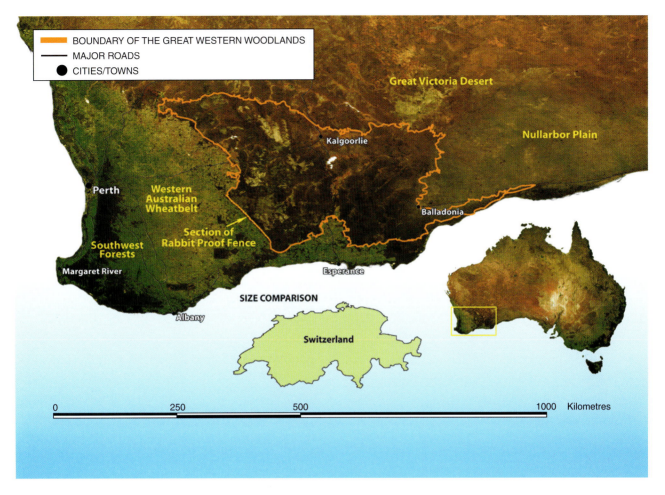

Figure 11.2: The boundary and relative size of the Great Western Woodlands. The boundary of GWW was established by researchers from Australian National University and was drawn using MODIS satellite data using a methodology that identified a characteristic 'Eucalypt' spectral signature that is found throughout the Great Western Woodlands (Berry 2002). The boundaries discriminate the eucalypt woodlands from the Mulga (*Acacia aneura*) country to the north, the treeless Nullarbor Plain to the east, the moist coastal heath to the south-east and agricultural land to the west and south. A map of Switzerland has been placed as an inset in the figure to indicate the relative size of the Great Western Woodlands.

area is crucial for conservation of the regions' biodiversity.

Over the last 150 years, more than 40 000 non-Indigenous Australians have come to live and work in GWW, many of them living in the towns of Kalgoorlie, Southern Cross, Balladonia and Norseman. The region is recognised as one of the most mineralised landscapes on Earth and most of people who live in GWW work in the mining industry. There are currently 334 operating mines in GWW and while the most lucrative industry is gold mining, extraction of other minerals such as nickel and iron ore are also an important source of employment (Department of Environment and Conservation 2010).

Pastoralism is another major industry with ~20% of the GWW currently being managed as pastoral lease. Despite the fact that many of these leases have been managed for more than 150 years, it is widely recognised that the long-term viability of this industry in the region is questionable as many of these operations are now marginal. Some pastoralists are diversifying into tourism and increasingly pastoral leases are being held by mining companies, or transferred into the conservation estate.

Natural values

We assessed the region's known natural values by first considering peer-reviewed published papers, and non-refereed reports from government departments and environmental non-government organisations (ENGOs). We also conducted interviews with ecologists who have worked in the region as we recognised that some important surveys may never have been reported upon. We supplemented this literature review and set of interviews with a formal analysis of the data from the Western Australian Museum (Faunabase), the Western Australian Department of Environment and Conservation's Western Australian Herbarium (Florabase) and Birds Australia (Birds Atlas) and the results of the relevant biological surveys of the West Australian goldfields (How *et al.* 1988; Keighery 1995; Newbey and Hnatiuk 1984, 1988; see Judd *et al.* 2008 for a review of the methodology). The following is an overview of (i) plant diversity, (ii) vertebrate diversity and (iii) ecosystem diversity established by this review.

Plant diversity

The Western Australian Herbarium has records of 3314 flowering plant species from 119 families in the Great Western Woodlands and over 4200 different taxa (which is a list that includes undescribed species as well as subspecies, hybrids and varieties). It is estimated that almost half of these species are endemic to the south-west of Australia but this has not been formally established (Burgman 1988; Hopper and Gioia 2004). If these proportions turn out to be accurate, then it is a very large number relative to other Mediterranean regions of the world (Wilson *et al.* 2007) and shows that GWW belongs as one of the cornerstones to the south-west of Western Australia's 'biodiversity hotspot' (Myers *et al.* 2000).

Since the GWW is still largely intact there are fewer rare, threatened or endangered species in this region than are found in more disturbed regions of Australia such as the Wheatbelts of Western Australia, Victoria, New South Wales and Queensland (Watson *et al.* 2010). Nevertheless, GWW contains

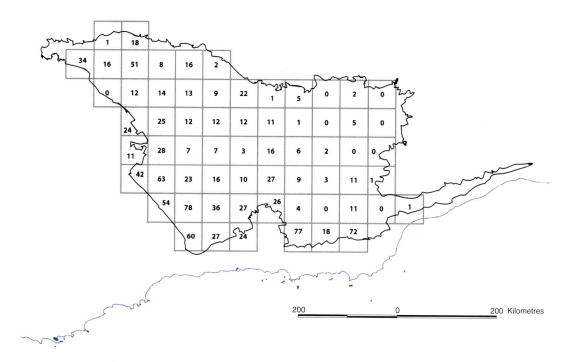

Figure 11.3: The distribution of the Declared Rare Flora and Priority plant species (*n* = 464) across the GWW landscape. A comparison of the cells shows that priority species are spread across the landscape. Many of these are by nature found only in tiny pockets and are threatened by a range of disturbances. It is likely that many of those in the western and southern part of the GWW are threatened because much of their original distribution has been cleared for agriculture.

many species that are listed as threatened by the Western Australian State Government. For example, 44 plant species are listed as 'declared rare flora' by the state government, because they are rare, in danger of extinction, or in need of special protection. A further 422 species have been listed as 'priority' species by the Western Australian Department of Parks and Wildlife. The 464 declared rare flora and priority species are distributed across the GWW, although the western and southern boundary appears to contain many more species per cell than recorded in other parts of the region (Fig. 11.3). This probably reflects sampling bias (that is, more sampling has been conducted in these cells) as well as the fact that this boundary abuts a landscape that has been heavily modified for agriculture (Judd et al. 2008).

Vertebrate diversity

A review of the Western Australia Museum (Biological Survey of the Goldfields and FaunaBase) and Birds Australia (Australian Ornithological Club Atlas Database) show that 49 mammals, 138 reptiles, 14 frogs and 215 species of bird have been recorded in the region. Our research, which is also based on consultants' reports, the notes of amateur enthusiasts and the habitats requirements of species published in the peer-reviewed literature, suggests there are likely to be an additional 51 species (nine species of mammals, 18 species of reptile, six frogs and 18 bird species) present in the GWW. While the level of vertebrate richness is not high on a global scale for mammals (Ceballos and Ehrlich 2006) and birds (Lee and Jetz 2008), the number of different reptile species makes the study area exceptional among the world's reptile communities (Bickford et al. 2010; Gasc et al. 1997). Different reptile communities are spatially and temporally variable, reflecting the heterogeneous vegetation communities and periodic rainfall events that occur in the region (Cowan and How 2004; Thompson and Thompson 2005).

The GWW contains many rare and vulnerable animal species. On the state government's rare and endangered fauna list there are 34 threatened vertebrate species that either exist, or are likely to exist, in the region. These comprise 15 mammal, 10 bird and nine reptile species. The Wilderness Society conducted a vertebrate fauna survey of the Honman Ridge–Bremer Range, and found 19 species that were of conservation significance in that area alone (Duncan et al. 2006). In addition, four species that are globally extinct were once found in the GWW (Pig-footed Bandicoot *Chaeropus ecaudatus*, Long-tailed hopping mouse *Notomys longicaudatus*, Crescent Nail-tailed Wallaby *Onychogalea lunata* and the Broad-faced Potoroo *Potorous platyops*) and a further six species are extinct in the GWW region and now only exist either on predator free islands, in enclosures designed to keep cats and foxes out of them or in very remote areas of Australia (Burrowing Bettong *Bettongia lesueur*, Banded-Hare Wallaby *Lagostrophus fasciatus*, Western Barred Bandicoot *Perameles bougainville*, Greater Stick-nest Rat *Leporillus conditor*, Plains Mouse *Peudomys australis* and Djoongari *Pseudomys fieldi*). With improved management, including well funded predator control and fire suppression strategies, it could be possible to re-introduce these animals back into GWW.

Ecosystem diversity

The GWW is dominated by eucalypt woodlands (woodland ecosystems resemble forests but with a more open canopy (Keally 1991). The dominance of woodland has lead to the GWW's name; however, the region also contains significant areas of mallee, shrublands and grasslands. An analysis of Australia's vegetation by the Australian Government in 2001 identified 23 'vegetation groups' across the continent (Commonwealth of Australia 2001). Of these vegetation groups, 14 are found in the Great Western Woodlands. The GWW contains more vegetation groups than iconic areas such as Kakadu and Cape York. With the exception of Tasmania (which contains 15 vegetation groups), there is no other area in Australia of equivalent size to the Great Western Woodlands that has as many vegetation groups. Four of the vegetation groups found in the GWW bioregion ('Eucalypt woodlands', 'Eucalypt open woodlands', 'Mallee woodlands' and 'Acacia forests and woodland') have been substantially cleared in other regions of Australia

(Table 11.1). In the GWW, these four ecosystems remain largely intact.

Conservation management

Historically, the environmental values of the GWW have been largely overlooked (Watson *et al.* 2008a, b). Unlike other parts of Australia, where so-called wilderness areas are more likely to have a higher proportion of protection (Pressey *et al.* 1994), less than 15% of the GWW region has been placed in protected areas. However, unlike other areas of Australia that have low levels of protection due to the fact that most land is privately owned or leased, the majority of the GWW (60%) is still designated as 'unallocated crown land' (Department of

Table 11.1. The land cover of each of the Australian Native Vegetation Assessment's major vegetation groups found in the Great Western Woodlands and a comparison of how much each type has been cleared since European inhabitation

Major vegetation group	Total Cover pre-1788 (km²)	Total cover 2001 (km²)	Proportion of 1788 cover now cleared (%)	Area in the Great Western Woodlands (km²)	Proportion of the Great Western Woodlands occupied (%)
Eucalypt woodlands	1 012 047	693 449	31.4	83 738	52.41
Mallee woodlands and shrublands	383 399	250 420	34.7	27 497	17.21
Low closed forests and tall closed shrublands	15 864	8749	44.9	10 501	6.57
Acacia shrublands	670 737	654 279	2.5	7499	4.69
Inland aquatic – freshwater, salt lakes, lagoons	–	–	–	6171	3.86
Other shrublands	115 824	98 947	14.6	5280	3.3
Hummock grasslands	1 756 962	1 756 104	0.1	3322	2.08
Heathlands	47 158	25 861	45.1	3051	1.91
Acacia forests and woodlands	657 582	560 649	14.7	3001	1.88
Casuarina forests and woodlands	73 356	60 848	17.1	2279	1.43
Chenopod shrublands, samphire shrublands and forblands	563 389	552 394	2	2277	1.43
Cleared, non-native vegetation, buildings	–	–	–	2243	1.4
Naturally bare – sand, rock, claypan, mudflat	–	–	–	2036	1.27
Acacia open woodlands	117 993	114 755	2.7	643	0.4
Eucalypt open woodlands	513 943	384 310	25.2	156	0.1
Callitris forests and woodlands	30 963	27 724	10.5	49	0.03
Eucalypt low open forests	15 066	12 922	14.2	44	0.03
Total	5 974 283	5 201 411	–	159 786	–

Environment and Conservation 2010; Watson *et al.* 2008a). This means that state and federal governments have significant opportunities to expand the protected area estate in the region. It also means that significant threats to the region's natural diversity, including inappropriate fire regimes, feral animals such as cats and foxes and weeds are not being managed across most of the landscape (Department of Environment and Conservation 2010; Watson *et al.* 2008a).

The conventional approach to expanding the protected area system typically focuses on achieving (as efficiently as possible) a minimal target level of protection for representative samples of each mappable class of ecosystem (usually defined in terms of major vegetation types), and/or populations of particular target species (typically threatened plants and animals) (Watson *et al.* 2008b; Woinarksi *et al.* 2007). We suggest that this type of planning exercise does not work in, or acknowledge, such diverse and largely intact and functional landscapes as the GWW (Soulé *et al.* 2004). We believe that pursuing minimalist targets will lead governments to fall into a 'minimalist trap' (*sensu* Sanderson *et al.* 2006) as the conservation targets they set are not necessarily based on a robust scientific understanding of what is needed for the long-term conservation of biodiversity (Brooks *et al.* 2004; Rodrigues and Gaston 2002; Svancara *et al.* 2005). Conservation programs based on inadequate conservation targets have been found to create a false sense of security about the conservation status of a region's biodiversity (Soulé and Sanjayan 1998). The end result of such exercises is often a landscape fragmented by contrasting isolated land uses, with conservation areas scattered throughout a landscape largely transformed by modern land use activities (Watson *et al.* 2011; Woinarksi *et al.* 2007). Generally, these conservation areas are isolated and few in size and number, surrounded by disturbance, and in locations chosen for their low economic value (often because these are infertile, too steep, or otherwise of low productivity) (Ford *et al.* 2001; Recher 2004).

Regardless of the size and location of protected areas, there is also a need to ensure the whole landscape retains its ecological functionality. Although it has become mandatory to assess the local environmental impact of large, site-specific projects such as mining, the environmental impacts of economic activities in the area are almost always discounted. The current environmental impact process ignores cumulative environmental impacts outside the immediate local area of the economic activity. Similarly, the widespread and pervasive impacts of activities such as infrastructure corridors and their cumulative effects are not considered. We believe these are symptoms of an *ad hoc*, episodic approach to land use and planning, with a narrow focus on site-specific, acute development pressures. Such an approach will ultimately result in the incremental erosion of the region's outstanding biodiversity and natural values.

We believe the conventional approach to conservation planning will not protect the values and future of the GWW because it fails to incorporate the landscape-wide ecological processes that sustain biodiversity (Soulé *et al.* 2004). These processes include managing 'eco-hydrology' (defined as: the relationships between water – both surface and ground water, vegetation, wildlife, and landform at local and regional scales) (Mackey *et al.* 2007); trophically important species (e.g. pollinators) (Soulé *et al.* 2003); ecologically appropriate fire regimes (Woinarksi 1999); and understanding temporal and spatial movements of species (Soulé *et al.* 2004). Management of these processes needs to occur across the entire landscape, not just within the minimalist reserve system, to ensure the conservation of the region's biodiversity and other environmental values. This is especially relevant in an era of rapidly changing climate; many ecologists now recognise that the conventional park system will not ensure the conservation of species as they attempt to move in order to find more favourable climate conditions (Hughes 2000; Mackey *et al.* 2008).

In light of the problems of the traditional approach to conservation, scientists have proposed that managers of regions such as GWW adopt the approach to land management and land use that has been developed in sparsely populated and largely natural extensive landscapes in North America (e.g. see Mackey *et al.* 2007; Soulé *et al.* 2004). This approach recognises that conservation can only ever be successful in Australia when it occurs across all land tenures and when different stakeholders work together with a common aim of biodiversity conservation. The central component of this approach is to identify and conserve the key large-scale, long-term ecological processes and interactions that drive and enhance 'connectivity' between ecosystems and species and maintain biodiversity at all scales (Soulé *et al.* 2004). In this context, connectivity draws attention to large-scale phenomena and processes that contribute to the maintenance of landscape ecological function and the diversity of native flora and fauna (Mackey *et al.* 2007; Soulé *et al.* 2004).

Developing a collaborative approach to land management

As a first step as to the practical management of this landscape, a process was developed that ensured all stakeholders were given equal voice to encourage their support. Defining and naming the 'Great Western Woodlands' led to the identification of stakeholders, and allowed them to be brought together to work as a group. From the earliest meetings of what is now known as the 'Woodlands Initiative', it became apparent that there was a shared frustration around the previously described *ad hoc* approach to planning decisions across the area and across sectors. The commitment of the different sectors, including mining, conservation, Traditional Owners, local government, science, pastoralism, and those who simply see themselves as 'community', to working together has allowed a land use planning process to begin.

Consultants determined what the stakeholders had in common and, more importantly, identified the contested issues (Del Marco 2011). Although the study found that there was agreement across all sectors that the area had high conservation values, future planning will require regional objectives for conservation be well defined. To achieve a comprehensive land use plan, the group decided these objectives should be relatively few and elegant, easy to measure and track, and capture the key aspects of the Woodlands' values, such as intactness and ecosystem diversity.

Defining the exercise as a 'land use planning' approach, rather than a 'conservation planning approach' was a subtle difference that, to date, has enabled stakeholders with diverse interests to come to the table and feel that their interests will be recognised and incorporated into any agreements (Del Marco 2011). The challenge now becomes clearly defining the conservation objectives by which the land use plan will be framed, raising and managing funding for the work, and, most importantly, finding a statutory model that allows the on-ground conservation work to be achieved.

Given the size and complexity of the area, and the paucity of information at a sub-regional scale, it is envisaged the land use plan will be introduced in phases. The early phases are likely to be based around a regional land use policy, codes of practice for those working in the GWW, and criteria for assessment of any new land use proposal for the area. Subsequent stages, informed by greater knowledge of the region's ecological values, processes and threats, may set sub-regional conservation objectives, and ultimately lead to the zoning of all lands within the area.

There are obviously difficult issues that have arose in this process, such as the conservation sector's concern about cumulative impact of mining and exploration and the irreplaceability of ecosystems (Pressey *et al.* 1994), and an industry expectation that all areas should be open to exploration. By naming contested areas, this systematic planning process identifies them as 'priorities' for more detailed data collection and survey work. This in turn allows more targeted investment of limited funds (Bode *et al.* 2008). Historically detailed survey work has been on individual mine sites, or areas of known conservation value, with little attention to regional or sub-regional context. Lack of standardised collection protocols has made comparison of survey data difficult or impossible before now.

What are the opportunities for the mining sector?

The mining sector wants certainty of access and security for investment. Many companies have recognised that the land use approach previously described, which is based on a clear framework, is easier to negotiate, less costly and more expedient than the current approach required by state and federal government. Within the GWW, the mining sector has acknowledged that some areas are of such high conservation value that they should be excluded from mining, just as the conservation sector recognises areas of high prospectivity, such as the greenstone belts (which are often highly mineralised). This is the beginning of a discussion around the establishment of different landscape zones for different primary uses.

This approach, which moves beyond the strict reservation of small areas (and theoretically should exclude mining) to one of conserving landscape function, requires the development of a continuum. A basis of this is the identification of areas which need to be afforded high level protection, and areas in other parts of the landscape that will have reduced protection, with management occurring alongside more intensive mining, exploration or pastoral activities. The concept of 'zoning' areas for multiple use and conservation management was tested in late 2010 in the Mt Manning area, where the state government announced four new reserves totalling ~526 000 ha, or 3.3% of the Great Western Woodlands. While the tenures of 'Conservation and Mining' and 'Conservation Park' created in this instance afford little statutory protection from a conservation perspective, it was a welcome move by the state government to the extent it clarified and gave certainty that these areas are to be vested and managed, at least in part, for their conservation values rather than remaining unmanaged as 'Unallocated Crown Land'.

Working at a landscape scale has shown the potential to provide significant offsets for mining operation, where 'bond funds' can be used for wider landscape management. This also creates an opportunity to address legacy issues from mining's past, such as uncapped drill holes and old grid lines, without blame. Practically, the mining community can see that working with others sectors using a standardised approach for data collection allows knowledge gaps to be targeted, which will save time and money. In a recent example, one of the mining companies in the GWW funded a bushfire threat analysis for most of the area, scaling up from what began as a mine site threat analysis. Such datasets covering the whole of the GWW are a rare and valuable contribution to the planning process. There is significant potential for stakeholder collaboration on the detailed vegetation mapping now beginning for the GWW.

Engagement of Traditional Owners

The intact nature of the GWW presents a rare opportunity for stakeholders to collaborate to maintain and enhance the environmental values of this region. This approach is not new. For tens of thousands of years, the Aboriginal people have implemented ecological management across the region. As we now focus on what is required to manage ecological processes, and the various legal and political processes that best allow us to achieve this, Traditional Owners should (and will) be involved in decisions affecting future ownership and management.

There is an important role for Traditional Owners in managing and conserving biodiversity rich areas like the GWW. For example, significant knowledge and expertise can be used to manage threatening processes such as ecological deleterious fire regimes (Watson *et al*. 2008a). Addressing issues of biodiversity loss, contraction of species, feral animals, weeds, changed fire regimes, pollution and overgrazing, are all high priorities. In addition, protection of Aboriginal culture and heritage sites will be important criteria in developing a new management framework.

Active engagement in developing projects that promote biodiversity conservation can provide economic, social and health benefits for local Indigenous people and communities. Elsewhere in Australia, Indigenous Land and Sea management groups are also now evolving into enterprises that carry out environmental services work, providing the opportunity for significant employment and opportunity for Traditional Owners. For example, in West Arnhem Land, an agreement with Conoco Philips to abate 100 000 tonnes of carbon emissions each year through customary fire management is protecting vast areas of savannah woodland, and returning A$1 000 000 per year for 17 years to Traditional Owners (May 2010).

The Woodlands Initiative has recognised that an appropriate ownership and/or management structure for the GWW has to include the Traditional Owners. This might, for example, define their role in managing conservation outcomes, protection of heritage and culture and the development of training programs that incorporate Indigenous knowledge and practice.

Conclusion

We have synthesised all the readily available information on the plant, vertebrate and ecosystem diversity of the Great Western Woodlands region. Our findings highlight the GWW's biological diversity has five distinct features. First, the high number of different major vegetation groups in the region when compared to other bioregions of Australia makes it worthy of national significance. Second, the extraordinarily high numbers of plant species makes this region worthy of global recognition. Third, the floristic composition and structure of these groups changes rapidly across small spatial scales across the landscape and creates special conservation problems. Fourth, while the degree of vertebrate richness is not high on a world scale for mammals and birds, the number of different reptile species makes the study area exceptional among the world's reptile communities. Fifth, the significant 'Wallacean' (i.e. knowledge on species' geographic and temporal range in the region) and 'Linnaen' (i.e. knowledge on how many species actually live in the region) shortfalls for all the major species groups mean that more research is required to understand the key values of the system and to optimise management strategies.

The GWW offers a unique opportunity to study a highly distinctive, and in some places close to pristine, functioning and globally-significant ecological region. The challenge now and in the coming decades is to maintain the natural values of the Great Western Woodlands, protect the ecological processes that sustain these values, and repair any environmental damage that has already occurred. The initial progress of the 'Woodlands Initiative', comprising major stakeholders in GWW, appears to shows that a collaborative approach to landscape management is perhaps the most significant way to achieve significant conservation outcomes in the Great Western Woodlands.

Acknowledgements

We thank Brendan Mackey, Sandy Berry, Rob Lesslie, Helene Marsh, Henry Nix, Hugh Possingham, Harry Recher, Michael Soulé, Regina Souter, Jann Williams, Richard Hobbs and John Woinarski, for discussions over recent years through the WildCountry Science Council that helped inform this book chapter. We also thank Barry Traill, Keith Bradby, Michael Looker, Steve Hopper, Ian Keally, Ian Hereford, Karl Brennan, Anthony Espositto, Trudy O'Connor, Andrew Del Marco, Jessica Chapman, Peter Robertson, Peter Price, Rob Howard, Phil Drayson, Vanessa Bray, Jim Williams, Ray Ciantar, James Fitzsimmons and many others for their input with respect to particular aspects of the project.

References

Berry SL (2002) Estimating mixtures of leaf functional types using continental-scale satellite and climatic data. *Global Ecology and Biogeography* **11**, 23–39. doi:10.1046/j.1466-822X.2002.00183.x

Bickford D, Howard SD, Ng DJJ, Sheridan JA (2010) Impacts of climate change on the amphibians and reptiles of Southeast Asia. *Biodiversity and Conservation* **19**, 1043–1062. doi:10.1007/s10531-010-9782-4

Bode M, Watson J, Iwamura T, Possingham HP (2008) The cost of conservation. *Science* **321**, 345–346. doi:10.1126/science.321.5887.340a

Brooks T, da Fonseca GAB, Rodrigues ASL (2004) Species, data, and conservation planning. *Conservation Biology* **18**, 1682–1688. doi:10.1111/j.1523-1739.2004.00457.x

Burgman MA (1988) Spatial analysis of vegetation patterns in southern Western Australia: implications for reserve design. *Australian Journal of Ecology* **13**, 415–429. doi:10.1111/j.1442-9993.1988.tb00990.x

Ceballos G, Ehrlich PR (2006) Global mammal distributions, biodiversity hotspots, and conservation. *Proceedings of the National Academy of Sciences of the United States of America* **103**, 19374–19379. doi:10.1073/pnas.0609334103

Commonwealth of Australia (2001) 'Australia's natural resources 1997–2002 and beyond'. Australian Native Vegetation Assessment. National Land and Water Resources Audit.

Cowan MA, How RA (2004) Comparisons of ground vertebrate assemblages in arid Western Australia in different seasons and decades. *Records of the Western Australian Museum* **22**, 91–100.

Del Marco AP (2011) 'Towards a Land Use Planning Model for the Great Western Woodlands'. Unpublished report to the Woodlands Initiative, Perth, Western Australia.

Department of Environment and Conservation (2010) 'A biodiversity and cultural conservation strategy for the Great Western Woodlands'. Department of Environment and Conservation, Perth.

Duncan S, Traill B, Watson C (2006) 'Vertebrate fauna of the Honman Ridge-Bremer Range district, Great Western Woodlands, Western Australia'. The Wilderness Society, West Perth.

Ford HA, Barrett GW, Saunders DA, Recher HF (2001) Why birds have declined in the woodlands of Southern Australia declined? *Biological Conservation* **97**, 71–88. doi:10.1016/S0006-3207(00)00101-4

Gasc JP, Cabela A, Crnobrnja-Isailovic J, Dolmen D, Grossenbacher K (1997) *Atlas of amphibians and reptiles in Europe*. Muséum National d'Histoire Naturelle & Service du Petrimone Naturel, Paris.

Hopper SD, Gioia P (2004) The southwest Australian floristic region: Evolution and conservation of a global biodiversity hotspot. *Annual Review of Ecology Evolution and Systematics* **35**, 623–650. doi:10.1146/annurev.ecolsys.35.112202.130201

How RA, Newby KR, Dell J, Muir BG, Hnatiuk RJ (1988) Biological survey of the eastern Goldfields of Western Australia. Part 4. Lake Johnston-Hyden study area. *Records of the Western Australian Museum* (Suppl. 30).

Hughes L (2000) Biological consequences of global warming: is the signal already apparent? *Trends in Ecology & Evolution* **15**, 56–61. doi:10.1016/S0169-5347(99)01764-4

Judd S, Watson JEM, Watson AWT (2008) Diversity of a semi-arid, intact Mediterranean ecosystem in southwest Australia. *Web Ecology* **8**, 84–93.

Keally I (1991) Management of inland arid and semi-arid woodland forest of Western Australia. In *Forest Management in Australia*, 18–22 September, Perth (Eds FH McKinnel, Hopkins, ER and Fox JED) pp. 286–295. Institute of Foresters of Australia, Perth, Western Australia.

Keighery GJ (1995) Biological survey of the Eastern Goldfields of Western Australia. Part 11. Boorabbin-Southern Cross study area. *Records of the Western Australian Museum* (Suppl. 49).

Lee TM, Jetz W (2008) Future battlegrounds for conservation under global change. *Proceedings. Biological Sciences* **275**, 1261–1270. doi:10.1098/rspb.2007.1732

Mackey BG, Soule ME, Nix HA, Recher HF, Lesslie RG, Williams JE, Woinarski JCZ (2007) Applying landscape-ecological principles to regional conservation: the WildCountry Project in

Australia. In *Key Topics in Landscape Ecology*. (Eds J Wu and RJ Hobbs) pp. 92–208. Cambridge University Press.

Mackey BG, Watson JEM, Hope G, Gilmore S (2008) Climate change, biodiversity conservation, and the role of protected areas: An Australian perspective. *Biodiversity* **9**, 11–18. doi:10.1080/14888386.2008.9712902

May K (2010) 'Indigenous cultural and natural resource management and the emerging role of the Working on Country program'. CAEPR Working paper No. 65/2010, Canberra.

Myers N, Mittermeier RA, Mittermeier CG, da Fonseca GAB, Kent J (2000) Biodiversity hotspots for conservation priorities. *Nature* **403**, 853–858. doi:10.1038/35002501

Newbey KR, Hnatiuk RJ (1984) Vegetation and flora. Biological survey of the eastern goldfields of Western Australia. Part 2. Widgiemooltha-Zanthus study area. *Records of the Western Australian Museum* (Suppl. 18).

Newbey KR, Hnatiuk RJ (1988) Vegetation and flora. Biological survey of the eastern Goldfields of Western Australia. Part 4. Lake Johnston-Hyden study area. *Records of the Western Australian Museum* (Suppl. 30).

Pressey RL, Johnson IR, Wilson PD (1994) Shades of irreplaceability – towards a measure of the contribution of sites to a reservation goal. *Biodiversity and Conservation* **3**, 242–262. doi:10.1007/BF00055941

Recher HF (2004) WildCountry. *Pacific Conservation Biology* **10**, 221–222.

Rodrigues ASL, Gaston KJ (2002) Rarity and conservation planning across geopolitical units. *Conservation Biology* **16**, 674–682. doi:10.1046/j.1523-1739.2002.00455.x

Sanderson J, Fonseca GAB, Galindo-Leal C, Alger K, Inchausty VH (2006) 'Escaping the minimalist trap: design and implementation of large-scale biodiversity corridors' in *Connectivity Conservation* (Eds Crooks K, Sanjayan M). Cambridge University Press, Cambridge. pp. 620–648.

Soulé ME, Sanjayan MA (1998) Conservation targets: do they help? *Science* **279**, 2060–2061. doi:10.1126/science.279.5359.2060

Soulé ME, Estes JA, Berger J, Del Rio CM (2003) Ecological effectiveness: Conservation goals for interactive species. *Conservation Biology* **17**, 1238–1250. doi:10.1046/j.1523-1739.2003.01599.x

Soulé ME, Mackey BG, Recher HF, Williams JE, Woinarski JE, Driscoll D, Dennison WC, Jones M (2004) Continental connectivity: Its role in Australian conservation. *Pacific Conservation Biology* **10**, 266–279.

Svancara LK, Brannon R, Scott JM, Groves CR, Noss RF (2005) Policy-driven versus evidence-based conservation: A review of political targets and biological needs. *Bioscience* **55**, 989–995. doi:10.1641/0006-3568(2005)055[0989:PVECAR]2.0.CO;2

Thackway R, Lesslie R (2006) Reporting vegetation condition using the Vegetation Assets, States and Transitions (VAST) framework. *Ecological Management & Restoration* **7**, S53–S62. doi:10.1111/j.1442-8903.2006.00292.x

Thompson SA, Thompson GG (2005) Temporal variations in reptile assemblages in the Goldfields of Western Australia. *Journal of the Royal Society of Western Australia* **88**, 25–36.

Watson AWT, Judd S, Watson JEM, Lam A, Mackensie D (2008a) *The Extraordinary Nature of the Great Western Woodlands*. The Wilderness Society, Perth. p. 72.

Watson JEM, Fuller RA, Barr L (2008b) Why are we still using a 'one size fits all' philosophy for systematic reserve planning in Australia? *Pacific Conservation Biology* **14**, 233–235.

Watson JEM, Judd S, Mackey B, Bradby K (2010) Ecological restoration in Gondwanalink (south-western Western Australia) using the WildCountry principles: a convergence of thought and action. In *Connectivity Conservation Management: a global guide* (eds GL Worboys, WL Francis, M. Lockwood), Earth Scan, pp. 105–115.

Watson JEM, Grantham H, Wilson KA, Possingham HP (2011) Systematic Conservation

Planning: Past, Present and Future. In *Conservation Biogeography* (Eds R. Whittaker and R. Ladle), Wiley-Blackwell, Oxford, pp. 136–160.

Wilson KA, Underwood EC, Morrison SA, Klausmeyer KR, Murdoch WW, Reyers B (2007) Conserving biodiversity efficiently: what to do, where, and when. *PLoS Biology* **5**, 1850–1861. doi:10.1371/journal.pbio.0050223

Woinarksi JCZ (1999) Fire and Australian birds: a review. In *Australia's Biodiversity: Responses to fire: plants, birds and invertebrates*. (Ed. AM Gill) pp. 55–111. Environment Australia, Canberra.

Woinarksi JCZ, Mackey BG, Nix H, Traill BJ (2007) *The nature of Northern Australia: natural values, ecological processes and future prospects.* ANU Press, Canberra.

INDEX

Aboriginal and Torres Strait Islander people   35
   cancer risks   45
   ecological management and   259–60
   native species and   170–1
   traditional land ownership case 1971   161, 162–3
   see also tradition land owners; *various entries under* indigenous
acacias   83, 84, 164, 165, 167, 168, 169, 170, 172, 180, 186, 187, 188, 199, 200, 201, 205, 206, 207, 255, 256
access certainty   259
accounting systems, ecosystems and   32
acid litter   187
adaptive management (AM)   5, 19–20
Africa   30, 48, 172
African spoonbill (*Platalea alba*)   36
agile wallaby (*Macropus agilis*)   176
agriculture   170, 253, 254
air-blow seeding   165
air humidity, Brazil   60
Alcoa   19, 112–14, 146
   orchid reintroduction and   144–5
   Western Australian forest restoration   111–39
alkaline lakes   36
Alligator River   35
alpha particles   45
aluminium concentrations   3, 161, 188–9, 194, 195, 196, 207
Alyssum bertolinii   239, 240
Amazon rainforest   58
Amitermes spp.   173
amphibians   176
analysis of similarity (ANOSIM)   82, 83, 88, 89
angiosperms   60
animal dispersal   127
animal husbandry   37
animal recolonisation   171, 205–7
ants   133–4, 166, 171–2, 190, 191, 205, 208
Appalachian coalfields   5, 7–27
approximate original contour (AOC)   10, 12
aquifers   38, 59, 64
arbuscular mycorrhizas (AMF) recolonisation, Gove   179, 181–4, 205
archaeological heritage, Brazil   59–60
arid zone   79, 84, 89, 90
Arnhem Land   161, 162, 172, 260

arrested succession   12
arsenic   41, 232, 233, 236
Asteraceae   83, 85, 87, 234, 238
Atlantic rainforest   56, 59, 66
Australasia, minerals industry   231–49
Australia   4, 5, 30, 34–6, 48, 65, 143, 154, 180, 187, 241, 242, 243, 244, 258
   ultramafic soils   241
Australian bustard (*Ardeotis australis*)   175
Australian Government   36, 255
Australian tropics, bauxite mine restoration in   159–227
autumn olive (*Elaeagnus umbellata*)   11
avoid–minimise–mitigate management   100

backfill   18
Bahi wetlands   31, 33, 36–8, 40, 41, 44, 45, 47
banded-hare wallaby (*Lagostrophus fasciatus*)   255
banded ironstone formations (BIF)   4, 5, 55, 56, 57, 59, 60, 62, 65, 79, 81, 84, 90, 95, 97, 98, 99, 105
basal areas, Gove restored sites   167–9
Baudin's black cockatoo (*Calyptorhynchus baudonii*)   133
bauxite mining   5, 56, 124, 162, 180, 189, 190, 196
   ecosystem restoration and   159–227
   jarrah forest and   111–39
   orchid recolonisation and   144–9
   photo survey of   115
Belo Horizonte   57, 59
Berkheya coddii   238, 239, 240
best practice, mine site restoration   19–20, 204–5
beta diversity   63, 84, 88, 90
beta particles   45–6
bioconcentration   233
biodiversity hotspots   58, 254
biodiversity restoration   13, 17
biodiversity threats   135, 259
   minerals exploration and   242–4
biodiversity values   9, 15, 16, 31, 151
   Brazil   60–6
   Kollyanobbing   97
   Mount Jackson   104
   native forests   18
   photoextractors and   237–8
   restoration of   125–31

biogeographic patterns, Yilgarn Craton survey   84–9
biological surveys
    Great Western Woodlands   253
    Yilgarn region   105
biomass production   238, 239, 241
bio-ore   239
biospeleological studies, Brazil   63
birds   171, 205, 206, 255, 260
    Jarrahdale mine   132
    recolonisations (Gove)   175
Birds Australia   253, 255
black-banded rainbowfish (*Melanotaenia nigrans*)   41
black flying fox (*Pteropus alecto*)   176
black locust (*Robinia pseudocacia*)   12
blue-green algae   36
blue lady orchid (*Thelymitra crinita*)   145, 147, 148, 150
bond funds   259
Bonferonia correction   88
Brachychiton spp.   165
Brassiceae   237, 239
Brazil   4, 5, 55–77
Breaking New Ground   22
broad-faced potoroo (*Potorous platyops*)   255
broad-leaved species   186
Broken Hill Proprietary   97, 99
brolga (*Grus rubicunda*)   175
bulk soil density   193, 208
bull banksia (*Banksia grandis*)   125
burial sites   43
burrowing bettong (*Bettongia lesueur*)   255
bush foods, uranium mining and   43
bush potato (*Eriosema chinense*)   170

Caatinga   58, 59
cactus (*Arthrocereus glaziovii*)   63
cadmium   41, 232, 233, 234, 236, 238, 239, 241
Caetité   56, 58, 62, 65, 71–7
Caladenia spp.   153
calculable costs   32
callitris woodlands   165, 256
camping sites   43
cancers   45
cane toad (*Bufo marinius*)   176
cangas   56, 57, 59, 60, 61, 62, 63, 65
canopy closure   134, 168, 169, 175, 176, 177, 186, 187, 188, 198, 199, 201, 203, 205, 206
canyons   58
Cape York Peninsula   162, 189, 255
capital costs   31

caprock   114
Carajás   56, 58, 59, 60, 62, 64
carbon concentrations   83, 177, 187, 188, 190, 194, 195, 196–8, 199, 200, 201, 202, 208
carbon dioxide emissions   198, 236, 260
carbon isotope values   198–9
carbon sequestration   9, 11, 15, 16
carbon-to-nitrogen concentration   188, 197, 200, 201, 202, 208
carnivorous pitcher plant (*Nepenthes reinwardiana*)   235, 237
carpet python (*Morelia spilota imbricate*)   133
carrion removal   133
Casuarina forests   256
catchment protection   112, 120
cation exchange capacity (CEC)   195, 196, 208
cattle (*Bos primigenius*)   176
caves   60, 61, 63, 64–5
ceremonial sites   43
Cerrado   56, 59
char   196, 202, 206, 208
chemical contamination   35, 44
chemically assisted phytoextraction   237
chemical soil properties
    Gove   194–6
    litter   187–9
chenopod shrublands   83, 87, 256
Chile   22, 64, 80, 97, 231, 234
Chloris gayana   165, 169
choke layer   194
chromium   56, 231, 234, 241, 244
chronosequences   166, 167, 169, 173, 174, 180, 185, 187, 188, 189, 190, 196, 197, 198
chuditch (*Dasyurus geoffroii*)   132, 133
clays   162, 193, 194, 207, 208
cleaning   116–17
cleared land   256
Cliffs Asia Pacific Iron Ore Pty Ltd (Cliffs)   95–7, 98, 103, 105, 106, 107
climate   151, 152, 161–2
climate change   160, 204, 257
clonal orchid species   130, 145, 146, 147, 153
coal industry   3, 5, 9, 20
coarse woody debris (CWD)   133, 178–9
coastal plain   112, 253
cobalt   231, 233, 234, 236, 239
cockatoo nests   133
collaborative management, Western Australian biodiversity and mining   251–63
column cacti (*Pilosocereus pachyclados*)   61

Commission of Geologic and Palaeobiologic Sites (SIGEP) 59, 60
commodity values 32
common brushtail possum (*Trichosurus vulpecula*) 132
compliance 20, 119, 121
concentration levels, metals 237
condition monitoring, Windarling mine 101–3
conservation concern species, Yilgarn Craton survey 85–6
conservation management 258
    Brazil 56, 58
    Great Western Woodland 256–8
    Western Australia 84, 90–1
    Yilgarn region 105–6
constant head disc permeameter 193
construction species 170
contamination
    deserts and 38
    Kakadu National Park 35
    uranium mining 41, 43
contingent valuation 32
Convention on Biological Diversity 30, 65
copper 3, 233, 234, 236, 239, 241
copper flower (*Polycarpaea spirostylis*) 243
copper leachates 231–2
Coptotermes acinaciformis 173, 174, 175
Corymbia spp. 165
cost–benefit analysis 31
    Alcoa plant propagation and 128
    dieback management 119
    ecosystem values and 32
    reseeding 130
    uranium mining 47–8
    Yilgarn region study 103
cowslip orchid (*Caladenia flava*) 145, 148, 150
crescent nail-tailed wallaby (*Onychogalea lunata*) 255
Cretacean iron-caps 56
critically endangered species 142
Cryptostigmatid mites 134
Cuba 234, 242
culturally significant sites 42–4
cultural resources 35
cuttings 128, 130
cycling tracks 112
cyclones 161, 210
Cyperaceae 85, 87

Darling Ranges 112, 120, 208
Darwin stringybark (*Eucalyptus tetrodonta*) 5

data collection 19, 20, 259
death adder (*Acanthophis* sp.) 176
debris slopes 11
Declared Rare Flora and Priority species (DRF) (Western Australia) 83–4, 99–100, 104, 253, 254, 255
decomposers 134, 172, 184, 186, 187, 206
deep ripping 128, 204
dendrogram 87, 89
Department of Environment and Conservation (DEC) (Western Australia) 82, 91, 107, 122, 124, 146, 253
Department of Industry and Resources (Western Australia) 82, 124
Department of the Interior (US) 16, 19
Department of Water (Western Australia) 120, 122, 124
desert ecosystems 29, 33, 38–9, 47
desiccation tolerance 60, 62
detoxification 233
dicotyledonous plants 177, 178
dieback disease (*Phytophthora cinnamomi*) 114–19, 135
dingo (*Canis lupus familiaris*) 176
direct impact mining 63
direct return of topsoil 128, 129, 164
Discorea spp. 171
disease control 114, 181, 210, 204
disturbance tolerant orchids 153, 154
disturbed environments 145, 152
Djalkmarra billabong, destruction of 35
djoongari (*Pseudomys fieldi*) 255
doliniform depressions 58, 61
dormancy breaking (seeds) 129
double soil stripping 128, 164, 204
drainage 11, 58, 231
drought 181, 238
drylands 37
dry sclerophyll open forest 125
dry season 185, 198
dust impacts 98–9, 189

earthworms 171, 200, 206
Eastern Coalfields (USA) 8, 9
echidna (*Tachyglossus aculeatus*) 176
eco-hydrology 257
ecological costs
    bauxite mining and 160
    environmental values and 66, 100
    mining and 29–51
    phytomining and 239–40

ecological tolerance, orchids   152–4
ecosystem development
    Gove   203
    Great Western Woodland   255–6
    litter system and   184–9
ecosystem engineers, termites   173
ecosystem function analysis (EFA)   17, 160–1, 209
ecosystem irreplaceability   258
ecosystem reclamation approach (ERA)   16, 17, 18–22
ecosystem restoration
    Appalachian coalfields   7–27
    Gove bauxite mine   159–227
ecosystem values   4, 31, 32–3, 35, 46
    ants and   171
    ironstone sites   55–6
    mining and   3–6
ectomycorrhizas (ECM) recolonisation   178, 179–81, 205
edaphic factors   82–3, 84, 89–90, 152, 232, 234, 235, 238, 243
edible species   170–1
electrical conductivity   83
emergency response plan (ERP)   122
employment   4, 97
environment, Gove mine and   160, 161
environmental ban, Koolyanobbing area   80, 82
environmental costs   4, 20
environmental impact assessment (EIA)   39, 40
environmental impact statement (EIS)   16, 17, 65, 257
environmental incident reporting, Alcoa   123, 124
environmental licences, Brazil   66
environmental non-government organisations (ENGOs)   253
environmental non-information   98
Environmental Protection Agency (EPA) (US)   16, 20
Environmental Protection Authority (EPA) (Western Australia)   100, 105
environmental protection, uranium mining and   47–8
Environmental Review and Management Program (ERMP) (Aloca)   113, 120, 121, 122, 124
Environment Protection and Biodiversity Conservation Act 1999 (EPBC)   30, 133
Ephelotermes melachoma   173, 174
ephemeral species   129, 130
Epipactis spp.   144
Equisetum ramosissimum (Equisetaceae)   231–2
equivalent cylindrical diameter (ECD)   191
Eremophila   60, 83
Eriochilus spp.   60, 153

erosion control   13
Eucalyptus   83, 149, 160, 162, 165, 167, 168, 169, 170, 172, 174, 177, 180, 186, 188, 199, 200, 201, 202, 204, 205, 206–7, 209
Eucalyptus miniata   162, 165, 167, 170, 172, 175, 180, 199
Eucalyptus tetrodonta   160, 162, 165, 167, 170, 172, 175, 177, 178, 180, 199, 205
eucalypt woodlands   79, 81, 179, 253, 255, 255, 256
Euclidean distance matrix   83
Euphorbiaceae   85, 170
Europe   144, 152, 154, 231, 234
evapotranspiration   60, 235
exchangeable acidity   195
excluder plants   233, 234, 235
exotic species   112, 114
extinct species, Great Western Woodland   255
extremely rare orchids   152

Fabaceae   83, 85, 87, 170
fallow period   204–5
fauna corridors   132, 133
fauna reintroduction   135, 253
feral species   176, 257, 259
    controls   133
    Kakadu National Park   35
ferns   61
ferriferous formations, Brazil   56
ferruginous geosystems   58–9, 63, 65–6
fertilisers   163, 165, 166, 196, 204, 206, 208
fertility   199
fine earth concentrations, Gove   192–3
fine litter mass   186–7, 200, 201, 202, 206
fire exclusion   166, 168–9, 171–2, 176, 203
fire regimes   133, 146–7, 161, 184, 202, 208, 209–10, 252, 257, 259, 260
fisheries   19, 34–5
Five Year Mine Plan   120
floods   9, 17, 41, 42
flora   5, 84, 253
    recolonisation   125–6, 135, 181
    Yilgarn Craton surveys   79–107
flow paths   13
food chain   44, 46, 238
food plants   163
forest ecosystems   12, 15, 16, 19, 112, 180, 184, 189, 203–4
forest islands   60, 61
forest productivity   9, 13
forest red-tailed black cockatoo (*Calyptorhynchus bankssi naso*)   133

forestry reclamation approach (FRA)   12, 13, 14, 15, 16, 17, 18
forest site index   13, 14, 16
fossils   60
fox (*Vulpes vulpes*) predation   132
fragmented landscapes   257
freshwater aquatic systems   256
freshwater snail (*Amerianna cumingi*)   41
freshwater turtles   34
frill neck lizard (*Chlamydosaurus kingii*)   176
frogs   176, 255
fruit bat   175
fungal colonisation   151, 152, 174, 187
future developments
    Gove mine site   202, 203, 209
        mining sector   259
        nickel phytomining   239–40

gallery systems, termites   175, 190, 191, 200
genetically modified organisms   238
geographical information systems   116
geographical separation, species composition and   89
geomorphic landscape design   17–18
germination requirements, jarrahs   129–30
gibbsite clay   195
glaciation   59, 60
Glomus spp.   182
goanna (*Varanus* spp.)   176
gold   3, 36, 60
Goldfields Region (Western Australia)   251–63
Gove bauxite mine   159–227
grading   11
grasses   9, 10, 11, 16, 17, 160, 164, 165, 169, 172, 175, 177, 178, 185, 187, 188, 199, 200, 201, 202, 205, 206, 207
grass-trees (Xanthorrhoeaceae)   130
gravel concentrations   191–2, 194, 207
greater flamingo (*Phoenicopterus ruber roseus*)   36
greater stick-nest rat (*Leporillus conditor*)   255
Great Western Woodlands (GWW), biodiversity management of   107, 251–63
Greenspan turbidity sensors   121
Grevillea spp.   83, 165, 169, 172, 201
ground cover   38, 63, 65, 184–6, 203
groundwater   12, 120, 121, 124, 235

hair sedge (*Tetraria capillaris*)   131
hand seeding   165
hardwoods   8, 9, 10, 15
harvestman (*Canga renatae*)   63

haul roads   114, 124, 133
Havel site vegetation types   126
hayland/pasture   11, 12
heathlands   256
heat treatment (seeds)   129
heavy metals   41, 44, 45
Helena and Aurora Range   80, 84, 90
hematite   56, 59
Hemerocallidaceae   85, 87
hemi-accumulators   234
Hemiptera   134, 135
herbaceous stratum   13, 160, 169, 185, 190, 199, 201, 205, 206
herbicides   13
Heteropogon spp.   169, 173
hidden costs, mining and   29–51
high magnitude impact mining   63
high pressure acid leach (HPAL)   242
highwalls   10, 11
holistic ecosystem reclamation approach   20
human welfare
    ecosystem values and   32, 33
    uranium mining and   40, 41, 44–6
hummock grasslands (*Triodia*)   79, 256
Huntly mine   112, 120, 123, 125, 127, 132
Hybanthus spp.   240, 241
hydraulic hose failures   122
hydraulic lift   176
hyperaccumulators   232, 233–4, 236, 238–44

Illinois   7, 9
impact management strategies   100
Indiana   7, 9, 10
indicators   232–3
indigenous conservation officers   107
indigenous land and sea management groups   260
indigenous landowners   33–5
    radium mining and   42–4
    see also traditional land owners
Indonesia   142, 238, 240, 241, 242–3, 244
infiltration   185, 193–4, 200, 201, 208
infiltration sumps   124, 125
insect-eating plants   237
insect pollination, orchids   141–2
intergenerational equality   31
Intermediate Rainfall Zone (IRZ)   121, 122, 124
invasive species   11, 12, 35, 132, 172, 204, 206, 210
invertebrates   63, 133–5, 171
investment targets   258, 259
ion pumps   236

iron caves   58
iron concentrations   3, 56, 64, 188–9, 194, 195, 207, 234
    Brazil   55–77
    Yilgarn region   95–107
iron oxides   60, 162, 193, 196
ironstone ranges   55–107
irreversible impact mining   63
island vegetation   97
isotopes   41, 45–6
itabirite   56, 59, 60

Jabiluka   30, 35
Japanese honeysuckle (*Lonicera japonica*)   11
Jarrah (*Eucalyptus marginata*) forest   5, 182
    bauxite mining and   111–39
    orchid recolonisation of   142–9, 152, 153–4
Jarrahdale   145, 146, 187
Joint Intermediate Rainfall Zone Research Program (JIRZRP) (Alcoa)   122, 124
J1 deposit, Mount Jackson   104

Kakadu National Park   5, 30, 31, 33–6, 40, 42–3, 44, 45, 47, 255
Kalgoorlie   97, 253
kaolinite   162, 193, 195, 207, 208
karstic systems   58, 59
Kentucky   7, 9
Key Biodiversity Areas (KBAs)   62
keystone species   165, 170, 172, 175, 199, 202, 204
Khaya spp.   163
king cricket (*Apterogryllus pedestris*)   190
Koolanooka Hills   90, 98
Koolyanobbing mine   80, 95, 97, 99, 105, 106, 161
Krakatau eruption   142–3

Lamiaceae   83, 85, 87
land clearing   114
land management   154, 258
landowners, biodiversity and   15–16, 31
landscapes, mining and   3–6, 114
landslides   142, 143
larch (*Eucalyptus marginata*)   115
lateral roots   178
lava flows   142, 143
lead   46, 233, 234, 236, 241
legumes (Papilionaceae)   11, 129, 170, 208
lespedeza (*Lespedeza cuneata*)   11
lesser flamingo (*Phoenicopterus minor*)   36
Leucaena leucocephala   163, 170
lichens   38

lilies (Anthericaceae)   130
lime   235
limestone quarry   143
litigation, mountaintop mining and   16
litter layer   18, 134, 152, 165, 166, 171, 174, 177, 180, 190, 196, 206–7, 209
litter recolonisation, Gove   184–9, 200–3
lizards   176
loams   194
local government   258
local seed collection   129
long-tailed hopping mouse (*Notomys longicaudatus*)   255
loss of containment (LOC) response   122, 123
lower slope forest   126

Macadamia neurophylla   241
Magela River system   34, 35, 41–2, 44
magnesium   41, 234
magnetite   56, 59
Malaysia   238, 240, 241, 242
malleefowl (*Leipoa ocellata*)   104
Mallee woodlands   255, 256
Malvaceae   83, 85, 87
mammals   131–2, 176, 205, 206, 255, 260
manganese   41, 56, 161, 232, 233, 234, 241
manmade habitats, orchid recovery in   144
mapping   116
mardo (*Antechinus flavipes*)   132, 133
market values, ecosystem values and   31, 32–3
marri (*Corymbia calophylla*)   125, 126
Maryland   7, 9
Mastotermes darwiniensis   173, 174, 175
mating flights, termites   173
medicinal plants   163, 170
medium roots   178–9
Melaleuca   150, 165, 175, 180
Melastomataceae   60
Melinis spp.   163
mercury   46, 236
metallic ores, Brazil   56–8
metallophytes   231–49
metallurgical coal   7
microbial biomass   184, 200, 209
Microcerotermes spp.   174
microchannels   63
microhabitats   152, 166, 260
    Brazil   60, 62
    disturbance of   142, 152
microorganisms   164, 205, 207

Midgley's grunter (*Pingalla midgleyi*)   34–5
mid-stratum vegetation   160, 167, 202, 203, 205
Midwestern coalfields   7, 8, 9, 10, 22
Midwest region (Western Australia)   96, 97, 98, 105
mignonette orchid (*Microtis media*)   147, 148, 150
milkweed (*Minaria monocoronata*)   63
Mimosaceae   60, 129, 241
mine operators   15, 16
mine pits   104, 112, 129, 135
minerals industry   22, 56
    Australasia   231–49
        metallophytes and   234–5
        rehabilitation   234
mineral soils   207
Minesite Drainage Book (Alcoa)   124
mine sites   100, 144, 152, 239
minimalist targets, Western Australian government   257
mining industry   258
    ecologically sensitive landscapes and   3–6
    exclusion areas   100
    future opportunities   259
    hidden costs   29–51
    history (Yilgarn region)   97
    Western Australia   253
Mining, Minerals and Sustainable Development (MMSD)   21, 22
Mirarr people   35, 36, 42–3, 48
monitoring   19, 20, 48, 114, 116, 253
    dieback disease control   119
    threatened species   133
    uranium mining   41–2, 44, 45
monoculture   189
Morro do Pilar   56, 58, 59, 62, 65, 71–7
mountaintop mining (MTM)   4, 5, 9, 16, 17
Mount Jackson   80, 84, 90, 96, 97, 98, 99, 100, 103–4, 105
Mount Manning Nature Reserve   90, 105
mudflats   36
Mudjinberri billabong   44
mulga woodlands (*Acacia aneura*)   79, 81, 253
mussels   41, 44
mycelial mats   180, 184, 187, 200, 201
mycorrhizal fungi   5, 142, 153, 154, 176, 181, 195, 205, 209
    orchid disturbed habitats and   149–50
    reintroduction of   151–2, 166, 178–80
Myrtaceae   60, 83, 85–6, 241

Nabalco Pty Ltd   162
Namibia   5, 31, 38–9, 45, 47
Namib-Naukluft National Park   31, 33, 38–9, 40, 41, 44, 45, 47
nara lemon (*Acanthosicyos horrida*)   38, 39, 44
Nasutitermes graveolus   173, 174
national parks   112
National Protected Areas System (Brazil)   64, 66
native rat (*Melomys burtoni*)   176
native species   13, 103, 165, 169, 243
native title   252
    see also traditional land owners
natural phytoextraction   237
natural resources policy, Brazil   63–5
natural steam channel design   18
nature reserves   105–6
nesting resources   133, 176
New Caledonia   233, 234, 240, 241, 242
new species, Yilgarn Craton survey   84, 87, 90
nickel   56, 231, 232, 233, 235, 236, 238, 240, 242, 244
nickel hyperaccumulators   234, 239–40, 241, 242
nitrogen concentrations   83, 187–8, 190, 197–8, 201, 205, 207, 208, 209, 237
noisy scrub-bird (*Atichornis clamosus*)   133
non-accumulating plants   233
non-arid landscape   252
non-marketable resources   48
non-metric multidimensional scaling ordination   88
non-saline soils, Gove   194
North America   152, 258
North Dandalup River   120, 121
Northern Land Council   48
Northern Territory   47, 161, 163, 173, 174
nursery-grown plants   127
nutrient deficiencies   166, 238, 243
nutrients   172, 179, 196, 206, 235

obligate soils   231
off-site legacies   4
Ohio   7, 9
on-ground management   107
open country birds   175
open forest   162, 163, 199, 203
open pit mining   63, 65, 97, 114, 143, 144
operator training   119
orange-footed scrub fowl (*Megapodius reinwardt*)   175
orchids (*Epipactis palustris*)   127, 141–58
organic matter   177, 184, 190, 196–8, 202, 207, 208, 235
oven dry litter mass   186–7
overburden   9, 11, 12, 18, 104, 114, 235
overgrazing   130–1, 259

overstorey invertebrates   134
oxides   162, 195, 208

pair-wise comparisons, soil chemistry   90
palaeosols   189
paleoenvironmental heritage, Brazil   59–60
Pantanal   58, 59
paper baits   174
pastoral industry   258
   Gove Peninsula   163
   Western Australia   253
Peixe Bravo   56, 58, 59, 60, 61, 62, 65, 66, 71–7
Pennsylvania   7, 9
peregrine falcon (*Falco peregrines*)   133
perennial species   83, 169, 181, 205
Persoonia spp.   129, 165
Perth   97, 112, 135
pH   83, 194
Philippines   240, 241, 242, 244
phosphorus concentrations   83, 97, 188, 189, 190, 195–6, 204, 205, 207–8, 234
phosphorus-limited soils   180, 181
Phyllanthaceae   86, 234, 238, 240, 242
phytomining   234, 235–8, 238–40
pig (*Sus scrofa*)   35, 44
pig-footed bandicoot (*Chaeropus ecaudatus*)   255
pig-nosed turtle (*Carettochelys insculpta*)   34
Pinus spp.   12
pioneer species, orchids   145, 146, 152, 153
pisoliths   189–90, 191, 192, 200, 207
pits   10, 11, 124
plains mouse (*Pseudomys australis*)   255
plan–do–check–revise cycle   119
plant-available phosphorus   190, 196
plant community development   167–9
plant ecology studies, Brazil   62
plant species   103, 260
   Alcoa mine restoration   127, 128
   Gove mine site   205–6, 222–7
   Great Western Woodlands   253–5
   hyperaccumulators   238
   nutrition   209
plateaus   56, 57
playa lakes   37
Poaceae   83, 86, 170
pollination   151, 175, 257
pollution   11, 64, 259
polonium   41, 43, 44, 45
ponds   114, 194
P1 protection areas   120, 121
population expansion, orchids   154

Portman Iron Ore Limited   80, 97
Portulacaceae   87
post-and-pillar mining   8
post-mining land use (PMLU)   11, 12, 13, 162–3, 166, 169–70
post-wet season   185, 186
potassium   83, 234
poverty, biodiversity legislation and   30
Prasophyllum spp.   144, 153
predators   134, 175
pre-mining process   163
priority conservation species   104
profile soil properties, Gove   190–1
profitability, sustainability and   21, 22
propagules   164
prospecting, metallophytes and   243–4
protected areas   65, 66
   Great Western Woodlands   256–7
provenance range (seeds)   129
Psychotria spp.   60, 233, 240, 241
Pterostylis sp.   145, 148, 150, 153
PT Vale Inco   242
public policies, natural resources   63–5
Pychandra acuminata   233, 240
pygmy goose   44

Quadrilátero Ferrifero (Iron Quadrangle)   56, 57, 58, 59, 60, 61, 62–3, 64, 66, 71–7
qualitative assessments   31, 32–3, 160
Queensland   240, 241, 243, 254
quenda   132, 133
quokka (*Setonix brachyurus*)   133

rabbit-proof fence   97
radioactive elements   40, 236
radionuclides   35, 40, 43, 44, 45
radium   41, 43, 44, 45
rainbowfish (*Melanotaenia exquisita*)   34
raindrop impact   184, 185, 189
rainfall
   Appalachians   9
   Gove   161
   monitoring of   120
   perennial species and   83
   Yilgarn Craton   80, 82, 103
Ramsar Convention listing   34, 35, 36, 41, 42
range centroids   88, 89
Ranger uranium mine   5, 33, 35, 41, 42–3, 44, 45
ranges
   dissimilarity   87–9
   irreplaceability of   90

rare earth elements   3
rare species   34, 105
    banded iron formations and   95–6
    Brazil   62–3, 65, 71–7
    Yilgarn region   97–8
recalcitrant species   129–31, 150, 152, 154
recreational land use   112
red beak (*Pyorchis nigricans*)   145–6, 147, 148, 149
red lists, Brazil   65
reforestation   5, 9–12, 18, 22
Regional Forest Agreement (RFA)   121
regional objectives   258
regulatory framework   15, 30, 47
rehabilitation areas   4, 5, 17, 234
    Appalachian coalfields   7–27
    bauxite mining   112, 114, 153
    Gove   169–70, 204–5
    management of   122, 150
    metallophytes and   244
    orchid density   145, 154
    sustainable mining   20–2
remnant species   132
reportable environmental incidents   117
reptiles   176, 255, 260
research   19, 20
    Great Western Woodlands   260
    metallophytes   244
re-seeders   129, 130, 166
residual liability   31
respread topsoil   144, 150
resprouters   129
rhizosphere   177, 238
rice   37, 44
Ricinocarpos brevis   99–100, 101
ridges   126, 189–90
Rinorea   240, 242
Rio Tinto Alcan   20, 159–227
ripping   189, 190, 194, 200
Rirratjingu people   163
risk-based approach   46, 119
roadside clearing   143
root systems   176–8, 191, 194, 196, 208
Rubiceae   60, 233, 234
runoff control   120
rushes   130
Rutaceae   86, 87

Sabah   240, 242
salinity   33, 36, 122, 124, 238, 243
salts   13, 37
Samson Reservoir   120, 121

sandy loam   192, 193
saturated infiltration rates   193
savannah   56, 60, 162, 187, 260
scarcity rent   33
Schedorhinotermes spp.   173
screening, metallophytes   244
Scrophulariaceae   86, 93
scrubland   10
secondary colonisers   146, 152, 175
sedges   130
sedimentation   124, 125, 135
seed banks   127, 128, 144, 150
seed dispersal   128–9, 144, 145, 150, 151, 164, 169–70, 172, 175, 190, 204, 206
seed mixture   129, 166
seed predators   175
seed production   18, 152
seed shortages   165
selenium   232, 233, 236, 241
self-pollinating orchids   147, 148, 153
semi-arid areas   33, 240, 241
semi-disturbed habitat   152, 153, 154
Senna   60, 83
Serpentine River   120, 121
sewerage plants   124
Shannon–Weiner diversity   132
sharp-nosed grunter (*Syncomistes butleri*)   34
sheoak (*Allocasuarina fraseriana*)   125
shoot-to-root ratio   232, 233, 238
shrublands   79, 81, 167–9, 175, 199, 200, 202, 256
single-insect pollination, orchids   142
single species action plans   36
sinker roots   178
smoke treatment   144
snakes   176
snottygobble (*Persoonia longifolia*)   125
social sustainability   20, 22
sodic soils   195
soil chemistry, Yilgarn Craton survey   89–90
soil density   12–13, 177, 191, 193, 208
soil depletion, phytomining   239
soil depths, termites and   175
soil development   166, 171, 172, 203, 206, 207–9
soil erosion   234, 237
soil nutrients   149
soil pH   13
soil profile development, Gove   191–3, 200–3
soils   11, 12, 30, 151, 152
    drainage   194
    invertebrates in   134
    litter system and   184–9

organisms in   205
pollution of   237
removal   164, 205
restoration   189–99, 236, 237
separation   117
stabilisation   12, 176
taxonomy   162
termites and   174–5
soil sampling   83, 183
soil substitutes   11
Sørensen dissimilarity matrix   82, 87, 88
sorghum   165, 169, 205
Soroako   240, 243
South Africa   234, 238, 239, 240
South African orchid (*Disa bracteata*)   144, 147, 148, 149, 150
South Alligator floodplain   42
South Dandalup River   120, 121
Southern Cross   95, 253
space-for-time substitution   166
specially protected species   133
species composition   48, 89, 167, 257
speleological ecosystems, Brazil   64–5
spiders   134, 135
spill management   63–4, 120, 122, 123, 124, 135
spore types   181–2
Stackhousia tryonii   240, 241
stakeholders   20, 107, 258
stem wood litter   169, 200
strategic conservation planning   39, 105–7
stream restoration   17, 18
stream zone heaths   126
strip mining   4, 5, 9, 239, 241, 242–3
subsistence economies   33, 44
subsoils   82, 164
sucrose density gradient centrifugation   181
Sulawesi   240, 241, 242–3
sulphur   97
supercolonies   172
superphosphate   165
surface drainage   117, 119, 121
surface mining   7–8, 16, 63, 65, 160, 165, 190
Surface Mining Control and Reclamation Act 1977 (SMCRA)   8, 10, 11, 12, 13, 14, 16, 17
surface protection   184, 209
survey methods, Yilgarn Craton   82–3
sustainable environment   4, 8, 20–2, 37
symbiosis   179, 180

tailings   64, 143, 174, 235, 236
Tanzania   5, 31, 33, 36–8, 44, 47

tap roots   178
Tasmania   252, 255
taxonomy
    mycorrhizal fungi   150
    Yilgarn Craton survey   83
Tennessee   7, 9
termites   134, 166, 171, 173, 174–5, 189, 190, 191, 194, 200, 201, 202, 205, 206, 208, 209
    Gove mine and   172–5
Tetratheca paynterae   84, 98–9, 100, 101–3
thallium   41, 46, 233, 236, 239
Thelymitra spp.   153
thorium   40–1, 43
threatened species   133, 255, 257, 259
    Brazil   65
    orchids   144
    Western Australia   83
    Yilgarn Craton survey   85–6
three uranium mines policy   29
tilapia   38, 44
timber resources   112, 114, 164
tine penetration   164
tissue culture   128, 130, 131
topsoil   10, 16, 18, 118, 119, 128, 129, 164, 169, 170, 176, 182, 204, 239
tourism   161, 253
toxic materials   11, 33, 40, 231, 232, 243
traditional land owners   161, 162–3
    Gove   203
    Great Western Woodland   251–63, 258, 259–60
    see also Aboriginal and Torres Strait Islander People; *various entries under* Indigenous
training   19, 20, 122
transient workforce   4
trapping   132
tree growth   9, 12, 13, 160, 167–9, 200
tree hollows   176
triple bottom line   21, 22
tropical ecosystems   161
    orchids and   141
    roots and   177
Tumulitermes sp.   173, 174
turbidity   120–2, 124, 135
Typha augustifolia   234, 236

ultramafic outcrops   231, 234, 235, 239, 241, 242, 244
unallocated crown land   256–7
unburned sites   168–9, 170
understorey   134, 164
uniform gradational change   87, 88

unimpacted ecosystems   32
United States Geological Survey (USGS)   58, 242
United States of America (USA)   5, 7–27, 240
unmined sites   11, 12, 135, 145, 171, 173, 176, 177, 178, 179, 181, 189, 192, 193, 194, 197, 198, 199, 202, 203, 205, 206, 208
   carbon concentrations   187, 188
   Gove Peninsula   162
   mycorrhizal fungi and   149–50
   phosphorus concentrations   195–6
   soil properties   190
upper stratum   160, 167, 202, 205
uranium mining   5, 29, 30, 31, 33, 35, 40, 41, 43
   Bahi Swamp   37–8
   environmental risk   39–46
Urucum   56, 58, 59, 60, 62, 64, 71–7

valley floor   16, 126
vascular plants   62, 82
vegetation damage, termites and   174
vegetation groups   255, 260
vegetation monitoring
   Gove Peninsula   162
   pre-mining   126–7
   Yilgarn Craton survey   84
vegetation recolonisation   164, 200–3
vehicle-cleaning station   118
vertebrate recolonisations   131–5, 175–6, 255
village stakeholders   37
vines   170, 201
Violaceae   234, 240, 241, 242
Virginia   7, 9, 12, 13, 14
volcanic eruptions   142, 143
vulnerable species, Great Western Woodland   255

walking tracks   112
washdown trailer   117
waste dump   143, 144
wastewater   124
water access, roots and   178
waterbirds   36, 44
water buffalo (*Bubalus bubalis*)   35, 176
Water Corporation (WA)   120, 122, 124
water quality   9, 11, 13, 15, 16, 22, 30, 114, 116, 120, 135, 151, 160, 176, 236
   bauxite mining and   120–2
   uranium mining and   41
water table   63

weathering   189, 194
weaver ant (*Oecophylla smaragdina*)   172
weeds   127, 128, 170, 257, 259
   orchids as   142, 144, 147, 149, 153
Weipa mine   162, 208
Welwitschia spp.   38, 39, 44
West Virginia   7, 9
Western Australia   5, 19, 29, 79–107, 111–39, 144–9, 189, 190, 208, 240, 241, 251–63
Western Australian government   105, 255
Western Australian Herbarium   82, 83, 91, 253
Western Australian Museum   253, 255
western barred bandicoot (*Perameles bougainville*)   255
western quoll (*Dasyurus geoffroii*)   133
wetland ecosystems   33–8, 41
Wheatbelt region (WA)   97, 112, 149, 254
white-lipped tilapia (*Oreochromis alcalica*)   36
white pine (*Pinus strobus* L.)   13, 14
whole-of-landscape planning   257
wilderness values   58, 114, 255
wildfires   133, 146–7, 161
   see also fire regimes
wildlife habitat   11–12
Willowdale Mine   112, 120, 127
Windarling   80, 96, 97, 98, 99, 100, 101–3, 105
wind dispersal, seeds   142, 143, 240
wind erosion   189
win–win encounters   6
woodlands   9, 97, 125, 126, 182–3
Woodlands Initiative   258, 260
wood products   11, 14, 163
   termites and   172–5, 191
woody litter mass   186, 202, 206, 207
woody pear (*Xylomelium occidentale*)   125
Woolwonga   34, 35
World Heritage listing   34, 38, 60
World Reference Base   162
Wundarling   104, 105

Xylochomitermes melvillensis   173, 174

yellow crazy ant (*Anoplolepis gracilipes*)   166, 172, 206
Yilgarn Craton, ironstone ranges   79–107
Yirrkala Dhanbul Community Association   163
Yolngu people   170, 210

Zamia palm (*Macrozamia riedlei*)   129
zinc   232, 233, 234, 236, 238, 239, 241